Chinese Research Perspectives on the Environment, Special Volume

Chinese Research Perspectives on the Environment

VOLUME 8

The titles published in this series are listed at *brill.com/cren*

Chinese Research Perspectives on the Environment, Special Volume

Annual Report on Actions to Address Climate Change (2013), Focus on Low-carbon Urbanization

Editors

WANG Weiguang
ZHENG Guoguang

Deputy Editors

CHAO Qingchen
PAN Jiahua
ZHUANG Guiyang

Deputy Managing Editors

ZHU Shouxian
HU Guoquan

BRILL

LEIDEN | BOSTON

This book is the result of a co-publication agreement between Social Sciences Academic Press and Koninklijke Brill NV. These articles were selected and translated into English from the original 《应对气候变化报告（2013)》 (*Yingdui qihou bianhua baogao* [*2013*]) with financial support from Innovation Project of the Chinese Academy of Social Sciences.

Brill Open Access options can be found at brill.com/brillopen.

Typeface for the Latin, Greek, and Cyrillic scripts: "Brill". See and download: brill.com/brill-typeface.

ISSN 2212-7496
ISBN 978-90-04-28001-4 (hardback)
ISBN 978-90-04-28002-1 (e-book)

This book is printed on acid-free paper and produced in a sustainable manner.

Contents

List of Figures and Tables

Figures

Tables

CHAPTER 1

Low-carbon Urbanization: China's Strategic Choice in Response to Climate Change

By the Preparation Group of the General Report

Abstract

At present, Chinese economic and social development is undergoing an important strategic transformation, and urbanization is invested with new historical missions: expanding domestic demand and fostering new growth points. How to improve the quality and efficiency of the development of urbanization in order to blaze the path of new-type urbanization characterized by intensivisim, intellectualization, environmental protection and low-carbon is the important issue that China is facing at present. Low-carbon cities have become the focus of countries across the world in order to reduce resource and energy consumption, change the traditional developmental mode and seek emerging urban competitiveness. In order to improve the quality of urbanization and follow the path of new-type urbanization, China needs to rely on the city as a spatial carrier to develop a low-carbon economy. Low-carbon urbanization is

* The General Report was prepared by Pan Jiahua, Zhuang Guiyang, Xie Haisheng, Zhang Xiaomei, Zhou Zhenge, Zhang Ying, Xiong Na, et al. and was finalized by Pan Jiahua and Zhuang Guiyang. Pan Jiahua: director, researcher and tutor of doctoral candidates at the Institute for Urban and Environmental Studies, Chinese Academy of Social Sciences (CASS); research field: world economy, economics of climate change, urban development, energy and environmental policy, etc. Zhuang Guiyang: researcher and tutor of doctoral candidates at the Institute for Urban and Environmental Studies, CASS; research field: low-carbon economy and climate change policy. Xie Haisheng (teacher at Beijing Language and Culture University), Zhang Xiaomei and Zhou Zhenge: doctoral candidates at the Graduate School, CASS; research field: sustainable developmental economics. Zhang Ying: associate researcher at the Institute for Urban and Environmental Studies, CASS; research field: development and application of the climate change economics model. Xiong Na: post doctorate at the Institute for Urban and Environmental Studies, CASS; research field: low-carbon economy and regional sustainable development.

 | DOI 10.1163/9789004280021_002

not only an essential element of the construction of new-type urbanization, but also an important way to achieve the improvement of the quality of urbanization and the control of greenhouse gas emission. Despite defined implications, low-carbon urbanization still shares the following features with new-type urbanization: highlighting the scientificity of the development idea, the coordination of the development process, the intensivism of the development model, and the people-centeredness of the development objective. Low-carbon is the fundamental rigid constraint of the new-type urbanization. What really matters is not the speed of urbanization but preventing high-carbon lock-in. Only by maintaining a proper urban scale can "urban diseases" be avoided effectively. Most important of all, it is necessary to enhance the urban low-carbon competitiveness in low-carbon urbanization, so that industry provides support to the urban low-carbon transition. In order to actively and steadily promote the construction of low-carbon urbanization, it is absolutely necessary to scientifically formulate a low-carbon urbanization developmental plan aligned with the major function area, make a reasonable layout, and promote balanced development among large, medium and small cities. Besides, urban low-carbon competitiveness also needs to be strengthened by taking full advantage of industrial support. A series of measures such as the construction of a low-carbon demonstration community and the enhancement of adaptive management and carbon budget management should be taken to create a low-carbon resilient city and deal with climate change, thus building a beautiful China.

Keywords

low-carbon – new-type urbanization – lock-in effect – practice approach

Urbanization is an irresistible trend in economic and social development, an important sign of social progress, and an important strategic task in the future as shown in the outline of the "12th Five-Year Plan." Currently, Chinese economic and social development is undergoing an important strategic transformation with a fast promotion of industrialization and urbanization. How to improve the quality and efficiency of the development of urbanization in order to blaze the path of a new-type urbanization characterized by intensivisim, intellectualization, environmental protection and low-carbon is the important issue that China is facing at present.

1 Historical Background and Practical Significance of Low-carbon Urbanization

1.1 *The Rapid Urbanization Process in China*

Urbanization[1] is the population in a country or a region shift from rural to urban areas; the rural areas are gradually evolving into urban areas and there is a gradual increase in the proportion of people living in urban areas. The 21st century is the city's century. Urban economy is dominant in the current economy and is the developmental trend of the future economy. Joseph E. Stiglitz,[2] an American economist, believes that Chinese urbanization is one of the two themes that deeply affect human development in the 21st century.

Before the reform and the opening-up, the developmental pattern of Chinese cities presented a stable trend; since the reform and the opening-up, Chinese urbanization has experienced a rapid development which may be divided into two stages. The first stage was a period that started from the implementation of reform and opening-up and went until the 1990s, during which urbanization was mainly promoted by the development of small cities and towns; the second stage started from 1990 and is still going on now, a period during which the big cities have been rapidly developing, which is reflected in the sharp increase in the proportion of the city population in the urban population, and the rapid expansion of the scale of cities. From 1978 to 2012, the Chinese urbanization rate increased to 52.6% from 17.9%, with an annual average growth rate of 1.02 percentage points, revealing not only the fast promotional speed but more importantly the unprecedented scale of expansion. During the "12th Five-Year Plan" period, China will enter the phase of double transformation of urbanization and city development. The urbanization rate

1 The word "urbanization" was first used by Western scholars. In a narrow sense, urbanization can be interpreted as a process of population shift from rural areas to cities and towns. As "urban" includes cities and towns, the town population is relatively small in many countries and even doesn't reach the supposed town population scale; "urbanization" often means simply the process of population shift to "city." For most Chinese towns, their population basically equals that of small foreign cities. The population shift to both "city" and "town" is the Chinese characteristic of urbanization. To highlight the difference, the Chinese population shift to urban areas is mostly described by the word "Urbanization."

2 Quoted from the speech by an American economist J.E. Stiglitz in Beijing "Urban Development Senior Round Table" on July 23, 1999. Quoted in Zou Deci, *Some Understandings on Chinese Urbanization Problems*, published in the *Urban Planning Forum* 2004(3).

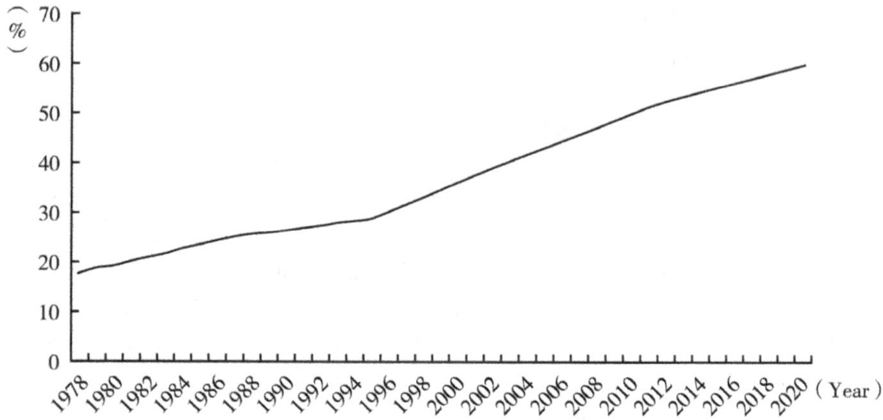

FIGURE 1.1 *The trend of change in the rate of Chinese urbanization (from 1978 to 2020).*

is estimated to enjoy an annual average increase of 0.8 to 1 percentage point; it was up to 55% in 2015 and will be about 60% in 2020[3] (see Figure 1.1).

The fast and large-scale promotion of urbanization exerts a far-reaching influence on current China and on the world. Urbanization provides space for industrialization, changes the distribution of Chinese population and industry, and drives the development of productive forces and the accumulation of social wealth. At present, China has ended the era dominated by a village-type society and is starting to enter the new urban era dominated by an urban society. Nowadays, the population of Chinese cities and towns exceeds 710 million. Advanced manufacturing industries and modern service industries gather in cities and towns, so that cities and towns have become the fundamental power that supports the rapid development of the Chinese economy.

However, since the reform and the opening-up, the path towards urbanization that the expansion of urbanization has followed is an extensive one. On the one hand, the promotion of urbanization highlights speed rather than quality; the citizenization of the migrant rural population is low, thus causing mismatching of urbanization speed and quality; on the other hand, the promotion of urbanization results in considerable costs in resources and the environment, specifically reflected in high consumption, high emission,

3 Data source: data is obtained based on calculation using the Chinese statistical data application system. In the article, the urbanization rate is obtained by dividing the urban population by the total population. The rate of urbanization from 1978 to 2012 is objective and real, while the rate of urbanization from 2013 to 2020 is obtained through an estimation based on an annual increase of 0.9 percentage points.

high expansion and low efficiency of resource allocation. Fast urbanization is accompanied by high consumption and emission. Some local governments and departments only focus on governmental images and economic growth, extort money from investment on people's livelihood projects and neglect the quality of cities and towns, causing urban-rural segmentation, the phenomenon of "thousands of cities look the same," an improper urban system, inefficient development of the urban space and urban sprawl, finally resulting in increasingly serious problems in the urban society, resources and the environment.

In the course of its fast urbanization, the Chinese government has never stopped to make plans and efforts. The 18th CPC National Congress report proposed that: We should propel the strategic adjustment of the economic structure, firmly follow the new distinctive Chinese path towards industrialization, informatization, urbanization and agricultural modernization, push forward deep integration between informatization and urbanization, sound interaction between industrialization and urbanization, mutual coordination between urbanization and agricultural modernization, and promote a synchronous development of industrialization, informatization, urbanization and agricultural modernization.

In December 2012, the Central Economic Working Conference suggested that we should establish a scientific and reasonable urban pattern, make a scientific layout for all forms of cities, small towns and urban agglomerations to closely connect with the regional economic development and industry layout and adapt to the carrying capacity of resources and environment; we should focus on promoting the citizenization of the migrant rural population; we should integrate the idea and principle of ecological civilization in the whole process of urbanization, thus following the path of new-type urbanization characterized by intensivisim, intellectualization, environmental protection and low-carbon. Early in 2013, Premier Li Keqiang pointed out that the focal point of the promotion of urbanization lies in the urbanization of population, the key is the improvement of the quality of urbanization, and the goal is to benefit the common people and enrich the peasants. It is a must to focus on the increase of the internal bearing capacity, avoid the artificial "city building," and strive to integrate industrial development with urban construction, thus gradually integrating the migrant workers into the cities and towns.

1.2 *Resource and Environmental Challenges in the Course of Urbanization*

In the process of rapid urbanization, the primary problem is the sharp increase in urban energy consumption and a large amount of pollutant emissions. The acceleration of urbanization means a tremendous need for urban infrastructures and residential buildings, so that the total energy demand will surely rise rapidly. A large quantity of steel, cement and electricity required by the construction of urban infrastructures mainly relies on the rapid development of domestic, high energy-consuming industries. Besides, with the intensification of urbanization and the increase in income, changes in the consumption pattern of urban residents also apply enormous pressure on the urban environment; for example, the surge of private sedans, the expansion of the residential areas and the upgrading of household appliances all exert an impact on traffic jams, resource shortage and environmental pollution in various cities, to some extent.

Rapid urbanization has speeded up the construction of urban infrastructures and public facilities, and promoted economic development. However, the original intention of urbanization was more than simple "building a city," the increase of the proportion of urban population and the expansion of administrative district areas. What is important is to achieve the significant transition from "rural" to "urban" in the industrial structure, employment mode, living environment and social guarantee, and that such a transition is harmonious and sustainable. In the process of such a transition, it is necessary to narrow down the rural-urban division and strive to achieve the equalization of rural-urban basic public services, to achieve overall urban and rural development, eliminate the rural-urban dual structure and the urban internal dual structure, thus attaining the goal of joint development and common prosperity. In the future, the urbanization rate in China will continue to rise, the rural population will continue to shift to an urban population, and a new round of investments will increase rapidly. The surge of consumption is just around the corner. While bringing a massive workforce to the cities, urbanization also leads to a huge demand by a new urban population for urban public services, further resulting in a series of issues such as child education, health care and old-age care.

However, the unfinished urbanization has triggered serious problems, mainly shown in the following aspects: The first is the insufficient bearing capacity of urban resources, for it is difficult for urban energy resources, natural resources and various social guarantees and services to bear the rapid expansion of the urban scale and population; the second is the frequent occurrence of environmental pollution, water pollution and gas pollution, which cause serious damage to the health of city-dwellers. The problems caused by these two aspects are "urban diseases," which go against the original intention of urbanization: "Better City, Better Life." Apparently, they are not the expected results of urbanization. Ideal urbanization is supposed to be resource-saving, environmentally friendly, comprehensive, coordinated and sustainable.

1.2.1 Mutual Constraint and Co-existence of the Energy Crisis and the Carbon Emission Reduction

Energy is the engine for economic growth. The energy demand remains high due to the fast development of the Chinese economy. The years 1990 and 2012 witnessed a total Chinese energy consumption of 990 million and 3.62 billion tce (tons of coal equivalent), respectively. Moreover, total Chinese energy consumption has exceeded total production since 1992, and the energy gap had reached 280 million tce by 2012. As fossil energy is a non-renewable resource, the energy security has been highly concerned, worldwide. China is facing a more serious energy crisis. In the course of the fast development of the Chinese economy, energy security has become the bottleneck that hampers the sustainable growth of the Chinese economy.

In 2012, total Chinese energy consumption amounted to 2,735.2 million toe (tons of oil equivalent), accounting for 21.9% of the global energy consumption. The enormous energy consumption results in a huge CO_2 emission in China; the total amount of carbon emissions came to 9.21 billion tons of carbon dioxide equivalent in 2012, accounting for 26.7% of the world's carbon emissions; the per capita carbon emissions were 6.82 tons CO_2 per capita, significantly higher than the average world level of 4.89 tons CO_2 per capita (See Table 1.1). As a large responsible country, controlling the emission of greenhouse gas to ensure sustainable development and relieve the international outcry is badly needed. The total carbon emission capacity is very limited. The *Work Plan for Greenhouse Gas Emission Control during the 12th Five-Year Plan* issued by the State Council of China in 2011 showed that the emission of CO_2 was around 1.9 tons per ten thousand yuan GDP (calculation based on the price in 2005), down 17% and 34% from 2010 and 2005, respectively. According to the goal as set out in the Work Plan, in 2020, the emission of CO_2 per unit of GDP is to fall by 40%–45% compared with that of 2005, and the ratio of non-fossil energy in primary energy is to reach 15%, which is the practical emission reduction pressure in the process of urbanization.

1.2.2 Coexistence of Short Natural Resources and Insatiable Practical Demand

As the city is an agglomeration of population and industries, it faces many challenges in terms of carrying the capacity of urban natural resources. By taking water resource as an example, China is a water-poor country. Among 600 cities, over 400 cities suffer from water shortage to different degrees, and over 200 cities suffer from serious water shortage. Unlike the water shortage in the traditional sense, the primary cause for water shortage in the coastal regions

TABLE 1.1 *Global energy consumption and the pattern of carbon emissions in 2012*

Country/region	Carbon emissions ($\times10^4$ tons of CO_2)	Energy consumption ($\times10^4$ toe)	Per capita carbon emission (tons CO_2 per capita)
The USA	5786.1	2208.8	18.43
France	383.4	245.4	5.84
Bangladesh	66.1	26.3	0.43
China	9208.1	2735.2	6.82
India	1823.2	563.5	1.47
The EU	3977.5	1673.4	7.88
Total	34466.1	12476.6	4.89
OECD	13899.8	5488.8	11.06
Non-OECD	20566.3	6987.8	3.55

DATA SOURCE: ENERGY AND CARBON EMISSION DATA COME FROM UK BP GLOBAL; THE POPULATION DATA COME FROM THE WORLD BANK. HTTP://WWW.BP.COM/EN/GLOBAL/CORPORATE/ABOUT-BP/STATISTICAL-REVIEW-OF-WORLD-ENERGY-2013.HTML. HTTP://DATACATALOG.WORLDBANK.ORG.

is the frequent occurrence of water pollution and the deterioration of water ecology. Now, many regions have had no clean source of drinking water, further resulting in an increasingly severe "water quality-induced water shortage." Sixty percent of the rivers running through cities fail to meet drinking water quality requirements due to pollution. The accumulated pollution of surface water and underground water makes it difficult to relieve the pollution problem of urban water sources in the short term.[4]

Besides, it is hard to address the practical need arising from the urbanization process. China's steel output accounted for 45.4% of the world's steel output in 2011, and the ratio reached 46.3% in 2012. The number of inhabitants who annually move from rural areas to urban areas exceeds 15 million, so that adequate housing, living facilities and relevant urban infrastructures must be provided. All of these mean that China will maintain a high consumption of building materials such as coal, steel and cement for a long time. With the gradual relaxation of the urban household registration system, huge capitals

4 Qiu Baoxing, *Challenges and Opportunities at Middle-and-late-stage of China's Urbanization: New Trends of Urban Planning Transformation, Urban Planning*, 2010 (3).

are required for education, medical care and social guarantees of the new urban population shift from rural areas, and in addition, the efforts on such aspects are far from enough.

1.2.3 Severe Environmental Pollution

Apart from water resource's pollution hereinabove mentioned, automotive exhaust emission has been one of the main sources of the atmospheric pollution in Chinese large and medium-sized cities. The monitoring data of the Beijing Municipal Environmental Protection Bureau in 2012 show that the nitric oxide and volatile organic compounds emitted by motor vehicles accounted for 58% and 40% of the total emissions in the whole city, respectively, and the PM2.5 arising from motor vehicle emission accounted for about 22.2% of the total PM2.5. Besides, since all Chinese cities feature compaction and denseness, the harm caused by light pollution and atmospheric pollution is more serious than that in cities in Western countries. The direct cause for this is the sharp surge in dense high-rises and the number of automobiles. Although the standard of motor vehicle exhaust emission has been improved in recent years, the number of private vehicles shows a rapid increasing trend due to an inadequate urban traffic strategy.

The fog and smog that swept across central and eastern China in early 2013 made people realize the seriousness of atmospheric pollution and the urgency of relieving it. The cause for atmospheric pollution is mainly associated with the urban industrial structure, the energy structure and the urban planning. Smog is mainly composed of three entities, i.e., sulfur dioxide, nitric oxide and inhalable particles. Oxynitride mainly comes from emissions from motor vehicles, electric plants and industrial enterprises, in which the automotive exhaust emission is an important cause. Smog can severely damage people's health; as the intense use, the dense concentration and the high-speed increase in the number of motor vehicles, it has become increasingly acute. Pollution from vehicle exhaust is becoming increasingly serious.

Waste is an accessory to city development. Hundreds of millions tons of waste are generated from city operation and people's consumption annually. The increasing amount of waste and the intolerable odor generated by it are difficult issues that are hard to be tackled in waste disposal. According to statistical data, 2/3 of the over 600 large and medium-sized cities are "overwhelmed" by waste, and 1/4 of the cities have no proper place to designate as a waste disposal site. In China, high-speed developing cities, especially large cities, are suffering from the "waste siege." Beijing has sounded the alarm of waste siege repeatedly, and Beijing's "waste seven ring roads" are being gradually formed.

1.2.4 Emergence of the "Large-urban Disease"

A city has three major functions, i.e., work, livelihood and rest, as well as the transport that connects the three functions together. With the increasing scale of cities, the excessive concentration of population, industry and transportation brings about a great number of disadvantages, known collectively as "urban disease." The main causes for large-urban disease are resource concentration in Chinese cities, regional disparity of social welfare and benefits, and blind expansion of city-planning. Serious urban disease is manifested in traffic jams, environmental deterioration, housing shortages and employment difficulties, which will increase the burden of the city, restrict the development of urbanization and trigger the physical and mental illnesses of citizens. The permanent resident population of Beijing and Shanghai are more than 20 million, far exceeding their resource-bearing capacity, so that it is unavoidable for Beijing and Shanghai to suffer from the "large-urban disease."

1.3 *Avoidance of the High-carbon Lock-in Effect in the Process of Urbanization*

Chinese achievements in economy are, to a large extent, attributed to the rapid process of urbanization. "Based on the current trend, the number of people living in an urban area will increase to 1 billion by 2030. At that time, China will have 221 cities with a population of 1 million or more (at present, there are only 35 cities with such a scale of population in Europe) and 23 cities with a population of 5 million or more. And by 2030, the urban economic value will account for 90% of the GDP."[5] For this, China needs to construct buildings of about 40 billion m^2 in the coming two decades. The surge in urban population imposes many challenges on the carrying capacity of urban areas. For instance, the sharp increase in the urban population not only requires a synchronous supply of social services, resources and energy, but also brings about urban environmental pollution. To meet the investment need, Chinese cities will consume 20% of the global energy, contributing to 1/4 of the growth in fuel oil demand worldwide at the most.

Ominously, high-carbon lock-in is gradually formed in the process of Chinese urbanization. During the "11th Five-Year Plan," the total amount of Chinese industrial energy consumption has increased at a significant rate, from 1.595 billion tce in 2005 to 2.4 billion tce in 2010, accounting for 73% of the total social consumption; the proportion of energy consumption of six high energy-consuming industries (the steel industry, the nonferrous metal

5 McKinsey Global Institute, "Preparing for China's Urban Billion," http://www.mckinsey.com/mgi/reports/pdfs/China_urban_billion/china_urban_billion_full_report.pdf, 2009.

industry, the building material industry, the petrochemical industry and the electric industry) of the total industrial energy consumption increased from 71.3% to around 77%. Such industrialization not only causes high energy consumption, employment difficulties, environmental pollution, but it also restricts the citizenization of migrant workers. It is the high-carbon lock-in effect of urbanization which creates endless troubles and is not based on people. Once such a mode has been formed, it is hard to change it in a short period of time. Avoiding the high-carbon lock-in effect is of vital importance in the development of Chinese urbanization, at least for the following reasons:

First of all, Chinese self-owned resources and environments are inadequate to support the developmental mode of urbanization characterized by high carbon and high emission. Even the abundant coal supply can only last 31 years based on the reserve-production ratio of proved reserves and current mining rate. Chinese urbanization features a large scale, a high speed and a long process, and faces a number of challenges such as land deficiency, energy crisis, water resource shortage and environmental deterioration; such challenges are, in the end, subject to the rigid restraint of the carbon budget.

Second, low-carbon urbanization is the necessary requirement in a period of low-carbon transformation for seeking new points of economic growth. It will generate hundreds of industrial departments and thousands of products characterized by low-carbon and low energy consumption, which will exert a pivotal impact on China's transformation and development. The Chinese low-carbon urban development has the problem of the reform of the upgrading of stock, as well as the problem of incremental optimization adjustment. In the light of the status of Chinese urbanization development, to boost the low-carbon development of the urban economy, the inevitable choice for China is to make a breakthrough in upgrading the traditional high-carbon industry, developing low-carbon emerging industry and building up the proportion of the tertiary industry, thus gradually realizing the low-carbon adjustment in the industrial structure.

Third, low-carbon is the rigid constraint in the road towards new-type urbanization with Chinese characteristics. New-type urbanization with Chinese characteristics means that the ideas and principles of ecological civilization are fully integrated into the whole process of urbanization to firmly take the road towards urbanization characterized by intensivism, intellectualization, environmental protection and low-carbon. Environmental protection, circulation, low-carbon and ecology are the fundamental requirements for the development of urbanization, important constraint conditions and the developmental limitation for the development of the national economy. However, environmental protection, circulation and ecology do not mean low-carbon, which is a deeper constraint condition and barrier than environmental protection and circulation.

1.4 *Building Low-carbon Resilient Cities Through Low-carbon Urbanization*

Climate change is a change in weather conditions due to a variation in the global atmospheric composition, directly or indirectly caused by human activities, besides natural climate change, after quite a long time of observation. In recent years, climate change has resulted in a higher frequency of catastrophic climatic activity, an acceleration of the melting of glaciers and snow, an imbalance of water resource distribution, and the threat to biodiversity. Moreover, the report by the Intergovernmental Panel on Climate Change shows that if the rise in temperature exceeds 2°C, it would exert a negative impact on any area in the world, especially in the developing countries; if the rise in temperature exceeded 4°C, it would cause irreversible damage to the global ecosystem and considerable losses to the global economy.

In the course of rapid urbanization in China, climate change has been occurring more frequently and seriously. Various extreme climate events, such as rainstorms in the north, dry and hot weather in the south, long-term smog in central and eastern cities and local heavy rainfall in cities, remind us that the urbanization process has a great influence on climate. In recent years, it is has been common for cities to encounter serious meteorological disasters, which has aroused wide public concern for urban public administrations, disaster prevention and mitigation. The meteorological disasters that frequently occur include waterlogging caused by rainstorms, drought and water shortage, hot weather, thunderstorms, smog and gales. With global climate warming, both the frequency and the intensity of extreme climate events are increasing, further resulting in more serious consequences for cities. With the acceleration of urbanization and modernization, the weakness of cities in the prevention of climatic disasters has become more evident.

A study shows that when city development reaches a certain scale, the local urban climatic features and ecological environment will change due to high intense human activities and the changes in the underlying surface of the city (the earth's surface that comes into direct contact with the lower atmosphere) and landform, such that any urban meteorological disaster is imprinted with human activities.[6] With global climate warming and the acceleration of

6 Zhou Jianming, *Introspection from Beijing Downpour, Ta Kung Pao*, November 6, 2012.

urbanization, the urban green space is decreasing and in addition, the heat island effect is becoming more evident, which causes more serious urban hot weather, threatens people's health, and results in a shortage of water and electricity, greatly affecting the production and life of urban inhabitants. Besides, problems in urban planning and distribution make the urban lifeline system, such as power supply, water supply, gas supply, communications and transportation, susceptible to meteorological disasters. Furthermore, the delay in and incompleteness of the construction of infrastructures such as an urban water drainage system, as well as large areas of armored concrete which has poor permeability, facilitate ponding and waterlogging.

At present, under the promotion of policies of energy conservation, emission reduction and low-carbon development, constructing a low-carbon city has become a common practice from the point of view of both idea and action in China. However, compared with emission reduction, China still lacks pertinent and viable policy planning with respect to strategic goals, action plans and governance mechanisms for cities in response to climate change. As a result, the concept of low-carbon resilient cities emerges at the right moment. Under the common goal in response to climate change, low-carbon resilient cities take two tasks into account, that is, the reduction of greenhouse gas emissions and the flexible treatment of climatic disaster risks, so as to increase the sustainable development capacity of the city.[7] A multi-goal decision process of low-carbon resilient cities requires urban administrators to adopt adaptive management[8] ideas and improve their urban governance mode.

The thought of Chinese urbanization development is to actively exert the agglomeration effect of regional central cities on accelerating local social and economic development, which makes many large cities (especially regional economic center megacities such as Beijing, Shanghai, Guangzhou and Shenzhen) undergo multiple pressures in terms of emission reduction, employment, economic growth and environment governance. In the meanwhile, against the background of climate change, extreme climatic events not only bring about more climatic risks to cities, but also make the city's vulnerability become more evident. Therefore, taking cities as the breakthrough point to carry out the demonstration and pilot of low-carbon resilient cities, contributes to the integration of mitigating pressures mentioned above and adapting

7 Zheng Yan et al., *Low-carbon Resilient City: Idea, Approach and Policy Selection, Urban Development Study*, 2013(3).

8 Adaptive management is a flexible decision mechanism adopted under various uncertain factors. Its epistemology is based on the complexity and uncertainty of the human ecosystem.

to climate change from ideas to action. Besides, building a low-carbon resilient city is the inevitable choice for China in order to follow the road towards urbanization.

2 Connotation of Low-carbon Urbanization and the Carbon Emission Features

Rapid urbanization means that the city will become the main concentration area for the emission and increase of greenhouse gas. In response to climate change and advancement of social development through a low-carbon economy, low-carbon cities will be the focal point for various countries in the reduction of resource and energy consumption, for changing the traditional developmental mode and for seeking emerging urban competitiveness. In order to improve the quality of urbanization and to follow the road towards new-type urbanization, China needs to rely on the city as a spatial carrier, by which to develop a low-carbon economy. Hence, the application of the idea of a low-carbon economy to guide urbanization, namely low-carbon urbanization, means that planning, design and construction of low-carbon cities will be carried out based on low-energy consumption, low pollution, low emissions, high efficiency and high output in the course of urbanization.

2.1 *The Relationship between Low-carbon Urbanization and New-type Urbanization*

In 2012, the Central Economic Working Conference took "promoting urbanization actively and steadily, focusing on improving the quality of urbanization" as one of the six tasks of the central economic work in 2013, revealing that urbanization is the important strategic direction for Chinese economic development. New-type urbanization means an intensive, intelligent, low-carbon and green road towards urbanization characterized by resource conservation, environmental protection, cost effectiveness, social harmony and urban-rural integration. This is based on the principle of urban-rural balanced development, reasonable layout, land conservation, perfect urban function and the radiation effect of big cities on small and medium sized cities and under the promotion of a market-oriented and government-guided urbanization mechanism to achieve good interaction among urbanization, industrization, informatization and agricultural modernization, as well as a rational layout and coordinated development among large, medium and small cities and small towns.

Low-carbon urbanization is an important part in the construction of new-type urbanization, as well as an important route to improve the urbanization quality and control the greenhouse gas emission. Compared with the new-type urbanization, low-carbon urbanization has a deeper connotation, but it shares common features.

2.1.1 Scientificity of the Development Idea

Low-carbon urbanization is a road towards new-type urbanization under the guidance of a scientific concept of development. It is necessary to follow the common rules of urbanization development, adhere to the people-first idea, respect people's aspirations and make the best use of the circumstances to steadily promote the process of urbanization.

2.1.2 Coordination of the Development Process

Low-carbon urbanization respects the objective rules of economic and social development, and constructs an urbanization system of coordinated development among large, medium and small cities and small towns through balancing urban and rural development, economic and social development and regional development, so as to advance the sound development of urbanization.

2.1.3 Intensivism of the Developmental Mode

Low-carbon urbanization is an ideal choice for China based on the rigid constraints of resource and the environment. To change the traditional extensive developmental mode, it is necessary to follow an intensive developmental road characterized by low investments, low consumption, low pollution and high efficiency.

2.1.4 The People-centered Idea of the Goal of Development

In the course of the advancement of low-carbon urbanization, it is necessary to adhere to the people-centered idea of the goal of development in regarding "people" instead of "objects" as the starting point and the aim of the work, and not to aim at urbanization only for the sake of urbanization.

The traditional developmental mode of urbanization has shown a positive correlation between the development of urbanization and carbon emissions. On the one hand, the road of extensive urbanization causes damages to the ecological environment and weakens the carbon sink function of the nature; on the other hand, the development of urbanization results in a further increase in carbon emissions. The focal point of a low-carbon economy is to use the market mechanism to construct a low-carbon energy system and industrial structure through the policy innovations, institutional design and technological progress, thus achieving a sustainable development of the economy, the society and the environment, which is exactly what is required by new-type urbanization. So, for new-type urbanization, it is a must to follow the road towards low-carbon development.

Low-carbon urbanization means that in the process of urbanization, with the principle of sustainable development, we should formulate a scientific low-carbon urban plan, establish low-carbon urban infrastructures and form a low-carbon energy consumption structure, develop a low-carbon economy, enhance the comprehensive governance of the urban ecological environment, reduce the impact of urbanization on ecological construction, and finally realize coordinated development in economy, society and the environment.

2.2 *Quality Management of the Urban Incremental Land and the Transformation and Upgrading of the Urban Stock Land*

The low quality of China's urbanization is manifested in a lag in the planning, poor implementation performance, resource waste caused by mass-demolishing and mass-construction and the phenomenon of urban villages, etc. Low-carbon urbanization needs to take the following two tasks into account, that is, the construction of new urban areas and the renovation of old urban areas. The former concerns the quality management of the urban incremental land, and the latter concerns the transformation and upgrading of the urban stock land. With rapid progress in the development of urbanization, the construction of new urban areas and the renovation of old urban areas have been the important parts of the current urban low-carbon development. The construction of new urban areas is an important spatial carrier to use low-carbon technology, implement low-carbon developmental ideas, foster low-carbon urban functions and speed up the creation of low-carbon urbanization. Lester R. Brown, an ecological economist and Chairman of the Earth Policy Institute of America, believes that the modes of urbanization and economic development can be classified into "A mode" and "B mode." The former is the developmental mode of the developed countries, led by America, which is mainly characterized by low-density urban sprawl, mechanization dominated by private cars, fossil-fuel-based energy consumption and a plethora of disposable products; consequently, America's population which contributes to 5% of the world's population, consumes over 1/3 of the world's energy.[9] The American developmental mode is a typical mode featuring high energy consumption, high carbon emission and unsustainability. Once it is formed, it is hard to be changed in a short period of time. Constant high carbon emission is called American-style high-carbon lock-in. According to international

9 Qiu Baoxing, *Urbanization Mode with Chinese Features—"C mode": Surpassing "A mode" Temptation and "B mode" Troubles, Urban Development Study*, 2009(1).

experience, urbanization is at the acceleration stage when the urbanization level reaches 30%–70%. In 2011, the Chinese urbanization level was 51.27%, at the middle point turning from the first half to the second half of the acceleration stage. The first half of the acceleration stage was that of volume expansion while the second half was that of quality improvement. Now, China is right in the middle period of urbanization, in which urban morphological flexibility is high. For newly-built cities in China, high-carbon lock-in should be avoided.

Low-carbon transformation of old urban areas is also an important part of low-carbon urbanization. With the increasingly rapid process of urban construction, the renovation of old urban areas has been one of the most urgent problems needing to be addressed, especially in the construction of metropolises and megalopolises. With the expansion of the urban scale, the renovation of old urban areas exerts a growing influence on urban development.[10] According to a foreign idea for urban renewal development, the renewal goal and content vary with different stages. From the history of the whole urban renewal, urban renewal undergoes, from large-scale reconstruction to small-scale, staged and progressive renovation; from beautification focusing on material space to comprehensive renewal in the economy, the society and the environment.[11] The international experience shows that the European city attaches great importance to urban planning, and strictly complies with quality standards to fulfill the construction in one step. On the contrary, in some Chinese cities, urban planning varies with mayors, and mass-demolishing, mass-construction activities are carried out in every plan. It can often be seen that roads are renovated and plaza tiles are replaced, which is rare in European cities.

During the "12th Five-Year Plan," the style of Chinese urban architecture will move towards one of green architecture with abundant connotations from a single energy-saving architecture, and toward large-scale urban district renovation from single building renovation. At present, there is still no mature and widely accepted practice for low-carbon and ecological reconstruction in existing communities.[12] With the rapid development of urbanization and industrialization, the available urban construction land resource is in increasingly short supply, and the need for land for urban industrial upgrading and

10 Zhang Pei, *Renovation of Old Urban Areas and the Development of New Urban Areas, Urban Problems*, 2000(1).

11 Lu Danmei, *Planning: Toward Stock Renovation and Old Urban Area Renewal—Three-old Renovation Planning Thought Exploration, Urban Development Study*, 2013(6).

12 Ye Ruike et al., *Promoting Energy Conservation and Carbon Reduction and Developing a Low-carbon Economy—Based on a Typical Survey on Business Buildings in Xiacheng District, Hangzhou City, Future and Development*, 2012(12).

economic structure transformation will be addressed gradually through renovation of old urban areas (renewal of old urban areas).[13] To promote intensive urban development, further improve policies and mechanisms relating to resource conservation and energetically promote technologies in urban energy conservation, material conservation, water conservation and land conservation, it is necessary to balance the renovation of old urban areas and the construction of new urban areas and actively guide the evacuation of the inhabitants of old urban areas and the agglomeration development of the industries in the new urban areas.[14]

2.3 *Features of Carbon Emission during the Process of Urbanization*

Low-carbon urbanization is finally measured based on the overall decline of carbon emissions, so it is necessary to get to know the features of carbon emission at various stages. In 1979, Ray M. Northam, an American urban geographer, summarized the common rules of the urbanization process of different countries and regions into a leveling S-shaped curve, and divided them into three stages: (1) the primary stage, in which the level of urbanization is low (the urban population makes up less than 30% of the total population), and the number and scale of cities are small; (2) the middle stage (the proportion of urban population exceeds 30%), in which the population and economic activities migrate rapidly towards the cities, the number of cities has sharply increased and the level of urbanization is rapidly increasing; (3) the late stage (the proportion of urban population exceeds 70%), in which the process of urbanization tends to be slow, even stagnated, and the urban-rural division has almost vanished.[15] Although not every country and region follows the same process of urbanization, this is a common rule based on the process of urbanization in most countries and regions around the world. According to Northam's Curve, though the overall level of Chinese urbanization has reached the middle stage, there is a great difference in the actual level of urbanization among different regions due to the vast area of China and the unequal development in the various regions. As a result, different regions may adopt different low-carbon urbanization strategies based on different stages.

---

13 Pan Yue et al., *Strategy Research on Old Urban Renewal under Change of Urban Planning Role, Well-known Cities of China*, 2013(3).

14 Wei Houkai and Zhang Yan, *Thoughts and Measures on Comprehensively Accelerating the Green Transformation of Urbanization in China, Economic Review*, 2011(9).

15 Northam, R.M., *Urban Geography*, New York: John Wiley & Sons, 1979.

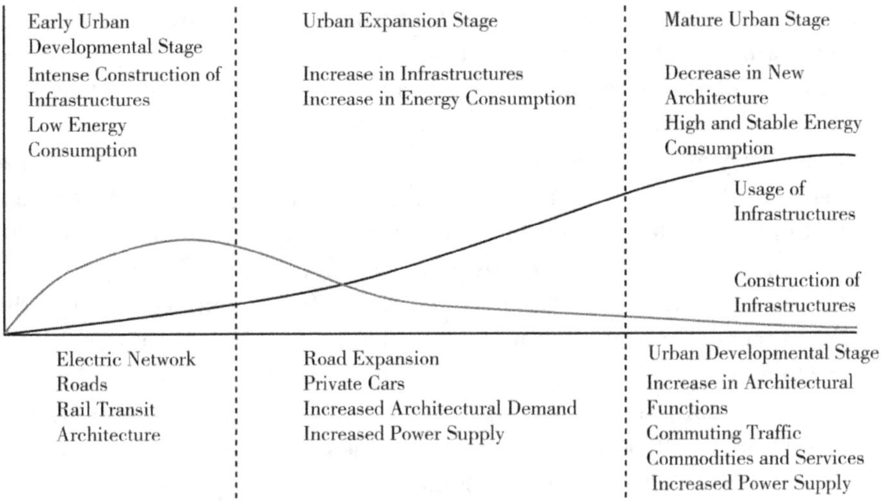

FIGURE 1.2 *The feature of carbon emission in the different urban developmental stages.*
DATA SOURCE: FERNÁNDEZ, J.E., "RESOURCE CONSUMPTION OF NEW URBAN CONSTRUCTION IN CHINA," JOURNAL OF INDUSTRIAL ECOLOGY, 2007, VOL. 11, NO. 2, PP. 99–115.

Carbon emission in the early stage of urban development mainly comes from building construction, public transportation, energy and water. As urban development reaches maturity, the size of urban areas will grow continuously, and the amount of energy needed to maintain normal operations of existing infrastructures will rise, too. In the different developmental stages, the city has different features in energy consumption. The urban morphology formed during the early urban developmental stage will determine the structure and features of carbon emissions in the coming years (see Figure 1.2).

At present, neither carbon emission reduction policies nor strategies are separable from the progression of social development. The population shift attendant to urbanization exerts great impact on energy and carbon emissions. The government can appropriately control the urbanization rate, and the growth rate of carbon emissions by treating urbanization as an opportunity for promoting low-carbon development, and it can also improve the efficiency of energy utilization and make the energy structure cleaner through the formulation and implementation of environmentally sound energy policies. Besides, the process of urbanization is also a process of the selection of lifestyle, and lifestyle directly affects energy consumption. To advocate an energy-saving lifestyle through policy guidance is also an important aspect in Chinese sustainable development. It is a must for China to go all out in its efforts to realize carbon emission reduction during the process of economic growth and urbanization, which is of primary significance in low-carbon transformation.[16]

Moreover, as the primary sources of carbon emissions, the consumption and tendency of primary energy in countries in various developmental stages are also worth exploring. As a whole, energy consumption in America, the representative of post-industrialization countries, tends to be stable; from 2001 to 2011, America's total energy consumption increased from 2.26 billion toe to 2.27 billion toe, with an increase rate of 0.44%. On the contrary, China, as the representative of emerging economies, had a sharp surge in energy consumption along with the rapid advancement of the process of industrialization; from 2001 to 2011, the Chinese aggregate economic volume surged to 47,310.4 billion RMB from 10,965.5 billion RMB, and the rate of urbanization increased to 51.3% from 37.7%. Moreover, China's primary energy consumption also increased constantly, and reached 2.21 billion toe in 2009, surpassing that of America (see Figure 1.3).

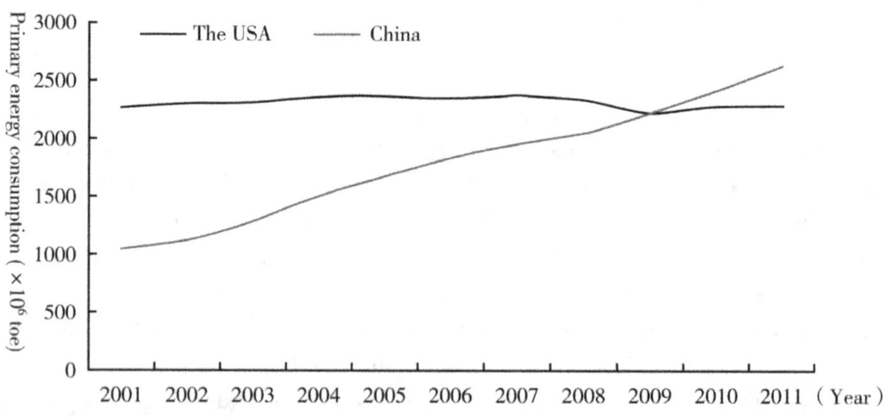

FIGURE 1.3 *Comparison of primary energy consumptions between the US and China (from 2001 to 2011).*
DATA SOURCE: BASED ON DATA FROM THE *BP STATISTICAL REVIEW OF WORLD ENERGY* (JUNE 2012), HTTP://WWW.BP.COM/LIVEASSETS/BP_INTERNET/CHINA/BP CHINA_CHINESE/STAGING/LOCAL_ASSETS/DOWNLOADS_PDFS/CHINESE_BP_STATSREVIEW2012.PDF.

16 Lin Boqiang and Liu Xiying, *China Carbon Emission under Urbanization Process: Influence Factors and Abatement Policies, Economic Research*, 2010(8).

Let's turn to the share of primary energy consumption in the two countries in worldwide primary energy consumption and their tendencies. The total energy consumption in America, the representative of post-industrialization countries, tends to be stable, but its proportion of worldwide energyconsumption is still large. For instance, the consumption of primary energy in America accounted for 24.0% of worldwide energy consumption in 2001; afterwards, the proportion dropped with each passing year. By contrast, the proportion of China's primary energy consumption in worldwide energy consumption rose continuously, from 11.0% in 2001 to 21.3% in 2011 (see Figure 1.4).

Aggregate energy consumption and change tendency apparently reflect the features of the developmental stages of industrialization and urbanization: the fast development stage of industrialization and urbanization of a country surely results in a sharp increase in energy consumption; with the completion of urbanization and the arrival of post industrialization, the technology of the country will advance constantly, the industrial structure will change dramatically, the consumption mode will also change, and the energy consumption will tend to be steady.

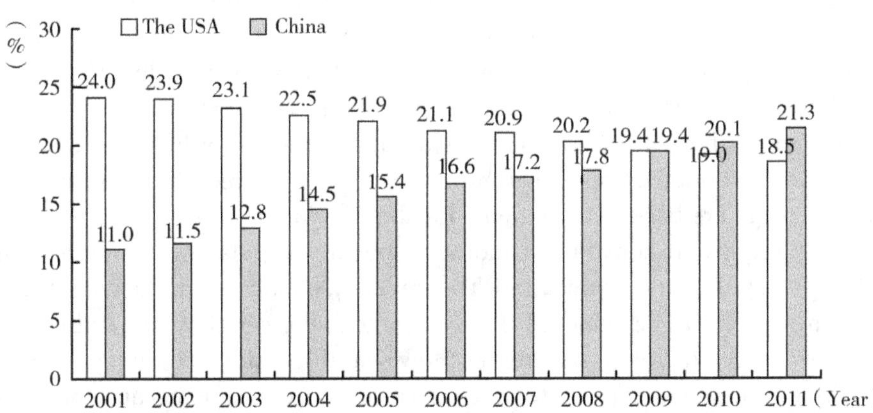

FIGURE 1.4 *Change and comparison of the proportion of primary energy consumption in the US and China in worldwide energy consumption (from 2001 to 2011).*

DATA SOURCE: BASED ON DATA FROM THE *BP STATISTICAL REVIEW OF WORLD ENERGY* (JUNE 2012), HTTP://WWW.BP.COM/LIVEASSETS/BP_INTERNET/CHINA/BPCHINA_CHINESE/STAGING/LOCAL_ASSETS/DOWNLOADS_PDFS/CHINESE_BP_STATSREVIEW2012.PDF.

3 Scientific Understanding of the Policy Implications of Low-carbon Urbanization

Urbanization is an inevitable trend of social progress, as well as a necessary need for people to improve their living quality. This means that urbanization is also the impetus of economic growth: expanding the domestic demand and fostering new growth points. Now, we do need to consider completing urbanization in the mode of low energy consumption, low emissions and low-carbon, rather than increasing urbanization.

3.1 *Low-carbon Requirement is the Final Rigid Constraint on Urbanization*

The fundamental objective of the development of urbanization is to address bread-and-butter issues. The city makes people's lives better, which has been fully manifested. Since urbanization is people-based, it is necessary to consider the number of people that local resources can sustain for survival, instead of blindly pursuing urban scale and economic scale. It is improper to focus only on urban construction.

All forms of development are ultimately bound by natural laws; anything that tries to go against them will only end in a costly failure. The development of urbanization obeys natural laws and the principles of ecological balance, and corresponds with the local environmental capacity, the climatic capacity and the resource carrying capacity, which are not only the requirements of the concept of scientific development but also the the evaluation criteria of low-carbon urbanization. Environmental protection, recycling, low-carbon and ecology are basic requirements for the development of urbanization, as well as important constraint conditions and developmental boundaries for the development of urbanization. Theoretically, circulation and environmental protection can be achieved through cyclic and green development only if investments and energy are infinite. However, the problem is that green and cyclic development is premised on low-carbon investments because of finite energy and investments. Some cities located at the upper reaches of the Yellow River and in short supply of water resources draw water from the Yellow River excessively for the construction of ecological cities, and some cities address plateau greening issues through high-pressure pumping, and even some southern cities build artificial lakes on the top of the mountains through pumping. All of these are achieved and maintained at the cost of a great deal of energy consumption, which goes against the laws of nature, and of course, low-carbon practices. Characterized by a large scale, rapid speed and long period of time,

Chinese urbanization faces a number of challenges such as the shortages of land, energy, water resource and environmental deterioration, which are finally subject to the rigid constraint of a carbon budget.

Chinese energy reserves can be described as "abundant in coal, lean in gas and deficient in oil," which determines that the energy consumption structure is dominated by coal, and that is the case. Besides, fossil energy is unable to support the current mode of urbanization development featuring high levels of carbon and high emissions because of its finiteness. Natural gas and oil are cleaner and more efficient fossil energy (compared to coal). In 2012, the Chinese reserve-production ratio of natural gas and oil was 28.9 years and 11.4 years, respectively; even for coal, with a relatively rich reserve, its production life is only about 31 years (see Table 1.2); this estimate is based on the reserve-production ratio of proven reserves and the current mining rate. China will continue to experience rapid industrialization for a long period of time, and energy security and carbon emission issues arising from energy consumption will become the bottleneck that will limit the advancement of Chinese urbanization.

The finitude of fossil fuel means that the operation of artificial ecosystems and the sustained development of human society cannot rely on fossil energy

TABLE 1.2 *Proven reserves and the reserve-production ratio of fossil energy in 2012*

Region	Oil		Natural gas		Coal	
	Proven reserves (100 million tons)	**Reserve-production ratio (year)**	**Proven reserves (trillion m³)**	**Reserve-production ratio (year)**	**Proven reserves (100 million tons)**	**Reserve-production ratio (year)**
The World	2358	52.9	187.3	55.7	8609.38	109
China	240	11.4	3.1	28.9	1145	31
India	8	17.5	1.3	33.1	606	100
The USA	42	10.7	8.5	12.5	2372.95	257
The EU-27	9	12.1	1.7	11.7	561.48	97

NOTES: DATA OF PROVEN RESERVES COME FROM OFFICIALLY PUBLISHED MATERIALS, AND MAY DIFFER FROM THE ACTUAL DATA BECAUSE OF DELAY IN UPDATING.

DATA SOURCE: BASED ON DATA FROM BP STATISTICAL REVIEW OF WORLD ENERGY, JUNE 2013 (HTTP://BP.COM/STATISTICAL REVIEW).

forever, and it also imposes ultimate constraint on the development of low-carbon urbanization.[17] Consumption of fossil fuel is the main source of carbon emissions. Placing a cap on total carbon emissions to protect the global environment establishes an objective and rigid constraints on low-carbon urbanization.

3.2 *High-carbon Lock-in Rather than Speed is the Problem*

Advancing towards new-type urbanization is the common rule followed by all countries in realizing modernization. The process of developed countries to achieve mature economies is the process of a gradual increase in the urbanization rate. According to the *World Urbanization Prospects*[18] issued by the United Nations in April 2012, the worldwide urban population will increase from 3.53 billion to 6.25 billion from 2011 to 2050, and the urbanization rate from 52.08% to 67.13%. For the developed regions, this ratio will increase to 86.26%, and for less developed regions, the ratio will be 64.08%.

In fact, all developed countries have undergon rapid urbanization during the course of modernization, and this stage is also the stage of rapid advancement of industrialization. Germany and Japan have grasped the historical opportunity of the stage of modern development to rapidly advance in the process of urbanization, and finally complete their task of urbanization. For instance, the Japanese rate of urbanization surged to 56.3% in 1955 from 37.5% in 1950, with an increase of 18.8% in five years, at an annual growth rate of 3.76%.[19]

Compared with the past rapid urbanization rate of France, Germany and Japan, China's urbanization rate is neither unique nor the highest, and cannot be simply considered as irrational. To move forward from an economic giant to an economic power, China must abide by the internal requirements of the rules of urbanization development in order to strive to advance towards new-type urbanization. The rationality of urbanization is actually measured by the quality and soundness of urbanization rather than the rate of urbanization.

Studies have shown that during the rapid development of urbanization, energy consumption and carbon emissions mainly derive from the construction of infrastructures. It can be said that such a process is unavoidable regardless of whether the rate of urbanization is fast or slow. What really matters is how to

---

17 Pan Jiahua, *Carbon Budget Management of the New-type Urbanization Road, Economic Study*, 2013(3).

18 United Nations, Department of Economic Social Affairs, "Population Division World Urbanization Prospects: The 2011 Revision," 2012.

19 Zhang Zhanbin, *Strategic Significance and Reform Difficulty of New-type Urbanization, Journal of the Chinese Academy of Governance*, 2013(1).

avoid the lock-in effect caused by carbon emissions. To prevent high-carbon lock-in, it is necessary to formulate a reasonable plan for the development of urbanization, which needs to integrate ideas and principles of ecological civilization into the whole process of urbanization, with carbon budget and constraints taken into account.

3.3 *Urban Disease may be Avoided in Case of a Proper Urban Scale*

The development and the rate of development of a city are closely associated with urban scale. Statistical analysis indicates that, in terms of population scale, the returns to scale will be negative if the number of inhabitants in a city is between 10,000 and 100,000. That is because, although a small urban scale leads to a small public burden, social services required by urban residents will be in short supply, and coordination among urban residents and the resulting efficiency, will be very low. Certainly, if the number of urban inhabitants is over 200,000, the returns to scale will turn from negative to positive. However, before the urban population reaches 500,000, the external costs for urban development will remain high, and the returns to scale will be lower than those of the cities with a population between one million and two million. If the urban population is up to and over two million, the external costs for urban development will be further increased, and the net scale yield of urban development will reach its highest point. With a further expansion of the urban population, the net scale yield will drop sharply. If the urban population exceeds 10 million, the net scale yield may be negative. Thus, neither a small nor a large urban scale is good, which is the rule[20] that must be obeyed in the process of promoting urban development and the process of urbanization.

Since 1990, in the process of China's urbanization, the population has gravitated towards megacities such as Beijing, Shanghai and Guangzhou, and the urbanization road that China has followed is a road on which megalopolises and megacities take the lead in development. Practice has proved that it is neither a scientific nor a low-carbon road towards urbanization. It has been demonstrated that the population, land area and real estate prices of megalopolises and megacities such as Beijing, Shanghai and Guangzhou have an irrational rise, which takes a large sum of financial capitals, occupies a great number of resources used for small urban development, aggravates the unequal development between rural areas and urban areas, among eastern areas, central areas and western areas, and among large, medium and small cities; this brings about traffic jams, the deterioration of the atmospheric environment and excessive energy consumption in urban traffic and in construction.

---

20 Liu Yingqiu, *Promoting Urbanization Actively and Steadily, People's Daily*, January 19, 2011.

Apart from Beijing, Shanghai and Guangzhou, other large and medium cities all pursue a large scale and high-rise buildings. For example, the Wuhan China Minsheng Banking Corp. Building is 336m tall, the Shenzhen Diwang Building is 383.95m tall, the Nanjing Greenland Square Zifeng Tower is 450m tall, and the Shanghai World Financial Center is 632m tall. On July 20, 2013, the commencement ceremony of the 838m-tall "Sky City Tower," 10m higher than Burj Dubai, currently the tallest building in the world, was opened in Changsha. Such a scale is sure to result in a huge waste of energy consumption of water supply, electricity supply and transportation. Thus, it can be seen that unscientific and irrational urbanization will make it impossible to achieve low-carbon urbanization.

An effective approach to avoiding "urban disease" is to develop urban agglomeration with a proper urban scale of large, medium and small cities, scientific division of labor, and perfect function and appropriate facilities, which not only helps relieve the pressure of large-city central urban areas, but it also promotes the development of medium and small cities, thus impelling a fast and better development of urbanization. Therefore, now and for a fairly long time to come, China should implement the suggestions made in the "12th Five-Year Plan" and the spirit of the Central Economic Working Conference, adhere to the principles of "overall planning, reasonable layout, improved functions and driving the development of small cities by large cities" and obey the objective rules of urban development to gradually form urban agglomerations with significant radiant effect, depending on large cities and focusing on small-medium cities, thus promoting a balanced development among large, medium and small cities and small towns.

3.4 *Optimization and Upgrading of Industrial Structures*

Urbanization is a natural historical process, and a result of marketization, industrialization and agricultural modernization. Although it can also promote marketization, industrialization and agricultural modernization, its effect is limited. Urbanization facilitates the expansion of the domestic demand and fosters new growth points, which will be achieved through industrial development, increase in employment and a rise in incomes.

Industry is the basis of urban development, and all forms of competitive industries are the vital support of urbanization. Low-carbon urbanization surely needs to be based on industrial support in order to achieve low-carbon transformation. The ideal urbanization mode is based on industrial development and premised on the increase of the employment, and coordinated with agricultural development. Optimization and upgrading of industrial structures include low-carbon transformation of traditional industries and the fostering of emerging low-carbon industries. The former can be achieved through the application of clean energy, the introduction of low-carbon technologies, the promotion of clean production processes and the recycling and reuse of waste; the latter can be achieved through the acceleration of the upgrading of the industrial structure and the expansion of the proportions of the tertiary industry, of modern service industry and of strategic emerging industries in industrial structure, etc.

In the low-carbon transformation of traditional industries, the emphasis should be placed on low-carbon and the ecological reconstruction of secondary industries, because the production value of the three industries was 10%, 46.6% and 43.4% respectively in 2011, showing that the industrial structure is at the stage of "secondary industry, tertiary industry and primary industry." China's current developmental stage and the features of the industrial structure in China determine that regarding the industrial developmental mode, it is a must to transform the traditional mode, follow new industrialization roads, apply such measures as clean energy, low-carbon technologies, advanced production processes, and recycling and reuse of waste to achieve energy conservation and emission reduction. As the tertiary industry has the potential to serve as a future urbanization motive force, we should pay attention to the development of finance, R&D, information, training and other modern service industries, to improve the level of compound industries such as the logistics industry and the tourism industry, increase the level of informatization of the logistics industry, and reasonably deploy and organize sources of goods and routes; we should impel the upgrading of the tourism industry, and reduce the burden and impact of the development of tourism in the ecological environment. For the primary industry, it is necessary to develop modern agriculture against the background of a "synchronized development of industrialization, informatization, urbanization and agricultural modernization" and focus on the control of energy consumption and on the prevention of non-point source pollution, to highlight the three characteristics of modern agriculture, i.e., organic, ecologically-sound and low-carbon.

First of all, emerging low-carbon industries are vigorously developed to achieve low-carbon transformation of industrial increment. To realize low-carbon urban development, it is a must to control high energy consumption and the excessive expansion of high emission industries, and strive to develop emerging low-carbon industries, especially strategic emerging industries such as energy conservation and environmental protection, new generation information technology, biology, high-end equipment manufacturing, new energy, new materials and new energy vehicles, thus radically reducing the amount of carbon dioxide emission.

Second, high-carbon traditional industries are transformed and upgraded to realize low-carbonization of industrial stock. Currently, traditional industries characterized by "high consumption, high emission, high pollution and low efficiency" are still dominant. Thus, transforming and upgrading the traditional industries play a critical role in the realization of low-carbon urban development. The transformation and upgrading of existing industries make up an important alternative approach to optimizing the urban industrial structure and promoting a low-carbon economic development.

Finally, the development of the modern service industry is accelerated so as to promote low-carbonization in the three-industry structure. Studies show that the average carbon emission intensity of industries in China is five times that of the service industry (Liu Xinyu, 2010). Thus, the effective approach to realizing the upgrading of the industrial structure and promoting low-carbon economic development is to energetically develop the tertiary industry, advance the upgrading of the urban industrial structure, increase the proportion of the service industry in the urban economy, reduce the proportion of industries and reduce the excessive reliance of urban development on the secondary industry that is the source of high carbon emission.

4 Basic Principles and Strategic Goals of the Promotion of Low-carbon Urbanization

Urbanization is the only way to achieve the development of modernization; it is also the important carrier and support of new industrialization, informatization and agricultural modernization. The 18th National Congress has specifically put forward the advancement of the strategic task of new-type urbanization, and defined the overall requirements on a synchronous development of industrialization, informatization, urbanization and agricultural modernization, which is the significant strategy proposed in order to conform to the transformation of economic and social development in the new historical period. In essence, new-type urbanization will pay more attention to the quality of urbanization and to the idea of "people first, industry-city interaction, layout reasonability, overall rural-urban development, environmental protection and low-carbon," thus low-carbon urbanization is the internal requirement of new-type urbanization.

4.1 *Basic Principles of Low-carbon Urbanization*

The development of urbanization involves a wide range of sectors, including the economy, politics, society, culture and nature, so it is a complicated, comprehensive and systematic engineering process. To realize scientific and sustained urban development, following basic principles should be observed in the course of the development of low-carbon urbanization.

First, a scientific plan for the development of low-carbon urbanization should be formulated. The scientificity of the plan determines the quality of the development of urbanization. As China has a vast area, the level of the development of urbanization varies with regions due to the influence of natural conditions, resources, climate and economic development. Thus, the low-carbon development plan should be developed based on local conditions.

Second, the developmental idea of "industry-based and energy conservation first" should always be observed. Industry is the important support for urban and regional development. New-type urbanization should rely on the comprehensive upgrading of urban industry and agriculture to further speed up and advance the process of urbanization. Similarly, the realization of low-carbon urbanization should also be based on the optimization of the industrial structure and energy structure, and strictly stick to the developmental idea of "energy conservation first" on the premise of an increase in the rate of energy utilization.

Third, institutional innovation should be encouraged to advance the construction of low-carbon urbanization. Internal impetus for low-carbon development and green transformation is stimulated through the innovation of urban administration and governance mechanisms. Through deepening reform and management of the key fields and links including the utilization of resources, public administration, investments and financing, and goal orientation, the intellectual management mode is promoted, and through the construction of new smart cities, an efficient institutional environment, which is conducive to low-carbon urbanization, is created.

Lastly, low-carbon value should be popularized to advocate a low-carbon lifestyle. Further intensification of the urbanization process, an increase in the income level of residents and a change in the consumption mode will bring more pressure to low-carbon development and transformation. At present, carbon emission mainly comes from high energy-consuming industries. But it can be predicted that the emission from traffic, building and consumption of urban residents will increase further. Therefore, to realize low-carbon urbanization, it is a must to launch a universal educational process for ordinary residents and the public and to develop a new low-carbon value in these people, to strive to attain the goal of emission reduction in the field of traffic, architecture and consumption closely associated with people's lives.

4.2 *Strategic Objectives of Low-carbon Urbanization*

4.2.1 The Overall Strategic Objective of Chinese Low-carbon Urbanization

The strategic objective of China's low-carbon urbanization must involve the schedule, path and approach of the overall realization of low-carbon development in the future. The development objective of low-carbon urbanization should focus on two core requirements, that is, low-carbon and urban development. From the perspective of time, it is necessary to set up short-term, medium-term and long-term goals, as well as regional goals that correspond with the national strategic objective and reflect the regional developmental differences.

From a national perspective, the overall strategic objectives of low-carbon urbanization must reflect two internal core requirements, i.e., low-carbon and urbanization, and in addition, the essential requirement of low-carbon urbanization must reflect urban habitability and sustainability. Therefore, in the establishment and attainment of the goal of low-carbon urbanization, the human developmental needs of urban residents must be taken into account. In the next short and long term, urban regions should offer urban residents a guarantee of basic needs, and pay attention to human development and needs while utilizing high efficient of energy. Based on the developmental process of urbanization of other countries, countries with a high level of development generally achieve an urbanization rate of over 80%. Thus, in the long term, the strategic objective of the development of urbanization in China should guarantee that 80% of the population will live in urban areas by 2050, and an absolute reduction of carbon emissions relative to the peak should be accomplished. According to such a macroscopic strategic objective, a strategic objective system for low-carbon urbanization involving six core factors (urban population, the level of human development, industrial structure, energy consumption, carbon dioxide emission and low-carbon economy) should be built. To be specific, the level of Chinese urbanization should maintain the trend of a stable increase in the future, and reach about 60% by 2020, about 70% by 2030, and 80% by 2050 (see Table 1.3).

Besides, low-carbon urbanization must make the needs of the human development of urban residents as its focus, and should consider the human development index (HDI) proposed by the United Nations Development Program (UNDP) as a measurement index, because this index and other

TABLE 1.3 *Overall strategic objectives of Chinese low-carbon urbanization*

Index	2020	2030	2050
Urbanization rate (%)	60	70	80
Human development index	0.75	0.80	0.85
Industrial structure	The secondary industry 47% The tertiary industry 45%	The secondary industry 41% The tertiary industry 54%	The secondary industry 35% The tertiary industry 60%
Energy consumption (100 million tons of standard coal)	48	60	75
Carbon dioxide emission (100 million tons)	120	115	100
Per capita carbon dioxide emission (ton/person)	9	8	7

sub-indexes related to resident's longevity, the level of knowledge and the level of wealth could make up a simple quantitative index, which reflects the impact of development on people's lives and choices that the development offers to every person. By quartering, UNDP classifies 186 countries and regions into four groups with respect to the level of human development, i.e., extreme, high, medium and low. At present, the HDI of countries at the forefront of the list of countries at a medium and a high level of development is approximately 0.7 and 0.8, respectively. Based on the HDI method updated by UNDP in 2010, the calculated Chinese HDI is 0.693, ranking 101st in the world,[21] and towards

21 Data come from the *Chinese Human Development Report 2013* jointly published by UNDP China Office and the Institute for Urban and Environmental Studies of the Chinese Academy of Social Sciences, which is slightly different from the 0.682 as set out in the *Global Human Development Report* in 2010, resulting in a slight difference in ranking. The main cause for this is the difference in the source of original data calculated and the method adopted. Data in the *Chinese Human Development Report 2013* come from the Chinese statistical yearbook and population census, while the data in the *Global*

the end of the list of countries at a medium level of development. Besides, in the future, the urbanization process should also pay attention to the basic developmental needs of urban residents. China is expected to be in the middle of the list of countries with a high level of human development around 2020, and be towards the top of the list of countries with a high level of human development or be in the list of countries at an extremely high level of development by 2030, and be at the middle of the list of countries with an extremely high level of human development by 2050. Based on a phased strategic goal of HDI, Chinese HDI is expected to reach 0.75 by 2020, 0.80 by 2030 and 0.85 by 2050, equal to the level of a country currently ranked around 25th in the world (see Table 1.3).

It is certain that low-carbon urbanization will promote optimization and upgrading of the industrial structure. In the coming ten years, China will still be in the process of rapid promotion of industrialization and industrial modernization. Emerging strategic industries will become the important force that stimulates industrial development. "Clean production," "energy conservation and emission reduction" and "elaborate manufacturing" will be the important features of Chinese industrial development,[22] and the proportion of tertiary industry will maintain the trend of a stable increase. According to to the prediction by the Institute of Industrial Economics of CASS, to better realize low-carbon urbanization, the proportion of secondary industry is expected to reach a peak and then decline, and the proportion of tertiary industry is expected to be constantly on the increase and reach around 60% in 2050.

With the sustained development of the economy and the intensification of the process of urbanization, Chinese primary energy consumption will keep rising in the future, but its growth rate will tend to be slow under the overall strategic objective of low-carbon urbanization. There are six typical foreign and domestic institutions that conduct studies on the future Chinese energy consumption and carbon emission scenario.[23] In the baseline scenario, the

Human Development Report come from the data provided by international organizations such as the World Bank.

22 Li Ping, Jiang Feitao, Wang Hongwei and Gong Shuxin, *Chinese Social and Economic Predication in 2030—Chinese Industrial Economy Development Prospects, Macroeconomic Study*, 2011(6).

23 Li Huimin and Qi Ye, *Comparison of Chinese Carbon Emission Scenarios in 2050, Advances in Climate Change Research*, 2011(4). Research institutions include the International Energy Agency (IEA), the Lawrence Berkeley National Laboratory (LBNL), McKinsey & Company, the United Nations Development Program (UNDP) China Office, the Energy Research Institute the National Development and Reform Commission, and the Tyndall Centre for Climate Change Research, etc.

total primary energy consumption in China is estimated to be 5.4–7.4 billion toc in 2050, while in the low-carbon scenario, the total energy consumption is very likely to be 4.6–5.7 billion toc. With such a rigid macro strategic goal, the total energy consumption in China is expected not to exceed 7.5 billion toc. Moreover, with the further adjustment of the energy structure, more and more clean and non-fossil energy will be put into use. So, carbon emission is expected to reach its peak around 2025, and afterwards, an absolute emission reduction is expected to be achieved. Under different scenarios created by different institutions, there is a great difference in the prediction of total carbon emission, that carbon emission is estimated to be 4.3–16.2 billion toc by 2050. To set a practical strategic goal, a middle predication result is used, that is, carbon emission in China is expected not to exceed 10 billion tocs by 2050. In combination with the population prediction, Chinese per capita carbon emission is expected not to exceed 9 toc, and then gradually decline to about 7 toc by 2050.

4.2.2 Regional Decomposition of the Strategic Objectives of Low-carbon Urbanization

The vast Chinese territory, the uneven distribution of natural resources and human resources, and the different socioeconomic developmental histories in the different regions, all contribute to the great differences in the level of economic development and the mode of social development in different regions, further resulting in regional differences in economic development, the level of urbanization, energy consumption and carbon emission. The realization of carbon emission reduction and the strategic objective of low-carbon development relies on efforts in low-carbon transformation, adjustments in the industrial structure and progress in technology in provincial levels. Based on a correlation analysis, three core factors, that is, carbon productivity, HDI and per capita CO_2 emission, are regarded as core indexes for determining the provincial level of low-carbon development.

Through clustering analysis, the level of low-carbon and of urbanization development in different provinces may be classified into four groups (see Table 1.4) based on carbon productivity, HDI and per capita CO_2 emission in 2008. The first group concerns the developed low-carbon development areas, specifically including Beijing, Shanghai, Tianjin, Jiangsu, Zhejiang and Guangdong, mostly in the eastern coastal area. Such provinces and cities enjoyed early rapid development by relying on regional advantages, and their current carbon productivity, HDI and per capita CO_2 emission all remain high. For provinces and cities in the first group, carbon emission will drop in the future, and in addition, most of the cities in the first group have been national

TABLE 1.4 *Low-carbon urbanization development goal of various Chinese regions*

Region	2015	2020	2030	2040	2050
First group (Beijing, Shanghai, Tianjin, Jiangsu, Zhejiang and Guangdong)	To attain the goal of the 12th Five-Year Plan and reasonably control the total energy consumption	To attain the goal of total quantity control of carbon emission; carbon emission reaches its peak, and the per capita CO_2 emissions are not more than 10 toc	Decreased by 10% from the carbon emission peak	Decreased by 20% from the carbon emission peak	Decreased by 30% from the carbon emission peak
Second group (Shandong, Hebei, Shanxi, Inner Mongolia, Liaoning, Jilin, Heilongjiang, Qinghai and Ningxia)	To attain the goal of the 12th Five-Year Plan	To attain the goal of a 40%–45% decline in carbon intensity and reasonably control total energy consumption	To attain the goal of total quantity control of carbon emission; the per capita CO_2 emissions are not more than 9 toc	Decreased by 8% from the carbon emission peak	Decreased by 15% from the carbon emission peak
Third group (Fujian, Henan, Shanxi, Chongqing, Yunnan, Sichuan, Hunan, Hubei, Jiangxi, Anhui)	To attain the goal of the 12th Five-Year Plan	To attain the goal of a 40%–45% decline in carbon intensity	To attain the goal of total quantity control of carbon emission; the per capita CO_2 emissions are not more than 8 toc	Decreased by 10% from the carbon emission peak	Decreased by 20% from the carbon emission peak
Fourth group (Guangxi, Hainan, Guizhou, Gansu, Xinjiang)	To attain the goal of the 12th Five-Year Plan	To attain the goal of a 40%–45% decline in carbon intensity	To attain the goal of total quantity control of carbon emission; the per capita CO_2 emissions are not more than 7 toc	Decreased by 5% from the carbon emission peak	Decreased by 10% from the carbon emission peak

demonstration areas of low-carbon development. Thus, the formulation of the low-carbon development goal and performance evaluation of such development areas should not impose high demands on the improvement of energy efficiency and tertiary industry proportions, should focus on energy structure and technology-intensive industry proportions, and analyze their reasonableness, so as to optimize the regional energy utilization structure and stimulate the development of regional low-carbon industries.

The second group includes Shandong, Hebei, Shanxi, Inner Mongolia, Liaoning, Jilin, Heilongjiang, Qinghai and Ningxia. Such provinces and cities have low-carbon productivity, high per capita CO_2 emissions and different HDIs, because they mainly regard energy industries and heavy industries such as the major pillar industries. With a further rise in the level of urbanization, continuous increase in the level of economic development and adjustment of the industrial structure, carbon productivity in such regions will surely improve further, and the potential for low-carbon development will be great. Therefore, in the future, the development of such regions should focus on the optimization of the industrial structure, not only increasing the level of human development and carbon productivity, but also controlling per capita CO2 emission increment.

The third group includes Fujian, Henan, Shanxi, Chongqing, Yun'nan, Sichuan, Hunan, Hụbei, Jiangxi, Anhui, etc. For provinces and cities in the third group, carbon productivity, per capita CO_2 emissions and HDI are all in the middle level. Such provinces and cities have great differences in resources, but are all at the stage of a rapid socioeconomic development, which is typically manifested in hastily getting high energy-consuming industries under way and the increasing proportion of these high energy-consuming industries in the regional industrial system. Hence, to impel the development of low-carbon urbanization of such regions, the first thing to do is to actively transform the developmental mode while guaranteeing economic growth, so as to form a low-carbon industrial system with regional characteristics and explore a low-carbon path suitable for local development.

The fourth group includes Guangxi, Hainan, Guizhou, Gansu, and Xinjiang. As such regions are rich in low-carbon resources, the carbon productivity is high but per capita CO_2 emissions are relatively the lowest, and the level of human development is not high due to the low level of economic and urbanization development. In the future, such regions are the key regions that have the most potential to achieve low-carbon urbanization. Energetically developing low energy-consuming and low-emission industries can prevent such regions from repeating the developmental path of high-carbon lock-in.

5 Practice Path of Low-carbon Urbanization

In order to actively and steadily advance urbanization and carry out low-carbon urban construction, it is necessary to work out a scientific low-carbon urbanization developmental plan that matches the comprehensive urban plans and special plans, so as to improve the quality of urbanization, make a reasonable layout and promote a balanced development among large, medium and small cities. As low-carbon industries and technologies are the commanding point in future urban competition, it is necessary to enhance the supporting role of low-carbon industries, increase the low-carbon competitiveness of cities, give full play to the role of low-carbon technology as the "No.1 productive force," strengthen the capacity for urban innovation; focus on energy conservation and environmental protection, create a clean and livable urban low-carbon life, actively conform to people's wishes to live a better life and appeals to enjoy a good ecological environment; build low-carbon demonstration areas, intensify adaptive management and carbon budget management, construct low-carbon resilient cities, cooperate in response to climate change, thus attaining the goal of creating a "Beautiful China."

5.1 *Planning First—Improvement of the Quality of Urbanization*

To actively and steadily advance urbanization and improve its quality, it is a must to adhere to the idea of planning first and following a low-carbon path to guide the urban development. As a whole, either the international need for reducing greenhouse gas emissions in response to climate change or the Chinese need to control energy consumption and reduce its energy consumption per unit of GDP, it is necessary to observe the principles of respecting the local natural environment, regional geographical conditions, the distinctive historical or cultural atmosphere and facilitate the production and living of local people; On the basis of strict implementation of the national regional developmental strategy and the national development plans for an overall functional area, focusing on the following continuously improved cyclic process: "understand the current conditions of urban carbon emission—study the scenario for medium- and long-term urban carbon emissions in the future—set the goal of carbon emission reduction—work out and implement the urban low-carbon developmental plan—evaluate and monitor the results of the reduction in carbon emissions—lay down the action plan,"[24] we should scientifically formulate the development plan for the development of low-carbon urbaniza-

24 Lei Hongpeng et al., *Chinese Low-carbon Urban Development—Strategy and Approach*, China Environmental Science Press, 2011.

tion that matches the comprehensive urban plan and the special plan, so as to guide urban development along a low-carbon path.

Implementation of low-carbon urbanization needs the planning of a major functional area as a guarantee based on the old plan. With regard to the imbalance of the development order of Chinese regional space and the low efficiency of land use utilization, the 11th Five-Year Plan put forward the idea of planning a major functional area, that is, the national land is classified into four types of functional areas according to the resource and environmental carrying capacity, the existing development density and future developmental potential, i.e., optimized development, critical development, limited development and prohibited development. Based on a functional orientation, the regional policy and performance evaluation are adjusted and improved, and the space development order is normalized to form a reasonable structure of space development. The 12th Five-Year Plan also examined the idea of a major functional regionalization and specifically developed a major functional area strategy based on three classifications, i.e., urbanized area, main area for the production of agricultural products and critical ecological functional area. Thus, it can be seen that the idea of a plan for a major functional area and its implementation will surely lead to the elimination of the blind pursuit of industrialization and urbanization prevalent in various regions regardless of conditions of resources and the real possibility; it will probably cause various regions to select their developmental directions based on assumed responsibilities in a wider range. During the course of the advancement of urbanization, it is important to cluster industries and population appropriately, based on the available resources and the environmental carrying capacity so as to form a differential but ordered pattern of urbanization,[25] instead of only pursuing rate and level.

The planning of a major functional area, as a major functional area strategy, should be considered as the basis and constraint of all kinds of planning, which is not only conducive to forming a differential but ordered pattern of urbanization, but it also provides a basis for implementing the low-carbon urbanization mode. Since the planning of a major functional area selects the regional developmental mode and path based on natural ecological conditions and carrying capacity, it is a strategic thought that radically removes the impact and damage of industrization and urbanization on ecologically sensitive areas, reduces the carbon emission in various regions, and prevents excessive development beyond the regional carrying capacity. Only if the planning

25 Hao Huayong, *Urbanization Transformation of Less Developed Regions, China Opening Journal*, 2013(2).

of the major functional area is put into practice, can the development order of national land be regulated and the coordination of functional localization in various regions be achieved, thus forming a differential and supplementary coordinated pattern of development. On such a basis, with regard to the various types of major functional areas, the implementation approach of carbon emission reduction that adapts to the intensity of development is formulated, and the collaborative effect of various regions in fixing carbon and reducing discharge is fully exerted, thus reaching the objective of overall energy conservation, carbon emission reduction, carbon fixation and carbon sink increase.

5.2 *Reasonable Layout—Balanced Development among Large, Medium and Small Cities*

To follow the low-carbon urbanization road, promote urbanization actively and steadily and achieve balanced development among large, medium and small cities, it is necessary to make a reasonable layout, construct a low-carbon pattern and space for urban development, strive for energy saving and profit increasing through space optimization and intensification, reduce unnecessary emission from traffic and buildings, to serve the urban low-carbon development.

Among the cities, it is necessary to make a reasonable layout, build a low-carbon urban developmental space, implement the regional overall developmental strategy and the major functional area strategy, actively promote the construction of urban agglomerations, comprehensively improve the level of the development of the eastern urban agglomerations and foster new urban agglomerations in the central and western regions with high resource and environmental carrying capacity, thus forming a regional pattern of development[26] characterized by complementary regional economic advantages, clear major functional localizations, efficient utilization of national land and a harmonious relationship between man and nature. Regarding the cities, it is necessary to include the construction requirements of the low-carbon urbanization developmental plan in the construction management system of the local regional urban planning, and include the key factors and indexes reflecting urban low-carbon development in the overall planning index system and regulatory plan index system while improving the coverage rate of the urban regulatory plan;

---

26 Hu Jintao, *Firmly March on the Path of Socialism with Chinese Characteristics and Strive to Complete the Building of a Moderately Prosperous Society in All Aspects—Report to the 18th National Congress of the Communist Party of China*, People's Press, 2012.

give full play to the leading role of low-carbon urban planning in the process of urban carbon emission reduction, build a land utilization mode and space pattern suitable for urban low-carbon construction, optimize the structure of urban morphology and the land use functional division, to achieve the goal of energy-saving and efficiency-increasing through spatial optimization and intensification, thus serving urban low-carbon development.

In terms of overall urban planning, it is necessary to pay attention to the construction of a corridor of public transport, improve the job-housing balance index, guide the job-housing balanced development; increase the proportion of mixed land use blocks, rationally use land resources, and weaken the lock-in effect of the land use mode; focus on the coverage rate of urban green areas, and guarantee an orderly connection among the green ecological systems. We should increase the coupling degree between urban center and public transit hub, improve the comprehensive supporting capacity of urban public facilities and efficiently utilize the surrounding lands under the guide of the transportation junction, and increase the proportion of green going-out of urban residents and the rate of the utilization of traffic clean energy. In terms of regulatory detailed plan, it is necessary to advocate the compact-oriented mixed layout mode and guide the construction of urban land development through the increase in various control indexes such as the utilization rate of rain water, the comprehensive energy efficiency of building, the proportion of green buildings in all constructions in the new areas and the utilization rate of renewable energy in buildings, so as to promote the transformation of the urbanization development model, relieve the urban carbon emission lock-in effect, and make urban development present the new-type urbanization features of "intensivism, intellectualization, environmental protection and low-carbon."

5.3 *Low-carbon Industries—Improvement of Low-carbon Competitiveness*

Low-carbon urbanization needs to be supported by industrial development. It is necessary to built a low-carbon industrial system on the basis of regional comparative advantages, formulate the low-carbon industry development plan, promote the adjustment of the industrial structure and improve low-carbon competitiveness, so as to provide substantial economy support and new economic growth points for urban low-carbon development.

To develop low-carbon industries and strengthen low-carbon competitiveness we need to regard the strategic adjustment of the economic structure as the focus of the transformation of the developmental mode, strictly control the excessive growth of industries stamped with "high energy consumption, high pollution and high greenhouse gas emission," and promote a balanced development between low-carbon industries and high-carbon industries. Besides, it is important to enhance the constraint effect of indexes such as energy conservation, reduction of carbon emission and environmental protection, try to take the mandatory material consumption standard and the level of clean production as a basis for the definition of outdated capacity, and make more efforts to eliminate the outdated capacity. Furthermore, it is necessary to further increase the access threshold of industries stamped with "high energy consumption, high pollution and high greenhouse gas emission," set up the approval mechanisms that connect new projects with energy and resource conservation, the carbon emission reduction of pollutant emission and greenhouse gas, and the elimination of outdated capacity, carry out the system of capacity equivalent or reduction replacement, and promote clean industrial production.

To develop low-carbon industries and strengthen low-carbon competitiveness it is necessary to control the development of industries stamped with "high energy consumption, high pollution and high greenhouse gas emission," and more importantly, to strive to realize the objective of improving the carbon productivity, adjust and optimize the industrial structure and the structure of energy utilization, enhance carbon emission management from the production source, production process and product usage, develop low-carbon industries based on local comparative advantages, actively promote the construction of industrial clusters with the agglomeration of industrial resource elements linked by industrial chains, and improve the industrial concentration. In the meanwhile, it is necessary to take the following actions: advance the deep integration of informatization and industrialization, strive to foster and develop emerging strategic industries such as high-end equipment manufacturing, new generation information technology, energy conservation and environmental protection, and new energy vehicles, energetically develop productive service industries and low-carbon service industries, promote the transformation and upgrading of the industrial structure against the background of climate change, actively establish a modern industrial system characterized by greening and low-carbon emission, implement the response to climate change and advancement of low-carbon development into the process of saving energy, reducing emissions, increasing the rate of energy utilization and the transformation and upgrading of traditional industries, impel the low-carbon transformation of industries, finally achieve win-win results in sustained industrial development and responding to climate change.

5.4 *Low-carbon Technology—Enhancement of Urban Innovation Ability*

Low-carbon technology is a commanding point in future urban competition, and the condition of its development is an important index that reflects future urban integrated competitiveness. In order to promote low-carbon urbanization it is necessary to enhance the urban innovation ability and develop the low-carbon technology.

To develop low-carbon technology it is necessary to work out the developmental plan of low-carbon technology, improve the innovation system of the enterprise-focused low-carbon technology based on "original innovation—integrated innovation—introduction, implementation, adoption and re-innovation," closely track the progress of low-carbon technology,[27] develop more core technologies with proprietary intellectual property rights and products with international brands, improve the manufacturing capacity and level of low-carbon equipment, give full play to the role (No.1 productive force) of low-carbon technology in promoting urban low-carbon development, thus making the "research—development—promotion" of low-carbon technology become an important means for responding to climate change and the transformation and upgrading of industries, and making the city become an important space and basic administrative unit that attempts to establish a low-carbon economic developmental mode, develop and promote zero-carbon or low-carbon technology, reduce greenhouse gas emission and facilitate sustained development.[28]

It is necessary to develop and apply technologies relating to source reduction, recycling and reuse, remanufacturing and zero emission by focusing on industrial parks and industrial agglomeration areas; it is also necessary to drive the innovation of low-carbon technology, improve the level of core and critical technologies, enhance the promotion and application of advanced low-carbon technologies, low-carbon processes, low-carbon equipment and low-carbon materials, and actively use low-carbon technologies to transform and upgrade the traditional industries. Furthermore, it is important to make great efforts to develop a series of significant, fundamental, cutting-edge and universal clean production processes and technologies and green raw materials (products) by focusing on industries stamped with high energy consumption, high emission, heavy pollution and resource-consuming, speed up the establishment of a clean production mode, guide enterprises to

27 WWF, "Booz & Company Analysis. Reinventing the City," March 22, 2010.

28 Lei Hongpeng et al., *Chinese Low-carbon Urban Development—Strategy and Approach*, China Environmental Science Press, 2011.

combine the transformation of low-carbon technologies with the upgrading of products, transform and upgrade the traditional industries by the utilization of new technologies, new processes, new equipment and new materials to attempt to improve the overall level of industrial technology. Moreover, based on different features and risks in different phases of the "research—development—promotion" of low-carbon technology, we should establish corresponding fiscal investment and loan systems and mechanisms, actively give full play to the role of the risk response of private-public partnerships and the guiding and motivating role of prices, finance and taxation policies in the "research—development—promotion" of low-carbon technology.

5.5 *Energy Conservation and Environmental Protection—Creating a Clean and Livable City*

Facing people's wishes to live a better life and appeals to construct a "Beautiful China," to follow the road towards urbanization with Chinese characteristics and to actively and steadily promote urbanization, it is necessary to adhere to the idea of people first, actively advance the optimization of the living environment, focus on energy conservation and environmental protection, and advance urban low-carbon development based on energy conservation and environmental protection, thus creating a clean and livable city.

To attain the goal of optimizing the living environment and creating a clean and livable city, it is a must to implement the double control[29] of the intensity and amount of energy consumption under the urban production and life system with carbon-based fuel as the main energy source, to form forcing mechanisms. By focusing on the increase of energy efficiency, we should promote energy conservation and usage in industry, architecture, traffic transportation, agriculture, villages, commercial and civil fields, and public institutions, implement key energy-saving projects, and achieve emission reduction through energy conservation.

To optimize the living environment and create a clean and livable city, it is necessary to pay attention to environmental protection. Within the cities, based on the construction of urban domestic sewage treatment facilities, the reduction of pollutant emissions by key industries, the prevention of agricultural pollution sources and the control of motor vehicle pollutant emissions, it is necessary to enhance the reduction of the principle pollutants, implement large environmental protection projects, and accelerate the development

29 The State Council, *Notice of the State Council on Printing and Issuing the 12th Five-Year Plan of Energy Development* (*G. F. (2013) No.2*), January 1, 2013, http://www.gov.cn/zwgk/2013-01/23/content_2318554.

of low-carbon architecture and traffic. Among the cities, according to the difference in major environmental functions in the various regions, to reach the objective of keeping water, air and soil environment in good condition, it is necessary to establish a national environmental function division in consideration of the planning of national major functional areas, implement a regional environmental protection strategy, reinforce the environmental governance in the optimized developmental areas and key developmental areas. In the critical joint prevention and joint control area, according to environmental capacity and climate capacity, we should strictly observe pollutant discharge standards, establish a regional environmental (air, water and soil) quality assessment system, develop the construction of multi-pollutant cooperative control engineering, promote the governance of fine particulate pollutants (PM 2.5) in the air, and enhance the environmental risk prevention. While protecting and expanding the ecological space, we should promote the equalization of basic public services of environmental protection.

5.6 *Low-carbon Life—Construction of a Low-carbon Demonstration Community*

In the face of the urgent need for energy conservation, emission reduction, response to climate change and improvement of the living environment under the constraint of resources and energy environment, and in accordance with the demand-oriented economic, social and environmental standards, the promotion of low-carbon urbanization needs to adhere to the idea of people first, construct low-carbon demonstration communities, actively advocate low-carbon life and constantly improve the level of human development.

Community is a basic social organization. We should give full play the guiding role of streets, neighborhood committees and community estates to arouse the enthusiasm of residents to participate. Through the construction of low-carbon demonstration communities featuring experiments, guidance and demonstrations, we should actively carry out the construction of a project and management system relating to low-carbon life, and innovate the low-carbon management methods and mechanisms in communities.

Family is a social cell. By centering on "food, clothing, housing and transportation," it is necessary to formulate a manual on household low-carbon life in terms of household water, electricity, gas, cooking, heating, travel, decoration, energy-saving seismic retrofit, indoor/community greening, waste separating collection, popularize low-carbon life knowledge, and actively promote the formation of "the consumption mode of energy conservation and ecological environmental protection." Besides, subject to demand-oriented economic, social and environmental standards, we can drive the consumption idea and the lifestyle of the community to transform to the low-carbon mode on the basis of families, and promote the transformation of production mode through consumption.

Low-carbon urbanization involves a consumption mode, lifestyle, concept of life, regional developmental rights, etc. To promote the transformation from a "high-carbon consumption mode" to a "low-carbon consumption mode" is an important part of urbanization. To realize the goal of transforming the consumption idea and the lifestyle and raising a low-carbon awareness, on the one hand, we should put emphasis on the performance of environmental responsibility by manufacturers, and guide those manufacturers to strengthen an awareness of environmental protection from the consumer side, and on the other hand, we should reduce unnecessary domestic waste and pollutant discharges to relieve the environmental pressure of a city on surrounding cities and towns.

Low-carbon urbanization is a project regarding livelihood. It helps to coordinate the urban-rural relationship, accelerate the construction of the new socialist countryside, promote the upgrading of rural-urban resident consumption structures and improve the living quality of rural-urban residents and their level of health. After "food, clothing and others" issues have been addressed through industrial economy, it is necessary to efficiently resolve the issues of "housing, transportation and education," which are in short supply, and then further improve "food, clothing and others" issues under the premise of resource conservation and environmental protection. We should promote industrial low-carbon transformation, and speed up the development of a low-carbon urban economy. The problem of a short supply of "housing, transportation and education" can be solved through the expansion of consumption investments. In the meanwhile, the application of an appropriate low-carbon technology should be widely advocated to create a harmonious situation of a good social development and environment, which is an important part of the building of low-carbon urbanization.

5.7 *Adaptive Management—Creation of a Low-carbon Resilient City*

To advance low-carbon urbanization, it is a must to coordinately consider the different needs for reducing greenhouse gas emissions and for dealing with risks and hazards arising from climate change in the urban governance and planning design, transform the conventional urban administrative mode and governance idea, and implement an adaptive management, so as to create a low-carbon resilient city.[30]

30 Zheng Yan et al., *Low-carbon Resilient City: Idea, Approach and Policy Selection, Urban Development Study*, 2013(3).

The adaptive management is a new environmental management idea that emerges against the background of the change in the global environment and climate. It refers to management in light of climate change and the corresponding risks so as to reduce the vulnerability of the ecological system and of the social economy and to strengthen the adaptive capacity. Adaptive management is a practical process of long-term, constant learning and experience summarization. First of all, it is required to evaluate current relevant policies and mechanism designs to improve the objectives, functions and structure of the existing decision mechanism and ecological system management, address problems and policy needs on different temporal and spatial scales, and enhance the building up of capabilities and mechanism guarantees in terms of information, technology and resources.

To conduct adaptive management and create low-carbon resilient cities, it is necessary to set up a collaborative governance mechanism in response to climate change for cities, consider the connection of multiple urban developmental objectives as far as possible, thoroughly consider job creation, social equity promotion, ecological protection and resource sustainability while reducing energy consumption and greenhouse gas emissions, and work out scientific and reasonable planning of urban spaces. While the integrity of the urban ecological environment is protected, unnecessary emission from traffic and architecture should be reduced, and compact urban design and construction should be carried out to relieve the "heat island" effect and promote energy conservation. On the basis of accurate positioning of the construction of resilience city—incremental adaptation or developmental adaptation, low-carbon resilient urban planning should be scientifically formulated. The key objectives and factors reflecting the idea of building low-carbon resilient cities are included in the system of the urban integrated planning index and the system of regulatory detailed plan index, to advance the construction of community demonstration and pilot projects of low-carbon resilient cities, and comprehensively consider the synergistic effect of architecture, traffic, ecological protection and greening, disaster reduction, disaster prevention and community governance.

5.8 *Carbon Budget Management—Coordinated Response to Climate Change*

The city, as the center of human production and life, is the main energy consumer and greenhouse gas source. To follow the road towards urbanization with Chinese characteristics and actively and steadily promote urbanization, it is necessary to respect the objective circumstances of the complex carbon emission relationship and interaction effect mechanism among urban internal and external production activities, consumption activities and relevant facilities; it is also necessary to get a clear understanding of the situation, formulate a plan as soon as possible and make a carbon budget, to prevent the high-carbon lock-in effect[31] caused by urban industrial development, urban infrastructure construction, large machinery equipment manufacturing and durable consumer goods production, eliminate too much reliance of the system of social production and consumption on fossil energy, and avoid the high cost arising from response to climate change in the future. By focusing on the enhancement of carbon budget management to promote urban low-carbon development, various fields, industries and departments should unite to deal with the challenge of climate change and to stimulate sustainable urban development.

To enhance carbon budget management requires adhering to the concept of human development with an aim to improving the level of human development and under the necessary condition of guaranteeing resource or energy environment security and sustainable use. Through the preparation of the list of urban greenhouse gas emissions and an analysis of the carbon emission scenario in the coming medium and long term, a scientific and proper carbon budget level should be determined. Focusing on the "initial distribution of a budget limit—budget adjustment—transfer payment—market mechanism—capital mechanism—measurement, report and verification (MRV) mechanism—compliance mechanism," a whole set of carbon budget schemes[32] that connects with the double control[33] of the intensity and amount of energy consumption should be formulated. The sustainability of the carbon budget schemes should be guaranteed by taking the allowable emission amount of the climate safety capacity as a global aggregate carbon budget and setting it as a rigid constraint. To ensure the fairness of the carbon budget scheme, the limited aggregate carbon budget is initially distributed to every urban resident on a per capita basis, to satisfy their basic needs. Based on historical carbon emissions and future demands, we made carbon budget transfer payment and designed a proper capital mechanism to ensure an efficiency configuration feature for carbon budget schemes.

---

31 Zhuang Guiyang and Xie Qianyi, *International Experience and Development Trend of Low-carbon Economic Transformation*, see *Annual Report on Climate Change Actions (2009): Approach to Copenhagen*, Social Sciences Academic Press (China), 2009.

32 Pan Jiahua and Chen Ying, *Carbon Budget Scheme: a Fair and Sustainable International Climate System Framework, Social Sciences in China*, 2009(5).

33 The State Council, *Notice of the State Council on Printing and Issuing the 12th Five-Year Plan of Energy Development (G. F. (2013) No.2)*, January 1, 2013. http://www.gov.cn/zwgk/2013-01/23/content_2318554.

CHAPTER 2

Global Achievement of 2°C Temperature Rise and China's Emission Peak Value

Jiang Kejun

Abstract

In recent years, a consensus has been reached regarding the establishment of a goal of globally controlling the temperature rise since industrialization within 2°C by the year of 2100. However, how to achieve this goal in the future is an issue requiring answers from research institutions in countries all over the world, and it was also the main topic of the Fifth Assessment Report of the IPCC (Intergovernmental Panel on Climate Change). Several recent research projects have emphatically utilized models to analyze the emission methods to use in order to achieve that goal. We used models to analyze the emission scenario of China's achievement of that goal within the global framework. This chapter gives the main factors of the scenario, and also discusses the feasibility of China's achievement of the emission scenario for that goal. The research shows that, in order to achieve the goal globally, China needs to make its CO_2 emission reach its peak value before 2025, and it is possible for China to achieve such a peak value of emissions. Many factors have provided a foundation for China to achieve that emission peak value in 2025, including the recent adjustment of its economic structure, progress in renewable energy sources and energy-saving technologies, investment potential in low-carbon development and a policy environment where China is moving forward in its low-carbon development. In the meantime, China's international negotiation strategies need to be adjusted accordingly, and China needs to be more proactive so as to lead the world in the reduction of global emissions.

Keywords

climate change – mitigation – model – scenario – China

* Jiang Kejun, researcher in Energy Research Institute of National Development and Reform Commission.

 | DOI 10.1163/9789004280021_003

1 The Goal of Global Climate Change Mitigation

The ultimate goal of the *United Nations Framework Convention on Climate Change* issued in 1992 was to stabilize the concentrations of greenhouse gases in the atmosphere at a level where the climate system will not be damaged. For a long time, scientific research has been trying to propose a concentration at which the climate system is not damaged, committed to establishing a threshold. However, due to the climate system and gradual changes in its effects, it is fairly difficult to propose a gas concentration or temperature threshold. So far, no consistent conclusion has been made.

In 2009, the goal of controlling the rise in temperatures within 2°C was written into the Copenhagen Accord at the G8 Summit in Italy. The proposed goal was mainly a political one, based on a large amount of research, but not based on threshold research. Anyway, a goal is required for policy-making. Much of the research around the world analyzed concentration goals of 450ppm, 550ppm and above (IPCC, 2007). Generally, a concentration level of 450ppm corresponds to a 2°C temperature rise. After a lot of political discussions, the goal of 2°C was finally proposed.

It is notable that, controlling the temperature rise within 2°C is just an average global goal. The temperature rise in some African countries may reach 3.5°C, anticipating widespread damage to Africa. At the Copenhagen Conference, African countries and small island countries made an appeal to control the global temperature rise to within 1.5°C, hoping to mitigate the effects of climate change on their local areas. Although the *Accord* suggested assessing 1.5°C before 2015, the main part of its text focused on 2°C, which was also the root cause of why many African countries and small island countries refused to accept the Copenhagen Accord. The Accord based on a 2°C goal had caused a gap among countries and also among environmental groups supporting different temperature rise indices.

Due to the lack of an emission scenario analysis based on 1.5°C, there is little research support now. The common goal in literature is to stabilize the concentration of CO_2 in the atmosphere. In the case of several greenhouse gases in the research, different concentrations of gas can be measured according to their radiation characteristics by means of the normal concentration of CO_2 or of radiative forcing. Another method is to make stabilizing the average global temperature the goal. The advantage of the radiation intensity index compared with the temperature index is that it is unnecessary to consider climate sensitivity in the calculation, and its disadvantage is that a unified radiative forcing level may correspond to a large range of temperature shock, and a temperature index is more directly related to climate change impacts.

The Cancun Climate Change Conference in 2010 and the Durban Climate Change Conference in 2011 reconfirmed the climate change control goal of 2°C. Cooperation on emission reduction promoted internationally has been constantly advancing, with faster and faster progress.

However, how to globally achieve the 2°C temperature rise goal in the future is an issue requiring answers from research institutions in all the countries around the world, and this goal is also the focal point of the Fifth Assessment Report of the IPCC. Several recent research projects have emphatically utilized models to analyze the emission methods that could be used to achieve this goal globally.

2 The Emission Scenario of Achieving the Goal of a 2°C Temperature Rise Globally

Existing research indicates that[1] there is a strong correlation between a normal concentration of CO_2 (radiative forcing) and the CO_2 concentration in 2100, since CO_2 is the most important influencing factor of radiative forcing. Based on this, to simplify scenario comparison and evaluation, we should stabilize the scenario (not only include researches on several greenhouse gases, but also include researches on CO_2 only).

When setting the climate change mitigation goal in the future, temperature rise control should generally be adopted. The IPCC report mainly discussed the temperature rise goal of 2100, so the emission reduction that is being discussed by people at present is fundamentally determined on the basis of the temperature rise goal given in the IPCC report.

Now, if temperature rise can be controlled within 2°C in 2100, each step is uncertain, including from 2°C temperature rise, to radiative forcing, to concentration, and to emission, which will lead to a wide range for the determination of emission methods. When achieving the same 2°C temperature rise goal in 2100, different model groups will have various emission methods. In order to have a greater possibility of achieving the 2°C temperature rise goal globally, it would be better to achieve the peak value of CO_2 emission before 2015, and then start decreasing. In case the peak value is reached in 2020 or 2025, it will be less likely that the 2°C temperature rise goal will be achieved. According to the global emission reduction goal being reached by 2050 as proposed by the European Union, it is only half possible that the 2°C temperature rise goal can be achieved in 2100. With the same emission reduction goal, European

1 IPCC, Fourth Assessment Report, 2007.

researchers and American researchers proposed their emission methods in the future respectively. The European research result is that global emission reduction peak value is achieved earlier, while the American research result is that the peak value is achieved later but decreases dramatically afterwards and is lower than that of the European research after 2050.

According to the above, there are many ways to reduce emissions in the future. To analyze the global emission reduction goal in the future, a global emission reduction scenario is also required. The recent Fifth Assessment Report of the IPCC analyzed and evaluated recent research and further proposed a global emission reduction scenario. Compared with the Fourth Assessment Report of the IPCC in 2007, the current model study proposed an emission scenario of a 2°C temperature rise goal and meanwhile, and most importantly, the 2°C temperature rise goal was confirmed to be feasible based on the research.

3 Key Factors for Achieving a 2°C Temperature Rise Goal Globally

When mentioning achieving the global temperature rise goal, we need to consider some key factors, mainly including global emission peak value, emission methods in all areas and the time of achieving the peak value. A political problem is burden distribution.

Currently, there are several emission methods that would allow the achievement of the emission reduction goal globally (e.g. 2°C temperature rise goal). The first is to rapidly achieve emission reduction: reach the global emission peak value in the near future and then relatively slow down the emission reduction, so as to achieve the emission reduction goal in the end. The second is to delay the time of globally mitigating and achieving the emission peak value, e.g. after 2020, and then carry out a rapid emission reduction (achieve emission reduction of more than 70% around 2050), and thereafter carry out a slow emission reduction. The third is to delay the time of globally mitigating and achieving the emission peak value, e.g. after 2020, and then carry out global emission reduction, to achieve a negative emission around 2070 (by means of biomass power generation or energy utilization, adopting CCS (carbon capture and storage) technology). In this case, in order to achieve the 2°C temperature rise goal globally, the first emission method is to achieve the emission peak value by 2015, which currently seems to have a slim hope of success. We will mainly use the last two ways to analyze China's emission.

At present, during the preparation of the Fifth Assessment Report of the IPCC, in the scenarios of achieving a 2°C temperature rise goal provided by the model groups, many of them are to achieve a global emission peak value during the five-year period 2020–2025. The sooner the emission peak value is achieved globally, the higher the pressure on China, so we should choose the scenario of achieving the emission peak value later from the global emission methods to analyze China's emission scenario. It can be said that China's emission scenario obtained by analyzing these considerations would be the most favorable scenario to China under the circumstance of achieving the global temperature rise goal.

By using the emission scenario of the IPAC model to analyze the goal, we get the emission path as shown in Figure 2.1.

When setting up the climate change mitigation goal in the future, controlling the future temperature rise is generally adopted. The IPCC report mainly discussed the temperature rise goal of 2100, and the current emission reduction discussed is basically determined according to the temperature rise goal given in the IPCC report.

Here, we use the climate model in the IPAC model of energy research institutes, i.e. the simple climate model MAGGIC of NCAR (National Center for Atmospheric Research, USA), to obtain global emission methods of achieving the future temperature rise goal. Figure 2.1 gives one of the results. The goal for this result is to control the concentration for the year 2100 at 450PPMe. Greenhouse gas emission input uses several greenhouse gas emissions of Global emission model of the IPAC methods.

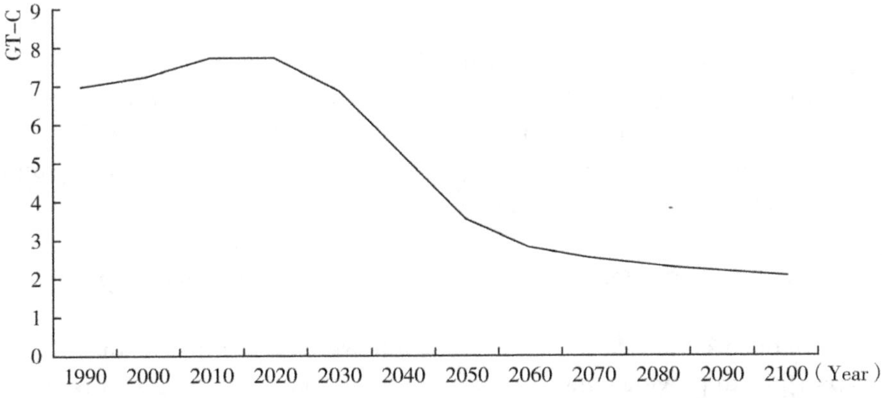

FIGURE 2.1 *CO_2 emission methods for globally achieving the concentration goal of 450PPMe.*

To simulate the future emission space of all countries or regions, we should adopt the form of burden sharing. Some mainstream opinions on burden sharing include the following points.

(1) Per capita emission convergence: This means that the annual per capita CO_2 or greenhouse gas emission converges to a target year to embody human rights. Generally, Year 2020, Year 2070 and Year 2100 are considered as the target years of the per capita emission convergence, with a difference for each research group. Before the final per capita emission convergence year, emission methods for all countries and regions are also important, so they need to be analyzed during the analysis of the per capita emission convergence.
(2) Accumulative emission per capita: It is believed that the accumulative emission per capita in the target years (Year 2050 and Year 2075) should converge, and Year 1850 and Year 1900 could be chosen as the starting years.
(3) Mixed method: With the overall consideration of payment capacity, per capita emission and emission reduction potential, the mixed method has been accepted by most people.

According to global emission reduction scenarios and the World Bank's prediction for the population (see Figure 2.2), we have analyzed the per capita emission convergence.

During the per capita allocation, there are several different main assumptions which lead to different allocations. The assumptions include the following aspects.

(1) Per capita consistent time: Year 2050, Year 2075.
(2) Developed countries reduce emission with the priority in the intermediate time slot, and developing countries deviate from the baseline.
(3) Per capita emission of developing countries may exceed that of some developed countries at some time.

Figure 2.3 gives the results of the burden analysis according to per capita convergence.

Developed countries must carry out intense emission reduction as soon as possible, to provide emission space for developing countries. Our research still assumes that developing countries will make efforts to reduce emission through international cooperation on the premise of sustainable development.

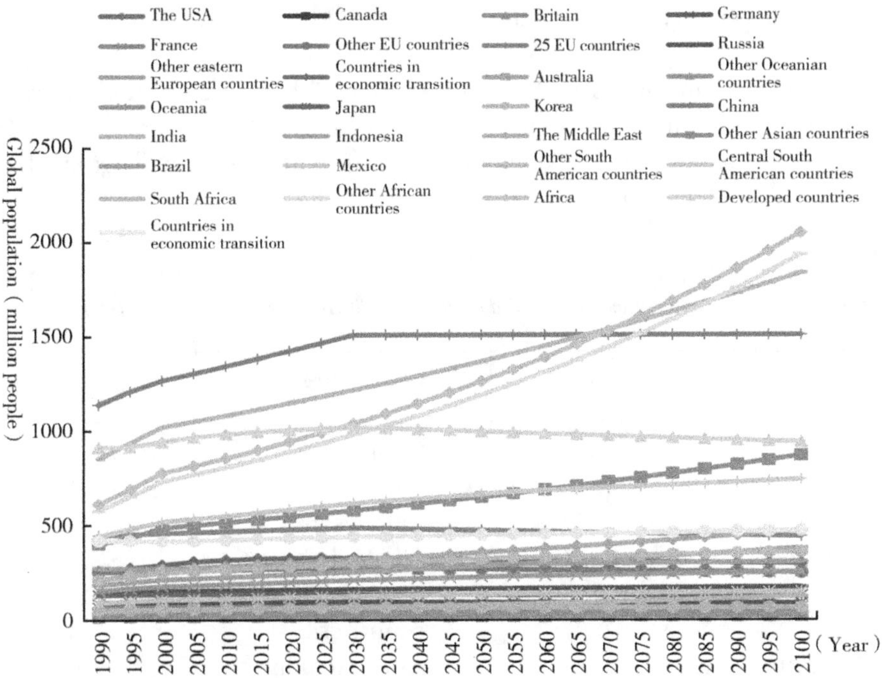

FIGURE 2.2 *Global population scenario of the IPAC model.*

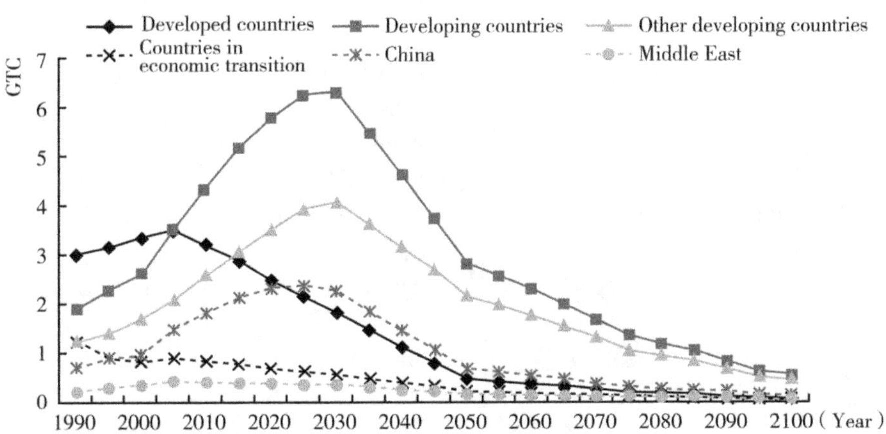

FIGURE 2.3 *Per capita convergence allocation plan of the IPAC model.*

4 Regional Emission Methods

Based on the mechanism of burden sharing, Table 2.1 gives the emission reduction amplitude of all countries and regions in 2020 compared with 1990. Through research on global emission scenarios of the IPAC model group, factors of technological advance are also considered here.

TABLE 2.1 *Emission reduction amplitude of all countries/regions compared with 1990*

Country/Region	2020	Country/Region	2020
The USA	16.63	China	–228.99
Canada	15.48	India	–323.19
Britain	33.72	Indonesia	–353.87
Germany	32.20	The Middle East	–75.25
France	27.80	Other Asian countries	–195.98
Other EU countries	23.87	Asia (excluding the Middle East)	–206.76
25 EU countries	27.74	Brazil	–138.92
Russia	43.75	Mexico	–88.99
Other Eastern European countries	47.63	Other South American countries	–121.91
Countries in economic transition	45.80	Central South American countries	–115.33
Australia	13.57	South Africa	–97.34
Other Oceanian countries	–18.72	Other African countries	–165.47
Oceania	10.48	Africa	–151.89
Japan	18.46		
Korea	–93.94	Global	–50.45

Unit: %

NOTE: A POSITIVE NUMBER REFERS TO EMISSION REDUCTION, WHILE A NEGATIVE NUMBER REFERS TO EMISSION INCREASE.

5 China's CO_2 Emission Peak Value

Based on the research of the IPAC global model, we used the IPAC-AIM model to carry out the analysis of the technical model according to China's ability to achieve the emission peak value by 2025. This research was carried out based on China's low-carbon scenario analysis which had already been carried out.[2]

Seen in Figure 2.3, China's CO_2 emission will reach its peak value around 2025, about 8.56 billion tCO_2e. But the premise is as follows: Developed countries are required to reduce emission by 30%–35% by 2020 compared with 1990, and other developing countries will also carry out strong emission reduction. This premise is possible but also seems to be difficult. That is, in order to achieve a 2°C temperature rise goal globally, China may need to reach the peak value before 2025. Our analysis above mainly considers space for China's emission reduction, but the space may be limited, so the situation is more severe than the IPAC strengthened low-carbon scenario. Based on different assumptions about the GDP, a decreasing amplitude of carbon intensity between 2020 and 2025 is 49%–59%, higher than the action objective issued by the government.

Based on research regarding the IPAC model, it is feasible to reach the CO_2 emission peak value before 2025. For this analysis, we use the IPAC-AIM model to further carry out utilization of renewable energy, CCS, energy conservation, etc. under the existing strengthened low-carbon scenario, in order to analyze the possibility of reaching the CO_2 emission peak value before 2025 (see Figure 2.4). The main factors are as follows.

China's economic structure will experience major changes before 2020. The output of most high energy-consumption industrial products will reach the production peak value before 2020, and then will begin to decline. Currently, the outputs of steel, building materials, nonferrous metals and chemicals have been able to meet the needs of China's rapid economic development for infrastructure and industrial production. As to building areas, airports, roads, railways, dams and others that have increased each year, the output is basically at the peak at present. In case of continuing expansion, China's per capita

2 Jiang Kejun, Hu Xiulian, Zhuang Xing, Liu Qiang, and Zhu Songli, *China's Energy Demand and CO_2 Emission Scenario in 2050, Climate Change Research Progress*, 2008–9, 4 (5), pp. 296–302; or Jiang Kejun, Hu Xiulian, Zhuang Xing, Liu Qiang, and Liu Hong, *China's Energy Demand and CO_2 Emission Scenario in 2050, China's Energy and CO_2 Emission Report in 2050*, Science Press, 2009.

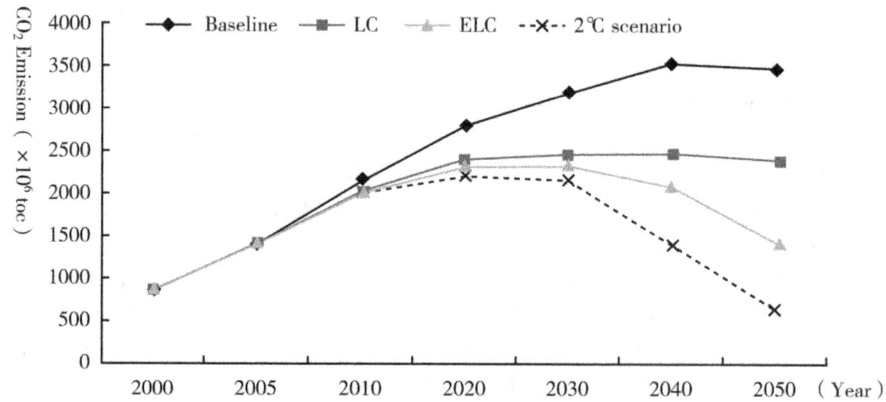

FIGURE 2.4 *China's CO_2 emission scenario of energy activities.*

building area will exceed the average level of Europe and Japan quickly and will approximately reach that of the USA in 2030–2040, but this is unimaginable for China with a huge population and a limited land area. The high energy-consumption industrial departments and related industries have consumed 70% of China's newly added energy in the past 10 years. If these departments stop increasing and consider improving their energy efficiency, their energy consumption may begin dropping in the next few years, so a big driving force of the increase in China's energy consumption will disappear.

Viewed from a policy environment, the development of China's low-carbon economy has had a good social environment and policy orientation. The Chinese government has announced a domestic action objective where the unit GDP CO_2 emission should decrease by 40%–45% by 2020 compared with 2005. China can do better: On the one hand, China needs to continue implementing its current energy-conservation, renewable energy and nuclear power policies in the next two Five-Year Plans, make efforts to promote low-carbon development and advocate low-carbon transportation and lifestyle; on the other hand, China could obtain more external support through such international cooperation as technology, carbon finance and carbon market.

In the past two years, rapid progress in renewable energy technology provides strong support for China to achieve the 2°C temperature rise goal. By 2011, fan cost was reduced to 3,200–3,500 yuan/kW and the installed photovoltaic power generation cost was reduced to 16,000 yuan/kW. In coastal areas, some wind power generation costs can compete with coal-fired power generation. In terms of end users, some photovoltaic power generation costs have also been competitive. China's renewable energy development goal of 2020 has improved continuously, e.g. installed wind power capacity of 3×10^7 kW by 2020 was planned for in 2006, 8×10^7 kW in 2008 and 150 million kW in 2010, and now an installed capacity of more than 200 million kW is being discussed. According to the pattern for the development of wind power generation at present, in spite of some difficulties in access to power grids for wind power, based on the conclusion in related research by the Chinese Academy of Engineering, as the power grid construction and the improved wind power access plan in the future, these difficulties will be solved quickly. Therefore, by 2020, an installed capacity of 250–300 million kW may be possible, as may be photovoltaic power generation. Recently, the photovoltaic power generation goal of 2020 has been raised to 2×10^7 kW, and relevant experts are discussing the goal of 5×10^7 kW, which may be further exceeded with good efforts.

So, in the 2°C scenario, the goal of utilizing renewable energy in 2050 has been raised for the future. The model considers the learning curve effect of the renewable energy and the declined cost of power generation technology of fossil fuels. Before 2020, with the continuously-increasing cost of fossil energy and the internalization of environmental externality, renewable energy has cost competitiveness compared to fossil energy power generation comprehensively. In this model, by 2050, the installed capacity of wind power generation will reach 860 million kW, and photovoltaic power generation 1.04 billion kW. In the meantime, the model further increases the application of solar hot water and heating, and also considers lots of distributed technologies to utilize renewable energy, such as solar air conditioning and photovoltaic power generation for heating.

The model further considers CCS technology. Due to a further increase in non-fossil energy, in the 2°C scenario, the installed capacity of coal-fired power generation respectively reduces from 760 million kW in 2020 and 630 million kW in 2050 in the strengthened low-carbon scenario to 710 million kW and 570 million kW. In the 2°C scenario, 410 million kW coal power will use CCS technology by 2050, when coal-fired power generation will focus on IGCC (integrated gasification combined cycle plant).

Meanwhile, low-carbon development will be comprehensively realized owing to the rapid development of national economic strength. By 2020, 2030 and 2050, China's GDP will have increased from 40.3 trillion RMB in 2010 to 130 trillion RMB, 290 trillion RMB and 520 trillion RMB, respectively (calculated at the current price). Due to a decrease in the cost of technology (the decrease is obvious even calculated at the current price), investments in low-carbon development in the future will be far more than the demand for low-carbon investments in the models proposed by the main model research institutions in

China at present. Adequate capital investments can comprehensively meet the requirements for low-carbon development in terms of low-carbon transportation and construction, achieving lower CO_2 emissions.

Based on the above-mentioned main factors, by using the IPAC-AIM technology model, it can be seen that it is possible for China to achieve the CO_2 emission methods of a 2°C scenario with clear policies and adequate investments, but the premise is a good policy design and enough capital investments.

6 Conclusion

It can be seen through analysis that, in order to achieve the 2°C temperature rise goal in 2100 from 1850, the global CO_2 emission load needs to reach the peak value before 2020. There are many ways of global sharing, but for China, in order to achieve the 2°C temperature rise goal globally, China's CO_2 emission of energy activities needs to reach the peak value by 2025 at the latest, and then begin to decline.

Rapid development of renewable energy technology in the past two years provides strong support for China's realizing the 2°C scenario. By 2050, renewable energy will have experienced large-scale development, and the installed capacity of wind power generation will be close to 900 million kW and that of photovoltaic power generation will exceed 1 billion kW.

CCS technology will be applied on a large scale in China so as to make fossil fuel power generation low-carbonized by 2050. CCS is a long-term technology. Lots of negative emission scenarios appear in research on the 2°C scenario in the recent IPCC report, making CCS application an important choice. China can take the lead in CCS technology. According to scenario analysis, 410 million kW coal power will use CCS technology by 2050, and at that time coal-fired power generation will have been based on IGCC.

In this case, China can make corresponding policies and institutional arrangements and propose a powerful climate change strategy, to guide social and economic development, industrial development and low-carbon technology development. At the same time, China also needs to significantly adjust its international negotiation strategies, in order to enter a more positive leading stage and urge all countries to achieve the global emission reduction goal together with us, which enables China to gain a good reputation in international morality, opens broad space for the exports of China's low-carbon technologies and promotes the formation of China's new economic development model.

CHAPTER 3

Progress of Negotiations on Funding Mechanisms of Climate Change and Suggestions on China's Participation in Constructing Funding Mechanisms

Zhang Wen and Pan Xun

Abstract

Funding mechanisms are the key to supporting developing countries in carrying out actions of mitigation and adaptation to respond to climate change. Since developed countries lack political sincerity regarding the performing obligations under the convention, they have been fighting against the developing countries concerning fund issues for a long period of time. With changes in global, political and economic situations, such problems as an insufficient scale of funds, unreliable fund sources, fund allocations to be improved, and the strong political complexion of funding mechanisms appear in fund issues of climate change. Therefore, improving and planning the funding mechanisms of climate change, fully recognizing and using international and domestic resources, establishing a diversified financing mechanism with multiple channels and guiding a balanced input of funds in the fields of mitigation and adaptation, are of important practical significance for developing countries to respond to climate change and achieve low-carbon development.

Keywords

climate change – funding mechanisms – mitigation – adaptation – financing

* Zhang Wen, a senior engineer at the Foreign Economic Cooperation Office, the Ministry of Environmental Protection, and a doctor in marine environmental science at Hong Kong University of Science and Technology; his research field is funding mechanisms of climate change conventions, funding mechanisms of environmental convention performance and international cooperation on environment; Pan Xun, an engineer and doctor at the Foreign Economic Cooperation Office, the Ministry of Environmental Protection.

 | DOI 10.1163/9789004280021_004

1 Introduction

Climate is the natural environment which humans rely on for their existence and it is an important basic resource for the sustainable development of economic society. Climate issues are seemingly environmental issues; the developmental space, industrial competitiveness, international political leadership and other interests behind the climate issues relate to fundamental interests of economic development for all countries, so climate issues are ultimately developmental issues.[1]

Usually, there are two aspects of measures and actions for people to use to respond to climate change: mitigation and adaptation. Mitigation emphasizes affecting the climate system so as to hinder the progress and the extent of changes; while adaptation aims at reducing the adverse effects caused by the climate system in order to control the risks of climate change. Thus, mitigation focuses on the reasons for climate change, while adaptation focuses on the results of climate change.[2]

Both mitigation and adaptation need a large number of funds, so fund issues become one of the core topics in international climate change negotiations and they become the key to implementing international cooperation arrangements for response to climate change. In recent years, cooperation on and competition regarding the funding mechanisms of climate change between developed countries and developing countries have shown new characteristics and evolutionary trends, so funding mechanisms of climate change will face both opportunities and challenges in the future. Considering that climate change negotiations have become a new stage in the competition of all countries, dealing with the topic of climate change funds scientifically and strengthening the research and prospective judgment on the funding mechanisms of climate change are of important practical significance for China to actively participate in the international cooperation on climate change, safeguard its current interests and future developmental rights, and promote the construction of an ecological civilization.

---

1 Xian Gang, Wu Yaling, and Zhu Liucai, *Economic Analysis on International Camp Changes in Climate Negotiations*, Environmental Economy, 2011, 12–14.

2 Jiang Dongmei and Zhang Mengheng, *National Funding Mechanisms of China's Adaptation to Climate Change*, World Environment, 2006 (5), 72–74.

2 The Situation of the Latest Progress and the Focus on the Funding Mechanisms of Climate Change

2.1 *The Progress Made in Negotiations on the Topic of Funds in the* Convention *and* Durban Declaration and Platform

The 18th *United Nations Framework Convention on Climate Change* conference of the parties held at the end of 2012 marked the end of the progress of negotiations of the "Bali Roadmap"; the emphasis of global climate change negotiations shifted to the course of *Durban Declaration and Platform*, which indicated that a negotiation timetable focusing on a new system of response to climate change internationally after 2020 would officially enter the 3-year countdown stage. The main mission of the *Durban Declaration and Platform* was to form a legal document or some kind of legal outcome suitable to all parties at the *Convention* before 2015, as the basis for all parties to strengthen the implementation of the *Convention*, to reduce and control greenhouse gas emissions and respond to climate change after 2020. The focus of the debate among all parties was to continue maintaining or fundamentally change the system of the existing *Convention* and the *Kyoto Protocol* (hereinafter referred to as the "*Protocol*"), and how to define and reflect the "common but differentiated responsibility" (CBDR) principle as well as other important principles such as fairness and the respective capacity.

The topic of funds was an important content of the *Durban Declaration and Platform* for strengthening the action arrangement after 2020 and improving the action strength before 2020. It is widely believed by the developing countries that the *Durban Declaration and Platform* should comprehensively and symmetrically consider the issues regarding funds, that the developed countries should provide funds for the developing countries to improve the action strength before 2020 and strengthen the self-mitigation actions after 2020, and that the *Durban Declaration and Platform* should specify how the developed countries should perform their fund commitments and strengthen fund arrangements. The developed countries believe that the issues regarding funds have been basically solved in the progress of the "Bali Roadmap" and that the core of future agreements is mitigation, so the contributions from the developed countries should not be included in the discussion. Although all parties will carry out intense negotiations on the basic framework, the delineation of the strengthening actions and other issues of future agreements, they will encounter serious divergences on the fund negotiation agenda and the specific realization form of funds, so it is less likely to make substantial progress in the topic of funds during the *Durban Declaration and Platform* in a short period of time.

2.2 *Latest Progress in the Funding Mechanisms of Climate Change Inside the* Convention

Based on the "common but differentiated responsibility principle" and the "fairness principle," the *Convention* requires parties of the developed countries to provide new and extra funds for parties of the developing countries, to support all funds or incremental costs incurred in the performance of parties of the developing countries, to guarantee a comprehensive, effective and continuous implementation of the *Convention*. The *Convention* also stipulates that the performance strength of the developing countries will depend on the extent of carrying out the *Convention* by providing funds and technical supports from the parties of the developed countries.

The Global Environment Facility (GEF) is the main funding mechanism of the *Convention* and is one of the main channels of funds for international cooperation on climate change; it provides grants to the developing countries to strengthen their abilities in responding to climate change and promotes the developing countries to make contributions to gain global environmental benefits. Since its foundation in 1991, the GEF has accumulatively provided more than 3.9 billion USD in the field of climate change to the developing countries, and enabled joint financing of more than 28 billion USD by various means. The climate change action supported by GEF is mainly mitigation, giving consideration to adaptation. At present, the GEF is in its 5th capital increase phase (from July 2010 to June 2014); the overall goal of the climate change field is to help the recipient countries transform into low-carbon development by means of market transformation, investments in environmentally friendly technologies and others. In June 2012, the Japanese Ms. Naoko Ishii was appointed as the 4th Chairman of the GEF at the 42nd Council. After taking office, on the one hand, Ms. Ishii actively promoted GEF's transformation reform and business innovation, proposed to carry out flagship projects in such fields as urban sustainable development, forest conservation and food safety, and strived to make GEF play a more important role in the protection of global environmental public products. On the other hand, Ms. Ishii tried to change the ways of cooperation with middle-income countries including China, set the disbursement limits for the countries and required them to increase the apportionment of the capital increase ratio. Since Ms. Ishii took office for one year, only two of China's national projects were approved, less than 4% of all project funds approved by the GEF in the corresponding period, with a sharp fall compared with China's previous fund proportion of over 10%. In April 2013, the GEF started the capital increase negotiations of the 6th phase (from July 2014 to June 2018), and the preferential strategies and details supported by funds in the climate change field are expected to be issued in February 2014.

The GEF is the main funding channel through which developing countries receive international cooperation and assistance regarding climate change under the *Convention*; the GEF usually pays great attention to large mitigation projects with global environmental benefits; due to the incapability of many small developing countries, most funds flow to large developing countries. In order to change this situation, the 7th Conference of the Parties (COP7) of the *Convention*, in 2001, decided to set up the Special Climate Change Fund (SCCF), the Least Developed Countries Fund (LDCF) and the Adaptation Fund (AF), to cover the gap.

The SCCF source concerns voluntary contributions from the developed countries or other countries; currently, the fields supported by SCCF include adaptation and technology transfer. Up to June 2013, the SCCF provided grants of about 179 million USD for 46 projects in 35 developing countries, and mobilized a joint financing of 1.559 billion USD through various channels. The LDCF is aimed at the least developed countries in Africa and other areas, supports the least developed countries in formulating the *National Adaptation Plan of Action* (NAPA) of climate change, implementing the urgent and preferential adaptation measures confirmed in the NAPA, and carrying out the construction of the related capacity of climate change and the related activities. Up to June 2013, the LDCF provided funds of 507 million USD for 161 projects in the least developed countries, and mobilized joint financing of 1.987 billion USD. China was not included in the LDCF support. At present, the funds available for the SCCF and the LDCF are 1.82 million USD and 8.38 million USD, respectively. Belgium, Germany, Norway, Sweden, Switzerland, and the USA have recently promised to donate 69.28 million USD and 129 million USD, respectively, to the SCCF and the LDCF, making the accumulative total amount of the two funds exceed 1 billion USD.

The AF is aimed at supporting specific adaptation projects and plannings of the developing countries; its source is mainly the income from 2% of the trading volume of emission reductions of CDM (Clean Development Mechanism) as well as voluntary donations from the developed countries. Although the package resolution approved at the United Nations Climate Change Conference held in Doha at the end of 2012 legally guarantees the implementation of the second commitment period of the *Protocol* in 2013, Canada, Japan, New Zealand and Russia have made it clear that they will not participate in the second commitment period of the *Protocol*, and in solving problems concerning the carbon emission balance in the first commitment period, only six countries, including Australia and Japan say that they will not use or purchase the emission balance in the first period of the *Protocol* to expand their carbon emission credit of the second period. Therefore, the emission reduction commitment locked in the second commitment period of the *Protocol* lacks strength. Currently, the Certified Emission Reduction (CER) price is continuously low and the market recovery is still unpredictable, so the AF source may experience adjustments; it needs further explorations and negotiations to decide whether the fund source will mainly be based on the CDM project or on voluntary donations from the developed countries in the future. According to the AF financial report issued by the World Bank,[3] up to December 31, 2012, the AF totally supported the adaptation projects of 28 developing countries, covering grants of about 198 million USD. Now, the funds available for the AF are 124 million USD. Considering that only 4–8 million USD will come from the income from CDM projects before 2020, even if the developed countries meet their grant commitments of 17.20 million USD on time, the funds available for the AF will be no more than 150 million USD.

Since the climate change issues have heated up, the imbalance of the GEF in undertaking responsibilities and controlling funds causes increasing dissatisfaction from the developing countries. The developing countries urgently hope to establish a funding mechanism exclusively for the *Convention*, and the developed countries are also willing to use the establishment of the new mechanism to strengthen the role of carbon market and private sectors in the climate financing system.[4] At the Durban Conference of the Parties in 2011, through hard negotiations during the course of three years, the Green Climate Fund (GCF) was finally launched, becoming a symbolic achievement of the *Convention* negotiations. The *Convention* required that the new multi-lateral funds to support the adaptations should be provided by the GCF, and that the funds would be used for mitigation and adaptation activities in a balanced way. Since new funds involve the core interests of all countries in the climate change issues, all the interested parties want to seize the chance of establishing regulations from the GCF to blend in institutional arrangements favorable to themselves. Although four meetings of the Board of Directors have been held since August 2012, the overall negotiations of the GCF are making slow progress. Centering on such important topics as funding business mode, financing mode and progress, private sector systems, and the establishment of an independent secretariat, the developing countries and the developed countries are carrying out heated discussions.

---

3 "Adaptation Fund Trust Fund Financial Report Prepared by the Trustee," http://fiftrustee.worldbank.org/webroot/data/AFTR1212.pdf.

4 Chen Lan, Wang Wentao, and Zhu Liucai, *Revolution and Development of International Climate Financing*, World Environment, 2012 (5), pp. 48–49.

2.3 *Steady Development in the Funding Mechanisms of the Climate Change Outside the* Convention

Since the funding mechanisms inside the *Convention* have such disadvantages as slow progress of negotiations, long preparation cycles of projects, complicated disbursement procedures, bad result-oriented, and lack of performance evaluation indices, both developed countries and developing countries tend to use the funding mechanisms of climate change outside the *Convention*, so the multi-lateral funding mechanisms outside the *Convention* have rapidly developed in recent years. For example, in July 2008, the World Bank officially approved setting up the Climate Investment Funds (CIF) according to the decision of the Group of Eight Ministers Conference. The CIF includes two sub-funds: the Clean Technology Fund and the Strategic Climate Fund, aiming at providing funds for the developing countries by means of grants, concessional loans, risk reduction tools and others, and helping the developing countries demonstrate and transform in clean technology, sustainable forest management, an increase in clean energy access, and the adaptation to the climate change. At its establishment, the CIF obtained grant commitments of 6.1 billion USD from the developed countries, and up to the end of 2012, the total funds for projects approved by the fund committee reached 2.78 billion USD. It is noteworthy that the CIF will provide concessional loans for adaptation projects of the developing countries, but, regarding the requirements of the *Convention*, the obligations of the developed countries are to provide funds for adaptation projects in the developing countries, not only concessional loans.

For the developmental tendency of the funding mechanism outside the *Convention* represented by the CIF, one possibility is to complement with the funding mechanisms inside the *Convention*, which can be seen in such aspects as launching the purpose of the CIF, key fields and relationships with the funding mechanism inside the *Convention*. The CIF does not establish direct cooperation with but complements the funding mechanisms inside the *Convention*. Another possibility is that as the CIF was entrusted to the World Bank to operate, which is conformed to the funding mechanisms based on multi-lateral banks and advocated by the developed countries under the platform outside the *Convention*, the developed countries may start all over again beyond the funding mechanisms inside the *Convention*, revolving the funding mechanisms outside the *Convention* into an important contribution channel and a tool for grasping international leadership in responding to climate change.

2.4 *Problems in Current Funding Mechanisms*

During the process of constructing the international cooperation system of climate change after 2012, as one of the core negotiation contents, the current funding mechanisms mainly faced with such problems as the scale of funds, fund sources and allocation, and the fund operation system.

First, the scale of funds of the current funding mechanisms is seriously inadequate. The developed countries are unwilling to fulfill their fund commitments, and continuously reduce the contribution standards and use multiple ways to "calculate" the amount of contributions. Both mitigation and adaptation need a lot of funds. According to the analysis of the *World Development Report 2010* of the World Bank, in the next 20 years, the developing countries need to spend 140–175 billion USD in mitigation each year, and 30–100 billion USD in adaptation each year.[5] In the *Speeding Up International Climate Financing*, the EU also points out that currently, the developing countries need more than 100 billion EUR for mitigation and adaptation each year.[6] The *Copenhagen Accord* requires the developed countries to provide 30 billion USD between 2010 and 2012 as rapid initial funds, and provide a long-term fund of 100 billion USD each year after 2020 for the developing countries to respond to climate change. Although the developed countries declare they have provided and made fund commitments of 29.4 billion USD for the developing countries through official development assistance, other official channels (such as cooperation with private sectors) and other ways, due to serious repeated calculations and distortions of statistical standards, the data of the official website of the rapid initial funds show that the developed countries have only delivered 3.6 billion USD, and the world development action agency also points out that the fulfillment ratio of the rapid initial funds has only reached 13%.[7] In the meantime, since the developed countries are less willing to contribute, the realization form and way of, and the management and usage method of the long-term funds of 100 billion USD still need long-time negotiations. In order to avoid unknown situations and a fatal "fund bluff" after the end of the "rapid initial funds" after 2013, the developing countries have collectively appealed to the developed countries to increase their scale of contributions for medium-term requirements between 2013 and 2015; the amount should be doubled at least, reaching 60 billion USD.

Second, the fund sources of the current funding mechanism are unreliable. The developed countries use multiple means to escape their contribu-

5 "World Development Report 2010," http://siteresources.worldbank.org/INTWDRS/resources/477365.

6 "Stepping up International Climate Finance: A European Blueprint for the Copenhagen Deal," http://ec.europa.eu/environment/climat/pdf/future_action/com_2009_475.pdf, 2010-09-27.

7 Zhang Wen, Wang Mou, and Lian Huishan, *Implementation Progress of Rapid Initial Funds of Climate Convention and New Trend of Environmental Performance of Developed Countries*, Ecological Economy, 2013 (3), pp. 29–32.

tion responsibilities, weaken or transfer their contribution obligations, leading to huge challenges for the operation of all funding mechanisms. On the one hand, the developed countries obscure the difference of their responsibilities and obligations with the large developing countries, requiring the large developing countries to increase their contributions, and attempt to transfer the contribution obligations to large developing countries; on the other hand, the developed countries greatly preach the role of market mechanisms and private sectors, intending to help the private sectors to expand the markets of the developing countries by using fund sources. Although private sectors have huge funds, due to their features such as the pursuit of profit, instability, and unpredictability, they do not meet the *Convention* requirements and have difficulties to provide sufficient and stable funds, so they can only be considered as a complementation of governmental public funds; meanwhile, developed countries attempt to take such innovative financing tools as the aviation tax, navigation tax and carbon-related tax as important medium- and long-term fund sources of multi-lateral funding mechanisms of climate change, and euphemistically incorporate the developing countries into the funding system to bring new obstacles to the economic growth of the developing countries, intending to consolidate and strengthen the leadership of developed countries in the process of responding to climate change globally.

Third, the operation efficiency and fund allocation efficiency of the current funding mechanisms need to be improved. The recipient developing countries complain that the disbursement procedures of the current funding mechanisms are complicated and the preparation cycles of projects are long; while the developed countries think that it is difficult to expect a greater influence of the current funding mechanisms, which is not favorable for them in using the funding mechanisms to exert a greater influence on the developing countries and does not facilitate them in controlling the operation of the international climate system. There are differences in the developing countries in terms of their level of development, their building up of a capacity for response to climate change and specific action requirements, as well as their abilities in using the international funding mechanisms of climate change and in developing the related projects, so funds mainly flow to large developing countries, and the policy implementation of allocating funds through performance also limits the efficiency of using funds of middle and small developing countries with relatively weak abilities. Since, currently, the funding mechanisms of climate change have parallel multiple channels, how to coordinate multiple funds inside and outside the *Convention* in subsiding fields and specific contents and guarantee a reasonable project fund supply, always present difficulties for the climate change funding mechanisms. Furthermore, multiple international institutions involved in the funding mechanisms act on their own, so it is hard to form effective resultant forces in order to mobilize and integrate the multi-channel funds to serve the projects.

Fourth, the current funding mechanisms tend to have political complexions. From the governance perspective, the current funding mechanisms gradually become a tool for the developed countries to penetrate their political intentions. During the negotiations of the GCF Board of Directors, the developed countries strived to make the GCF separate from the Conference of the Parties; in terms of the fund goal, they required the GCF to plan a positive role in helping the developing countries to implement the goal of emission reduction and promoting the *Durban Declaration and Platform* to reach an agreement on the global emission reduction, and they also strengthen the introduction of indices for the evaluation of the results-oriented projects and the performance of the projects. The developed countries escape their contribution responsibilities in the progress of capital increase, are unwilling to set a specific timetable, believe that priority should be given to business modes of funds and specify the business modes as a prerequisite for capital increase, and advocate that developing countries should also actively make donations. The developed countries advocate expanding the private sectors, reducing the investment risks of private sectors in the developing countries, and promoting the private sectors to play a more positive role in policy making, capacity building, technology transfer and other aspects. During the 6th capital increase negotiations of the GEF, the developed countries attempted to consider the GEF as a political tool, ignoring the fact that the GEF's purpose is devoted to improving the global environmental benefits, and proposed to make gradual policies towards the large developing countries and change their disbursement conditions. Good execution abilities and outstanding project performances of large developing countries not only produce wide global environmental benefits, but also become bases for the GEF to sustain in the international community. Politicization of source allocation will definitely shake the foundation of the GEF as a professional environmental agency, endangering its long-term development.

3 Suggestions on the Participation in Constructing the Funding Mechanisms of the Climate Change

3.1 *Insisting on the Position of the State and Protecting the Disbursement Rights*

The developed countries advocate the "accountability theory of carbon emission powers," and together with other developing countries, require China to undertake emission reduction obligations and contribution responsibilities.

China needs to stress and insist on maintaining the "common but differentiated responsibility principle," stick to its position as a developing country and should not undertake the binding contribution obligations. China should actively respond to the graduation issues of the upper- and middle-income countries in the multi-lateral funding mechanisms of climate change, and emphasize that the climate funds, especially the funds used to reduce emission, should reflect the global emission reduction effects, paving the way for China to continue reserving the disbursement rights. With an efficient utilization of funds, remarkable achievements, and obvious demonstration effects, China should continue to be provided with funding support.

3.2 *Considering the Interests of All Parties to Innovate the Cooperation Models*

China should promote and strengthen the external public opinion stressing that the "developed countries escape their contribution responsibilities under the *Convention*," urge the developed countries to accelerate the progress of the capital increase, enhance the propaganda of China's efforts, its emission reduction achievements and its contributions of funds in responding to climate change. China should further strengthen its alliance with large developing countries such as India and Brazil which have similar interest demands and the same positions, and pay attention to the climate benefits of the least developed countries and small island countries, seeking consensuses. Moreover, China should consider innovating the ways of South-South cooperation, and carry out cooperation on funds and technology with other developing countries by proposing to establish a "South-South Cooperation Fund for response to Climate Change" and other bilateral or multi-lateral channels, to jointly improve the capabilities of the developing countries for response to climate change, creating a harmonious external developmental area.

3.3 *Actively Participating in Making the Rules and Promoting the Establishment of Fair and Efficient Funding Mechanisms*

China should actively participate in the governance and operation of the GEF and the GCF, guide the making of rules and the developmental direction of the funds, and effectively improve its influence on international affairs; and then, based on the principles of promoting the development of the funding mechanisms of climate change featured with "fairness and reasonability, practicability and effectiveness, cooperation and mutual benefits" and of uniting the other developing countries, China should research and determine its medium- and long-term considerations on climate fund issues as soon as possible and undertake international responsibilities conforming to its developmental stage and economic strength under the framework of the climate change funding mechanisms.

3.4 *Changing the Traditional Ways of Cooperation and Exploring New Ideas of Cooperation*

As the cooperation between China and the climate change funding mechanisms enter the transformation period, China needs to continuously expand the new cooperation space by innovating cooperation content and form, gradually expanding pure cooperation on funds to cooperations on knowledge, capacity and talents. First, China needs to focus on jointly carrying out comprehensive and strategic research under the multi-lateral funding mechanisms with external intelligences; second, it should actively promote its system and mechanism innovation and capacity building by participating in multi-lateral funding mechanisms; third, it should systematically sum up its project experience, comprehensively evaluate the economic and social benefits of the projects, reach out to the world by virtue of the multi-lateral funding mechanism platforms, deepen the understanding of the international community in China's developmental concepts and experience, and strengthen the promotion of China's project implementation achievements on an international scale; fourth, China should pay attention to the cultivation of international talents, and actively try to carry out cooperation on the cultivation of talents with multi-lateral funding mechanisms in order to lay a solid talent foundation for solving new problems and meeting new challenges.

CHAPTER 4

Energy and Carbon Emission Issues in China's Urbanization

Hu Xiulian

Abstract

This chapter analyzes the status and trend of China's urbanization development; it discusses the pattern choices of urban development, the ever-increasing urban energy services and energy demands, the increasingly-prominent contradiction between urban development and environmental change and the transformation to low-carbon development, as well as other principle issues in China's urban development; it simply introduces the referential international experience to solve China's urban energy and carbon emission problems, and puts forward policy choices promoting China's urban energy and low-carbon development.

Keywords

urbanization – energy – carbon emission – policy choice

1 The Status of and Trend in the Process of China's Urbanization[1]

Since the reform and the opening up, China's urbanization has entered a rapid stage of development. The urbanization rate quickly increased from 17.9% in 1978 at the early stage of the reform and opening up to 52.57% in 2012; it increased by 16.35 percentage points in a 12-year period from 2000 to 2012. China's urbanization rate was close to the world average level in 2012.[2] China has taken only more than 30 years to experience the 200-year urbanization

* Hu Xiulian, a researcher at the Energy Research Institute of the National Development and Reform Commission, e-mail: huxl@eri.org.cn.

1 The article mainly discusses urban energy and carbon emission issues, so the word "urbanization" is used.

2 http://api.wordbank.org/datafiles/16_Topic_MetaData_en_EXCEL.xls.

 | DOI 10.1163/9789004280021_005

process of Western countries. The main driving factors of the acceleration of China's urbanization are its reform and opening up and its rapid economic development, especially its rapid industrial development; it is also related to the acceleration of the transformation of the land-use types due to low land resource costs, the entry into cities of landless peasants who provide a large amount of labor at a low cost, an extremely low emission cost of pollution in the process of urbanization, etc. In terms of the stage of China's development, industrialization has become a main boosting factor of China's urban development. At present, in large- and medium-sized cities of a certain scale, the secondary industry is still the main force driving urban development.

In the future, although the rapid development of urbanization will face many challenges, such as the construction of urban infrastructures (including urban residential housing, roads and pipe networks), employment, energy supply and consumption, air quality, environmental pollution, and an increase in CO_2 emissions, these challenges will not prevent the course of rapid development of urbanization. With continuous acceleration of China's social and economic development and industrialization process, the level of China's urbanization will improve constantly in the next several decades.

According to the results of the scenario analysis of the changing trend of China's future urbanization rate by related research institutes at home and abroad, China's urbanization rate by 2020 will rise to be 58%–63%, meaning that from now until 2020, about 200–250 million people in China will move from rural areas to cities to work and live. By 2050, China's urbanization rate will continue to rise to be 60%–79%,[3] reaching the current level of high-income countries. See Figure 4.1.

2 Main Issues on Energy and Carbon Emission in China's Urban Development

As a main gathering place of the social and economic activities of human beings and a spatial combination form of social productive forces with high concentration, cities are closely related to energy production, supply and demand conditions of the whole society. On the one hand, the particularity and complexity of urban operations make their energy demands more diverse, thus raising higher requirements for internal substitution and interconversion of urban energy systems; on the other hand, the utilization of urban energy

3 Fan Gang and Ma Weihua, *Low-carbon City Action: Policy and Practice*, China Economic Publishing House, 2011.

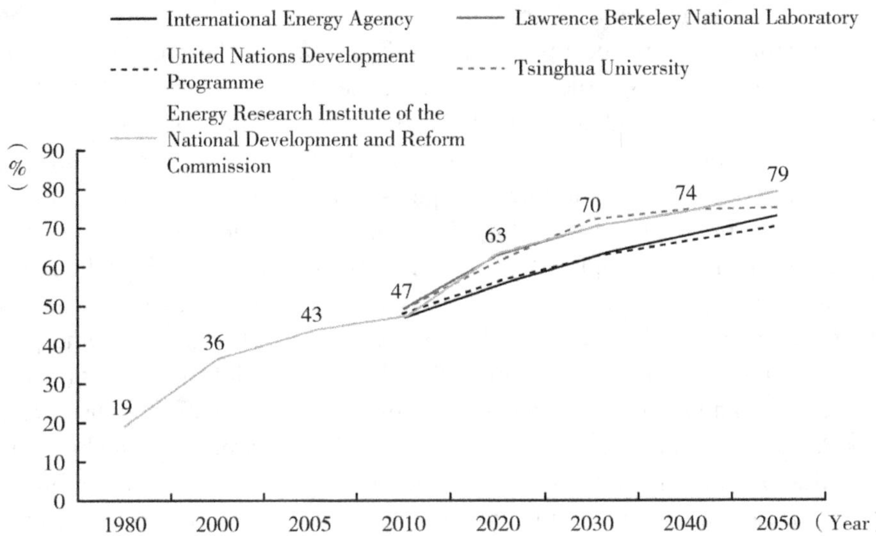

FIGURE 4.1 *China's future urbanization developmental trend analyzed by related research institutes at home and abroad.*

is closely linked to the quality of the urban environment, because a friendly urban environment needs an efficient and clean use of energy. Therefore, urban energy issues are a huge systematic project, requiring a long-term process of development.

China is experiencing urbanization. The pattern of extensive urban development of "high input, high consumption, high pollution" experienced by Western developed countries is worth thinking about. However, because one characteristic of China's developmental stage is the urbanization development driven by industrialization, the development of urban energy systems still faces many problems and challenges, such as complicated choices of urban development patterns, great potentials for the increase in energy service demands, ever-increasing energy demands, an insufficient supply of excellent energy, great potentials for energy conservation and emission reduction, resource constraints, increasingly-prominent contradictions among urban development, the environment and low-carbon development, reform and innovation abilities of urban modern energy systems that need to be improved, and weak intelligent management of energy systems.

2.1 *The Choice of Urban Development Patterns*

Cities are a gathering place of resource and energy consumption. With cultivated land accounting for 7% of that the world, freshwater resources for

7%, oil reserves for 4%, and natural gas reserves for 2%, China promotes the urbanization of people occupying 21% of the global population; so China's urbanization cannot copy the urbanization developmental pattern of Western developed countries, otherwise China's urbanization could not be supported.

Effects of the pattern of urban development on urban energy consumption are mainly shown in the following aspects: urban size and economic structure decide the urban demands for infrastructure construction, thus leading to production and consumption of a large number of raw materials and energy; the low-density and decentralized developmental trend of cities will cause a huge cultivated land area to rapidly diminish, thus leading to a low land plot ratio; urban size and spatial layout decide the distances for the transportation of urban residents in daily leisure and work; urban population, density and spatial layout decide the daily transportation trips; the scale of urban building (i.e. the total building area) will affect the total energy consumption of buildings. For example, in the past ten years, China's per capita urban building area has doubled, so has the energy consumption of buildings.

In recent years, when expanding the urban areas, China has had a low efficiency of land use. Take Shenzhen as an example, which has the highest efficiency of land use in the top 10 cities ranked by the economic aggregates in China, the GDP per square kilometer of land in Shenzhen is only 1/25th that of Osaka, Japan, and Paris, France, and 1/10 of Tokyo, Japan, and Seoul, Korea, obviously showing the extensive growth of urbanization.

Now, China is at a double-turning point of the adjustment of the economic system and the economic structure, and there are large gaps in different regions regarding social and economic development, so China's urbanization will be a long progressive process of social transformation. During the promotion of the development of urbanization, it is a severe challenge to choose a new urbanization development pattern, which should be consistent with the national situations in China, adapt to China's development, reflect the differences in different regions, and conform to a sustainable development trend of the society in the future.

2.2 *Ever-increasing Demands for Urban Energy Services and Energy*

2.2.1 Ever-increasing Demands for Urban Energy Services

Energy service demands are one of important factors that reflect the level of urbanization. In 2011, the average number of households possessing a family car for each 100 of China's urban households was 18.58, and 1.96 for the lowest income households taking up 10% of total urban households, but it was 170 for the USA, 61 for Japan and 27 for the world's average level in 1970. In 2011, the per capita living area of China's urban residents was 32.7 m^2, but that of the USA,

Denmark and Germany reached 60 m^2, 52 m^2 and 40 m^2 respectively, in 2007. Thus, there are great potentials and space for an increase in energy service demands of urban residents in China's urbanization process. Related research shows that, by the middle of the 21st century, the number of household appliances possessed by each of China's urban households will reach or exceed the current level of Japan or the European countries, the per capita living area will exceed 40 m^2, and the number of households owning a car per 1,000 households will also reach or exceed 400. At that time, the total living area of China's urban residents will increase by about 20 billion m^2 compared with 2010, and besides the public building area, the total building area will be doubled, reaching 90–100 billion m^2. The number of family cars owned will exceed 140 million.[4]

At present, 53% of the steel consumption and 48% of the cement consumption in China are used in building construction. According to sampling surveys of typical cases and combined with analysis results of consumption flow direction of such products as steel and cement in China each year, the construction of 100 million m^2 of buildings in China's urban areas needs to consume about 8.10 million tons of steel, 22 million tons of cement and 2.40 million m^3 of wood. If the per capita building area of urban residents decreases by 1 m^2, according to the calculation of 70% for China's urbanization rate and 1.07 billion people for the urban population by 2050, the use of steel and cement each year in China can have a drop of more than 80 million tons of steel and nearly 300 million tons of cement.

Highways, railways, airports and other transportation infrastructures are important material bases for the society to exist and for urban development. Currently, the density of roads in China is only 0.4 km/km^2 and the per capita length of roads is less than 30 km for each 10,000 people, respectively 40% and 50% of that of the developed countries. China's highway construction has become the greatest bottleneck restricting economic development in many areas. As a country with a large population and high travel demands, based on the current scale, China needs to at least double its highway mileage.

Seen from railway development demands, at present, the per capita railway mileage of the USA and Japan is respectively 6.9 km for each 10,000 people and 1.6 km for each 10,000 people, but China's per capita railway mileage is only 0.6 km for each 10,000 people. China's railway density for each 100 km^2 is about 1.54 km, while that of Germany has reached 10.6 km. In Europe and

4 Research group of the Energy Research Institute of the National Development and Reform Commission, *China's Low-carbon Developmental Road in 2050: Energy Demand and Carbon Emission Scenario Analysis*, Science Press, 2009.

Japan, trains have become the most convenient means of transport, and high-speed railways with a speed of up to more than 300 km/h are playing a key role in local transportation. As a large country whose national territorial area is almost the same as that of the USA and the EU, China has great demands for construction of both ordinary railways and high-speed railways.

Many domestic research results show that the output of China's main energy-intensive products such as steel, cement, glass, and ammonia synthesis will reach their peak values between 2020 and 2030. Up to 2050, China's building area and transportation will continuously increase, but the speed after 2030 will slow down compared with that before 2030.[5] In short, ever-increasing energy service demands in China's urbanization process will promote the acceleration of China's industrialization process and the sustainable growth of energy demands.

2.2.2 Ever-increasing Urban Energy Demands

The statistics for China's urban metabolic rate show that China's urban energy consumption accounted for 85% of total national energy consumption in 2010, and the per capita energy consumption and the per capita household electricity consumption of urban areas were 4.3 times and 5.03 times that of rural areas, respectively.

At present, in large- and medium-sized cities of a certain scale, the secondary industry is still the main force driving urban development. The rapid development of industry has caused a corresponding growth in energy consumption. The current industrial consumption accounts for more than 70% of China's terminal energy consumption. Even for developed cities such as Guangzhou, its industrial added value accounted for about 45% of the urban economic aggregate in 2011, while the industrial energy consumption accounted for up to 70% of the total urban energy consumption.

Many domestic research results show that, driven by ever-increasing energy service demands during China's urbanization process, China's urban energy demands will increase from 2.8 billion tce in 2010 to 3.6–4.0 billion tce by 2020. At that time, the per capita energy consumption of China's urban residents will have increased from 4.1 tce in 2010 to 4.3–4.8 tce by 2020. With the increase in the proportion of energy consumption of construction and transportation departments in the urban energy consumption structure, the

5 Research group of the Energy Research Institute of National Development and Reform Commission, *China's Low-carbon Development Road in 2050: Energy Demand and Carbon Emission Scenario Analysis*, Science Press, 2009.

guaranteed supply of such excellent energy as oil, natural gas and electricity will face huge challenges.

2.3 *Increasingly-prominent Contradictions among Urban Development, Environmental Improvement and Transformation to Low-carbon Development*

The global climate change deeply affects the existence and development of mankind, becoming a huge challenge for all countries. Cities are the most active part of the global economy. With the interweaving of carbon emission and urbanization, the construction and development of low-carbon cities have become important fields and a focus for responding to global climate change and promoting harmony between humans and nature.

The CO_2 emissions of China's cities account for 90% of national CO_2 emissions, and the SO_2 emissions for 98% of national SO_2 emissions; the CO_2 emissions of 287 cities at prefecture level and above takes up about 72% of the total national CO_2 emissions. Therefore, the contradiction among China's urban development, environmental improvement and transformation to low-carbon development is increasingly prominent. The process of urbanization faced various challenges to avoid high-carbon developmental patterns, which are mainly reflected in the following aspects.

2.3.1 Restriction of Resource Endowments

Limited by China's energy resource endowment, the structure of China's energy production and consumption has been coal-oriented for a long term; the proportion of coal in the total energy consumption has been hovering around 70% in recent 20 years.[6] When burning fossil fuels with the same calorific value, the CO_2 emission of coal is over 25% higher than that of oil and over 40% higher than that of natural gas. Due to resource endowment limits, the possibility for China to adopt relatively low-carbon or carbon-free oil, natural gas, water and electricity, nuclear power and renewable energy to largely replace coal in the near future is limited, and China is also under pressure to convert solid coal to liquid fuel in order to guarantee the safety of oil used for transportation.

In 2011, China imported 271 million tons of oil, accounting for 55.3% of China's oil consumption. Some cities in China which are in old industrial bases, have a large number of heavy chemical industries and are in resource-based areas, will face difficulties and challenges in low-carbon transformation.

---

6 The Department of Energy Statistics of the National Bureau of Statistics, *China Energy Statistical Yearbook*, China Statistics Press, 2012.

For example, the coal consumption of such cities as Handan, Chuzhou, Yulin, Altay, and Haixizhou accounts for about 80% of the total primary energy consumption. Since the "11th Five-Year Plan," although the energy consumption per unit of GDP of these cities has decreased year by year, it sometimes rebounds due to a poor foundation and a big population base. Thus, the structure of coal-based energy consumption becomes an important obstacle for these cities to improve the environment and transform to low-carbon development.

2.3.2 Restrictions of the Developmental Stage

With China's vast territory, differences in the development of various regions are great, and those at the level of urbanization are greater, so the primary task of some cities, in the long run, will be develop the economy. For instance, in some underdeveloped cities, such as Altay and Chuzhou, the overall carbon emission level is lower than the national level. The per capita energy consumption of Chuzhou City is only 44% of the national level, and the energy consumption per unit of GDP is only 88% of the average national level. According to data, Chuzhou seems to already be a low-carbon city. However, when further analyzing the reasons for the low level of carbon emissions in Chuzhou City, we can find that the city is still underdeveloped, and the proportion of agriculture with less energy consumption in GDP is up to 23.3%. Therefore, for underdeveloped cities such as Chuzhou, low-carbon transformation is limited by their developmental stage.

2.3.3 Great Potentials for Energy Conservation and CO_2 Emission Reduction

Now, there are no special statistics regarding urban energy consumption in statistical system for China's energy. According to related data in the Energy Statistical Yearbook and reference to related data estimates in the City Statistical Yearbook, in 2010, the national energy consumption was 3.25 billion tce calculated by means of specific coal consumption in power generation, and the total urban energy consumption was about 2.76 billion tce, accounting for 85% of the total national energy consumption. The total national power consumption was 4.2 trillion kWh, of which the total power consumption of municipal districts was 2.31 trillion kWh, i.e. the municipal districts occupying only 13.3% of the total national territorial area, consumed 55% of the power of the whole society. This shows that China's cities, especially the municipal districts, are regions with a concentrated amount of energy consumption.

Experts in related research institutes have compared and analyzed the indicators of cities in different areas of China between 2005 and 2010, such as energy consumption per unit of GDP, per capita energy consumption, power consumption per unit of GDP, energy consumption per unit of industrial added value, and per capita power consumption of residents in municipal districts, and the results show that the utilization efficiency of China's urban energy is characterized by "more in middle areas and less at both ends."

Take the energy consumption per unit of GDP as an example. In 2009, among 214 cities with an observed value, the average value of the intensity of urban energy consumption was 1.43 tce per 10,000 yuan, and Shizuishan City had the biggest intensity of energy consumption, equal to 6.75 tce per 10,000 yuan, 4.7 times the average value; Shanwei City had the smallest intensity of energy consumption, equal to 0.528 tce per 10,000 yuan, accounting for only 36.9% of the average value. It can be seen that there is a difference in the level of efficiency of energy utilization among cities.[7]

At present, the per capita energy consumption in the transportation and construction of China's cities is much lower than that in developed countries, but the per capita energy consumption in the urban production and the corresponding per capita CO_2 emission are much higher than those of the cities in developed countries. Now, the per capita CO_2 emission of China's big cities such as Shanghai, Tianjin, Beijing and Hong Kong exceeds that of New York, Tokyo, London and Paris.

What's more, the operational energy consumption in the industry, transportation and construction of China's cities is higher than that of developed countries. The average energy consumption per unit product of China's main industrial products is 15%–20% higher than the advanced international level, the heating energy consumption per unit area is 2–3 times higher than that in developed countries with the same climatic conditions; besides, there is a great gap in the average fuel consumption of China's passenger vehicles compared with those in Europe and Japan. Related research results show that, by 2020, China's cities may achieve about 1.7 billion tons of CO_2 emission reduction by applying various energy-saving and emission-reduction technologies, in the departments of industry, construction and transportation accounting for 62%, 24% and 14% respectively.[8]

---

7 Song Guojun, Ma Ben, et al., *Research on China's Urban Energy Efficiency*, Chemical Industry Press, 2013.

8 Research group of China's energy and carbon emission by 2050, *China's Energy and Carbon Emission Report of 2050*, Science Press, Sep. 23, 2009.

3 International Referential Experience for China to Solve the Problems Regarding Urban Energy and Carbon Emissions

Energy issues are important in the development of all cities at home and abroad. The practical experience of foreign cities such as London, New York and Tokyo shows, that energy characterized by diversification, environmental protection, high efficiency and low-carbonization is an irresistible trend in the future development of global urban energy. Particularly, some common experiences of foreign cities with a high level of urbanization is worth learning about, which is shown as follows.

(1) When formulating energy strategies in promoting urban development, we should follow the principle of laying equal stress on both conversation and development for short-term, medium-term and long-term energy strategies, and take measures regarding two aspects of supply and demand to guarantee the safety of the urban energy supply and improvement of the utilization efficiency of energy.
(2) We should consider the construction department and the transportation department as preferential and key fields for energy conservation and energy efficiency improvement in urban development, and adopt various targeted technical measures and polices.
(3) We should consider the development and utilization of new energy and renewable energy as an important part of urban energy development strategies in response to fossil energy that is limited and can cause pollution.
(4) Controlling and reducing CO_2 emissions becomes an important part of urban energy development strategies.
(5) The government's support in policy, law, taxation, finance and other aspects plays a convoy role in the implementation of the urban energy strategies.

At the global level, many cities have realized the important role of cities in mitigating climate change, and have started to take action. By developing low-carbon development strategies and making related action plans to promote urban development and transformation has become an important direction in modern urban development and has gradually become a new goal of urban brands, playing an increasingly key role in the competition of the soft power of cities.

Currently, most world-famous cities have issued their low-carbon developmental strategies and action plans. Particularly, such big cities as London,

Tokyo and New York have had a quick response to climate change, playing a leading role in urban low-carbon construction. Understanding the practical experience of the low-carbon development of foreign cities is helpful for China's cities to explore a new pattern suitable to China's urban low-carbon development by combining those experiences with China's own characteristics.

4 Policy Choices in Promoting China's Urban Energy and Low-carbon Development

4.1 *Actively Explore a New Road Towards Urbanization and Construct Low-carbon Cities*

China should comprehensively assess its urbanization policies of the past 30 years, and establish an overall strategy of a new road towards urbanization based on low energy growth and low-carbon emission as well as high environmental quality and high living standards. China's new road towards urbanization should make such issues in urban development as energy, environment and response to climate change as preferential fields with overall consideration, pay adequate attention to key fields with a rapid increase in the energy consumption of urban construction, transportation and households, and strengthen policy-making, to make China gradually move towards sustainable urban development.

Advocate a compact road towards the development of urbanization, and implement the developmental idea of compact cities to national urban planning. Properly improve the urban density, rely on megalopolis and central cities, develop urban agglomerations, urban belts and urban clusters, optimize the spatial functional layout, and strengthen the urban sustainable developmental capacity. Accelerate the formulation and implementation of the national urban system planning and national land use planning, and carry out low-carbon and low-pollution pilot demonstrations based on low-energy consumption.

4.2 *Identify the Developmental Potentials and Developmental Advantages and Set Down Developmental Plans and Strategies*

China's cities are developing rapidly, urban areas have expanded and large-scale infrastructures are under construction. In such a dynamic environment, policy makers and all interested parties regarding the cities should coordinate their actions, based on fundamental national and urban conditions, clearly understand and identify the status, the potentials, the capacity, the advantages and the necessary conditions of the development of cities, based on this, make five-year plans, short-term plans and long-term strategies which are suitable to the characteristics of each one of the cities and make cities gradually move towards a sustainable and low-carbon development.

Urban low-carbon developmental plans should be positioned between the five-year plans and the special plans of the national economy and social development, as a link and bridge between the two. The purpose of urban low-carbon developmental plans is to further refine the principles of and the guiding opinions on low-carbon development determined in the five-year plans, and combine and coordinate them with related special plans. The comprehensive low-carbon developmental plans are actually tools and carriers promoting the combination of urban low-carbon development demands and the main direction of urban development. Low-carbon developmental plans will affect the decision-making of investment projects, urban fiscal expenditures, land use, technology choices, and other developmental topics.

4.3 *Determine Key and Preferential Fields in Promoting Urban Low-carbon Development*

At different economic stages of development, key and preferential fields of the development of different cities are different. For most of China's cities, the industrial departments will be a key field with great potentials in the emission control and reduction of greenhouse gases for a period in the future. With the expansion of urban areas, the increase in the number of motor vehicles, the increase in building areas, and improvements in people's living standards, the absolute amount of emissions from construction and transportation departments will rapidly increase. Recently the emission reduction costs of construction and transportation departments have been higher than those of the industrial departments. Therefore, cities should specify the key and preferential fields of mitigating and controlling the greenhouse gas emissions in different periods during their low-carbon development, so as to aim at formulating and adopting a series of technological and policy measures to achieve the low-carbon developmental goal of key fields.

4.4 *Construct an Energy Supply System Supporting the Sustainable Development of Urban Areas*

Cities are the subjects of China's energy production and consumption, and they are also the chief culprits causing air pollution and greenhouse gas emissions. We should change the original energy structure and energy utilization methods of cities, closely combine the change in the national energy consumption structure with the improvement in urban fuel supplies, expand the consumption of oil and natural gas, achieve as soon as possible the goal of no direct combustion of raw coal in municipal districts, promote clean energy such as natural gas, solar energy and electric power, achieve clean and low-carbon urban energy, and improve the proportion of excellent energy supplies of all cities to the furthest extent.

The system of urban energy, including power supply, heat supply and air supply, should be considered in an overall way with the urban functional layout, the industrial structure and the transportation system. The construction of smart power grids can provide convenience for developing distributed energy systems and electric transportation systems. We should pay great attention to energy conservation in all industries in order to reduce the pressure of energy supply.

4.5 *Promote, Apply, Research and Develop Advanced Energy Utilization and Low-carbon Technologies*

Promoting, applying, researching and developing advanced energy utilization and low-carbon technologies in various fields of cities are the core factors in achieving sustainable urban development. Industry has the greatest potential for China's cities to improve energy efficiency and reduce CO_2 emissions. Efficient low-carbon technologies available to the industrial departments mainly include high efficient terminal electricity-consuming and gas-burning equipment, heat and electricity recovery technologies, material recycling and substitution technologies, non-CO_2 gas emission control technologies, lots of advanced process technologies, various technologies to improve energy efficiency, carbon capture and storage technologies in the production of cement, ammonia and steel, technologies for the utilization of new energy and renewable energy.[9]

Transportation is one of the basic industrial departments of China's national economy, a link of logistics and passenger flow in social and economic activities, and also a big oil consumer in China. Therefore, we need to implement the compact urban spatial planning to reduce traffic demands; give priority to the development of urban public transportation, and construct a reasonable traffic structure; promote the construction of an urban rail transit system, and develop bus rapid transit (BRT) systems with a large traffic volume; strengthen the R&D of transportation science and technology, and improve the proportion of clean energy.

Since most buildings have a long service life, low-cost, high-efficiency and low-carbon measures and technologies are important choices. In construction departments, energy saving and carbon emission reduction technologies with development potential in the future mainly include building envelope energy-saving technology, heating system energy-saving technology, cold

---

9 International Energy Agency, *World Energy Outlook 2010*, IEA, 2010.

supply system energy-saving technology, efficient lighting and illumination technology, efficient appliances, new buildings energy-saving design, and integrated design of commercial buildings. For spatial heating and water heating, efficient heat pumps, solar thermal systems and power-heat cogeneration systems using hydrogen fuel cells are all optional technologies.[10]

4.6 *Establish Complete Policy Mechanisms and Improve Various Supporting and Safeguard Measures*

Ways to achieve sustainable development of urban energy can be summarized as follows: choosing a reasonable way of life and consumption mode; carrying out multi-level structural optimization, such as optimization of the economic structure, industrial structure, business structure, product structure and energy structure; adopting technological progress in terminal energy consumption and in energy supply; providing safeguard and supporting policies for possible problems and obstacles in achieving the goals above, such as corresponding laws and regulations, various economic and taxation policies, an industrial policies of low-carbon development, strict energy efficiency standards, low-carbon product standards, various reform mechanisms, market mechanisms, policies promoting technological innovation, and policies for strengthening international cooperation and technology transfer. Therefore, establishing complete policy mechanisms as well as various supporting and safeguard measures is an important measure to achieve the sustainable development of urban energy.

4.7 *Strengthen Capacity Build-up and Carry Out Widespread International Cooperation*

Capacity build-up has a wide range, including capacity in policy-making, the formulation of low-carbon planning, the list formulation of urban greenhouse gas, the statistics, report and monitoring of the energy consumption and the load of greenhouse gas emissions, technical research, development and innovation, and implementation of policies and laws, etc. At different levels, long-term and constant capacity build-up for different subjects will favorably improve the quality and process of urban low-carbon development. Furthermore, learning from the practical experience of foreign cities in low-carbon development and carrying out widespread international cooperation and exchanges will be conductive to the sustainable and low-carbon development of China's urban energy systems.

---

10 Dai Yande, Hu Xiulian, et al., *Research on Technology Potential and Cost of China's CO_2 Emission Reduction*, China Environmental Science Press, 2013.

CHAPTER 5

Evaluation of the Progress of the Low-carbon Work in the Pilot Provinces and Cities

Ding Ding and Yang Xiu

Abstract

While striving to develop the economy and improve people's livelihood, the National Development and Reform Commission issued notices in 2010 and 2012 regarding the conducting of low-carbon pilot work in 42 provinces and cities in two batches, with an aim of effectively controlling greenhouse gas emissions. Since low-carbon pilot work was carried out in five provinces and eight cities in the first batch in the last three years, experimentation has actively been done on the refinement of the leadership mechanism, on the enhancement of the strategic plan formulation and guidance, on the determination of low-carbon development tasks and objectives, on the preparation of the list of greenhouse gas emissions, on the construction of a system for monitoring and assessment, and on the advocacy of ideas for low-carbon consumption and lifestyle, with good achievements obtained. The scope of the pilot work has been further expanded in the second batch. As the work in the first batch laid a good foundation, the pilot work in the second batch has been paying more attention to top-level design, so that the goal of quantitative emission control could be set in the plan for the pilot work. By summarizing the progress of the work in the two batches of pilot provinces and cities, it has been found that there is a lack of an indicator system for low-carbon assessment, and the data foundation is weak, resulting in a failure to meet the needs of low-carbon development. Therefore, it is necessary to actively promote the improvement in the method of the regional assessment of low-carbon development. In terms of policy and technology, low-carbon development is in urgent need of a top-level design, and supporting policies remain to be perfected, and more importantly, it is badly in need of enhancing input in the development and popularization of low-carbon technologies. Promoting urban low-carbon industrial development and park construction is an important aspect in low-carbon development. The low-carbon upgrading and transformation of traditional industries is the foundation of the

* Ding Ding, Ph.D., associate researcher, and director of the Department of Policies and Regulations, the National Strategic Climate Center; Yang Xiu, Ph.D., assistant researcher of the Department of Policies and Regulations, the National Strategic Climate Center.

 | DOI 10.1163/9789004280021_006

adjustment of the industrial structure, and modern service industries are the focus of industrial structural adjustment. Thus, it is suggested that great importance should be given to guidance mechanisms and policy measures in the low-carbon development of pilot cities, and the pilot program should stick to the idea of four combinations (the combination of government promotion and public participation, the combination of concept guidance and institutional innovation, the combination of independent innovation and technology introduction, and the combination of demonstration driving and scientific promotion).

Keywords

climate change – low-carbon pilot – low-carbon industry

Climate change concerns the survival and development of human beings, so it is a grand challenge that the world is facing. While striving to develop the economy and improve people's livelihood, China encounters a fresh new subject, i.e., effectively controlling greenhouse gas emissions and properly coping with climate change. Against the background of climate change, the National Development and Reform Commission issued notices on July 19, 2010, which determined the implementation of the first batch of pilot work in five provinces and eight cities;[1] it again issued a notice on November 29, 2012, determining the implementation of the second batch of national low-carbon provincial and urban pilot work.[2] This chapter evaluates the progress of that the low-carbon pilot regions are making by stages, in order to provide a reference for future work to achieve low-carbon development in an all-round way.

1 Summary of the Progress of the Work in the Low-carbon Pilot Provinces and Cities

To fully stimulate the initiative of all quarters of the society, gather experience in classified guidance in different regions and industries, and promote the

1 *Notice of the National Development and Reform Commission on the Launch of Low-carbon Provincial and Urban Pilot Work* (F. G. Q. H. (2010) No. 1587), July 19, 2010.

2 *Notice of the National Development and Reform Commission on the Launch of the Second Batch of Low-carbon Provincial and Urban Pilot Work* (F. G. Q. H. (2012) No. 3760), November 29, 2012.

implementation of greenhouse gas emission control action goals, the National Development and Reform Commission issued the *Notice on the Launch of Low-carbon Provincial and Urban Pilot Work* (F. G. Q. H. (2010) No. 1587) on July 19, 2010, to determine the implementation of pilot work in five provinces (Guangdong Province, Liaoning Province, Hubei Province, Shaanxi Province and Yunnan Province) and eight cities (Tianjin, Chongqing, Shenzhen, Xiamen, Hangzhou, Nanchang, Guiyang and Baoding); it issued the *Notice on the Launch of the Second Batch of Low-carbon Provincial and Urban Pilot Work* (F. G. Q. H. (2012) No. 3760) on November 29, 2012, to determine 29 cities as the second batch of the national low-carbon pilot cities.

1.1 *Work Progress and Achievements of the First Batch of Low-carbon Pilot Regions*

The criteria for selection of the first batch of pilot regions includes: the initiative of application and the degree of attention of leaders; work experience in low-carbon development; the role of experience gathering, demonstration and promotion of selected provinces and cities at different economic levels and developmental stages. In addition, regional balance is also an important factor that the National Development and Reform Commission needs to consider in order to select the first batch of low-carbon pilot regions.

According to the five requirements proposed by the central government (working out a low-carbon developmental plan, formulating supporting policies for green and low-carbon development, accelerating the construction of an industrial system characterized by low-carbon emissions, establishing a statistical and management system for greenhouse gas emission data, and actively advocating the idea of a green, low-carbon lifestyle and consumption mode), the first batch of low-carbon pilot regions has taken active actions, obtaining tremendous progress and accomplishments.

1.1.1 Refining the Leadership Mechanism

All pilot provinces and cities have formed leading groups for low-carbon work before their application for pilot region or for becoming a pilot region. Functions of the leading group are to perform uniform leadership, command, coordination and supervision in the low-carbon development of the whole province (city). To enhance the organization and coordination of the pilot work, quite a number of provinces and cities have set up a joint conference system, an incentive and restrictive mechanism, and a whole-process management system for major projects and demonstrative pilot units, thus promoting low-carbon construction in a comprehensive way. The establishment of a low-carbon leading group led by local major leaders reflects the great attention that the local government pays to low-carbon pilot work. Besides, most offices for the leading group are housed in the Development and Reform Commission of each province (city), which are responsible for the overall planning, co-ordination and promotion of the low-carbon development of the whole province (city).

1.1.2 Enhancing Strategic Planning Preparation and Guidance, and Improving the Planning of Low-carbon Development

Low-carbon development planning plays a critical role in ensuring that the pilot work is carried out steadily, orderly and effectively. From the response to climate change to the construction of low-carbon pilot regions, low-carbon pilot provinces and cities formed a planning framework of "1+4+x" from 2007 to 2011. In the planning framework of "1+4+x," 1 means that the response to climate change in the pilot regions is included in the local "12th Five-Year Plan"; 4 means four actions, that is, preparing a local climate change tackling scheme, fulfilling the local government's carbon emission reduction control index as set out in the *Work Plan for Greenhouse Gas Emission Control during the 12th Five-Year Plan*, working out the local pilot work implementation plan, and formulating a low-carbon developmental plan for pilot provinces and cities; x means various special plans developed by pilot provinces and cities in respect to energy conservation, clean energy development, a circular economy, construction and transportation on the basis of their own characteristics. For instance, Guiyang City constructs a low-carbon city mainly through "low-carbon transportation" and "green buildings";[3] Tianjin City promotes low-carbon urban development proceeding from the adjustment of its industrial structure, the acceleration of the formation of an industrial agglomeration area, the enhancement of the energy assessment of its investment project, and the implementation of industrial energy conservation, etc.[4]

1.1.3 Accelerating the Preparation of a Greenhouse Gas Emission List, and Starting the Construction of a System for Statistics, Monitoring and Assessment

Among the pilot provinces and cities, six provinces and cities (Guangdong Province, Hubei Province, Liaoning Province, Yunnan Province, Shaanxi Province and Tianjin City) also were the pilot provinces and cities to prepare

3 "Guiyang: Construction of a Low-carbon City through 'Low-carbon Transportation' and 'Green Buildings'," http://www.gov.cn/jrzg/2010-08/20/content_1684771.htm, 2010-08-20.

4 "Low-carbon City Pathfinder and Local Active Attempt," http://energy.people.com.cn/GB/12653517.html, 2010-09-07.

a Greenhouse Gas Emission List in 2005; they completed the report of the list by the end of 2011, and successfully passed a national inspection in 2012. The six provinces and cities successively conducted a 2010 province-level, and a 2005 and 2010 city (district, county)-level list preparation work. Among the pilot cities, Hangzhou City prepared a greenhouse gas emission list from 2005 to 2010,[5] and other cities such as Chongqing, Xiamen and Shenzhen are also actively studying and preparing their greenhouse gas list in 2005 and in 2010.

Moreover, pilot provinces and cities are actively arranging a work coordination mechanism, and implementing an index decomposition plan and assessment system according to the requirements on carbon dioxide emission control as stipulated in the *Work Plan for Greenhouse Gas Emission Control during the 12th Five-Year Plan* issued by the State Council in 2011. Guangdong Province[6] and Chongqing City[7] have decomposed the index of carbon emission to the subordinate administrative units, and other provinces and cities such as Guiyang, Shaanxi and Hubei are formulating index decomposition schemes.

1.1.4 Advocating the Idea of Low-carbon Consumption and Lifestyle

Low-carbon consumption and a low-carbon lifestyle involve many aspects, such as low-carbon transportation, low-carbon building, low-carbon community construction and low-carbon consumption. The first batch of low-carbon pilot provinces and cities (five provinces and eight cities) all view low-carbon transportation or green transportation as an important part of low-carbon pilot work. Special transportation-related plans have been introduced by various regions, whose main work includes: (1) giving priority to the development of public transportation; (2) developing a slow-traffic system and regarding slow travel mode such as walking and riding a bicycle as the focus of city transportation; (3) promoting new energy vehicles, in which, on the one hand, the new-energy vehicle industry is developed, and on the other hand, clean energy is promoted in the field of transportation; (14) developing intelligent transportation, which has been covered in the action plan of most pilot provinces and cities, but has not been put into practice yet. Besides, pilot provinces and

5 Zhejiang Provincial Development and Reform Commission, "Hangzhou Greenhouse Gas List Successfully Passing Inspection of National Experts," http://www.zjdpc.gov.cn/art/2013/2/5/art_10_498022.html, 2013-02-05.

6 "Guangdong Province Pioneering Control of Carbon Emission, and Decomposition of Carbon Index to 21 Cities," http://www.clinic.org.cn/site951/cjkx/2012-03-13/543193_1.shtml, 2012-3-13.

7 "Decomposition of Chongqing Carbon Emission Index to 38 Districts and Counties," http://www.cq.gov.cn/today/news/385384.htm, 2012-03-07.

cities are actively promoting low-carbon building, whose main actions include: (1) energy-saving reconstruction of existing buildings; (2) establishing energy-saving standards for new buildings; (3) enacting energy conservation standards and energy-use monitoring for public institutions; (4) utilizing renewable energy; (5) adopting energy-saving materials, products and technologies, etc. Moreover, pilot provinces and cities pay attention to press propaganda, and launch widespread publicity, education and social mobilization activities via newspapers, radio stations, TV channels and the Internet. With the construction of low-carbon communities and demonstration projects, the transition of residents' value and their lifestyle and consumption mode is expedited.

1.2 *Work Progress and Changes Regarding the Second Batch of Low-carbon Pilot Regions*

Apart from the five requirements that the first batch of low-carbon pilot regions observe, an additional requirement was specially developed for the second batch of low-carbon pilot regions, that is, the requirement of determining the work direction and principles. The second batch of low-carbon pilot regions should consider comprehensive coordination and sustainability as the fundamental requirement for conducting low-carbon pilot work. Based on the principles of comprehensively implementing the "five-in-one" overall layout (economic construction, political construction, cultural construction, social construction, and ecological civilization construction), they should further coordinate the relationship among resources, energy, the environment, development and the improvement of people's livelihood, reasonably adjust spatial distribution, actively innovate the institution and mechanism, and constantly improve policy measures, so as to expedite the formation of a new pattern for green low-carbon development and bring about a new situation for the creation of an ecological civilization.

As the work done by the first batch of low-carbon pilot regions has laid a good foundation in recognition and practice, comparatively speaking, the work plan developed by the second batch of low-carbon pilot regions is scientific, reasonable and comprehensive, with viable measures and a higher preparation level. Compared to the first batch, the work of the second batch of low-carbon pilot regions is characterized by:[8]

8 The low-carbon pilot work implementation plan of the second batch of low-carbon pilot regions.

1.2.1 Expansion of the Scope of the Low-carbon Pilot

The first batch of low-carbon pilot regions involves 12 provinces and direct-controlled municipalities, while there are 46 provinces and cities applying for becoming the second batch of low-carbon pilot regions, involving 25 provinces, autonomous regions and direct-controlled municipalities, of which 29 provinces and cities have been approved as the second batch of low-carbon pilot regions, involving 24 provinces, autonomous regions and direct-controlled municipalities. The expansion of the scope of the low-carbon pilot project shows the the idea of low-carbon development has become rooted in people's hearts, and has become a critical aspect for local governments to achieve transformation and development.

1.2.2 Emphasis on Top-level Design by Pilot Application Regions

Unlike the process of first determining low-carbon pilot regions and then approving their pilot work implementation plan in the application for the first batch of low-carbon pilot regions, in the application for the second batch, the pilot work implementation plan is first reviewed and modified to select the regions with a mature design of a work plan and a good work foundation as pilot regions. As a result, regions applying for becoming the second batch of low-carbon pilot regions have made great efforts in advance at fully demonstrating and modifying their work implementation plans, which remarkably strengthens the top-level aspect of the designs, as well as the comprehensiveness and guidance of the low-carbon developmental work.

1.2.3 The Objective of Quantitative Emission Control in the Work Plan

Based on the initial draft of the pilot work implementation plans submitted by pilot application regions in June 2012, most regions set the control objective not lower than that assigned by the national or provincial authority in their work plans. For instance, Beijing, Shanghai, Suzhou, Zhenjiang, Shijiazhuang and Guangyuan explicitly show a total volume target for greenhouse gas emission control, a peak time or per capita emission control target in their work plans; Beijing and Shanghai also set a total volume target for energy consumption control. After several rounds of review of the low-carbon pilot work implementation plans, various regions have brought forward an emission peak value on the basis of their own conditions.

1.2.4 Good Foundation of Preliminary Work

By fully drawing on the experience of the first batch of low-carbon pilots, various regions have carried out a great amount of work in capacity building, in fund guarantees and in the preparation of the greenhouse gas emission list before applying to be one of the low-carbon pilot regions, and have a good foundation for future work. In terms of emphasis on organization and leadership, the applying regions have formed the leading groups in response to climate change or of low-carbon development, with the local governmental leaders as group leaders and the leaders of relevant departments as group members, headed by the Development and Reform Commission of each province (city), so as to tackle climate change and achieve low-carbon development. In terms of input increase and low-carbon development fund guarantees, Wenzhou of Zhejiang Province, Ganzhou of Jiangxi Province, Jiyuan of Henan Province, Jinchang of Gansu Province and other cities have set up special funds for low-carbon city construction in different forms, to strengthen the support for low-carbon technologies' research & development and for capacity building projects. In terms of the preparation of the greenhouse gas emission list, a number of cities such as Beijing, Shanghai, Zhenjiang, Wenzhou, Chizhou, Hulun Buir, Huai'an, Ganzhou, Jilin, Zunyi and Jinchang are actively engaging in the preparation of the list, and in establishing and improving the statistics, computation and evaluation system of greenhouse gas emission.

2 Needs Analysis of the Policies and Technologies Promoting Urban Low-carbon Development

2.1 *Low-carbon Development is in Urgent Need of Top-level Design*

At present, low-carbon development is in urgent need of top-level design, and pilot work is prone to deviation. Low-carbon development is a fresh new type of work, as well as a systematic kind of engineering involving developmental mode, production mode, lifestyle, and interest adjustment of the various parties concerned. Now, China is undergoing fast urbanization and industrial development. It is really a great challenge to exercise a concurrent control over stock and increment for low-carbon development. One the one hand, fast urbanization is always accompanied by extensive construction of urban infrastructures including housing, roads, pipe networks and sanitary facilities and a huge consumption of steel and cement; the construction, utilization and maintenance of such facilities and buildings all require more energy than ever before, and will produce more greenhouse gas. On the other hand, the construction of urban infrastructures has long-term lock-in effect. So it is badly in need of top-level design to avoid a high-carbon mode in the future long-term development locked-in by the developmental mode of the current urbanization and industrialization.

2.2 *Support Policies for Low-carbon Development Remain to be Improved*

In terms of local guidance policies, the central government issued the *Notice of the State Council on Issuing the Work Plan for Greenhouse Gas Emission Control during the 12th Five-Year Plan* (G. F. (2011) No.41) which established the overall requirements, main objectives and control measures for the national greenhouse gas emission control work, and it also proposed the establishment of support policies, pilot construction norms and evaluation criterion, and an evaluation method to enhance the support, overall coordination and guidance of the pilot work. The No. 41 document first presented the CO_2 emission reduction index per unit of GDP in various regions during the "12th Five-Year Plan"; it viewed the index as a quantitative objective of low-carbon development, required local governments at all levels to decompose and put the index in place, and work out an implementation plan. Besides, there are also some other guidance policies (see Table 5.1).

Until today, not many guidance policies directly related to local low-carbon pilot work have been issued by the central government. As a whole, low-carbon pilot work is practical work from bottom to top. Under the idea of "carrying out implementation, summary, learning and promotion simultaneously," regions are encouraged to conduct "early and pilot implementation" locally to actively explore the mode and path for the realization of low-carbon development from bottom to top. Focusing on specific tasks assigned by the central government, pilot regions actively explore and improve supporting policies on the basis of order and control, special funds, financial policies, a carbon trading mechanism, governmental procurement and preferential service. However, no great support has been made for low-carbon pilot work, and supporting policies badly need to be improved. The road towards transformation for Chinese economic development in the future needs the experience obtained through low-carbon developmental pilot work, further forming beneficial guidance for low-carbon development of different provinces having different natural conditions and developmental foundations.

2.3 *Low-carbon Development is in Urgent Need of Developmental Technology*

Pilot work is facing bottleneck constraints due to a low technological level, an imperfect economic structure, a large population base and urban sprawl. As a consequence, low-carbon industries are low in economic aggregates, and are incapable of supporting the development of pilot provinces and cities; the level of low-carbon technology is also low, and is unable to meet the developmental needs of pilot provinces and cities; in addition, the idea of low-carbon

TABLE 5.1 *Guide policies released by the central government*

Type of policy	Name of document	Issuing authority	Date	Main content
Launching carbon emission trading pilot work	*Notice on Launching Carbon Emission Trading Pilot Work*	National Development and Reform Commission	October 2011	Agreeing to carry out carbon emission trading pilot work in Beijing, Tianjin, Shanghai, Chongqing, Hubei Province, Guangdong Province and Shenzhen
	Interim Measures for Administration of Greenhouse Gas Voluntary Emission Reduction Transaction	National Development and Reform Commission	June 2012	Ensuring that voluntary emission reduction work is carried out in an orderly fashion.
Enacting industrial developmental planning, and establishing a low-carbon industrial system	*Decision of the State Council on the Acceleration of Fostering and Developing Strategic Emerging Industries*	State Council	October 2010	Speeding up the cultivation and development of strategic emerging industries such as energy-saving environmental protection, new generation information technology, biology, high-end equipment manufacturing, new energy, new material and new energy vehicles.
	The 12th Five-Year Plan for National Strategic Emerging Industries	State Council	July 2012	Speeding up the cultivation and development of strategic emerging industries such as energy-saving environmental protection, new generation information technology, biology, high-end equipment manufacturing, new energy, new material and new-energy vehicles.
Preparing list	*Notice on Launching the Preparatory Work on a Greenhouse Gas List at the Provincial Level*	National Development and Reform Commission	September 2010	Requiring all provinces, autonomous regions and direct-controlled municipalities to launch preparatory work on the 2005 greenhouse gas emission list

consumption is not prevalent in the society, which restricts the sound development of pilot provinces and cities. Today, the structure of Chinese energy is still dominated by coal, causing low efficiency of energy utilization and difficult adjustment within the energy structure. China is undergoing rapid development of industrialization and urbanization now, so some high-consuming industries are still pillar industries supporting economic development. To reach the objective of carbon emission reduction while maintaining economic development, it is absolutely necessary to play the vital role of industrial structure adjustment and optimization in low-carbon development. Energy saving-based industrial structure is adjusted mainly in the following two aspects. First, a great effort on technological innovation and promotion has been made to realize "low-carbon" in high-carbon industries; second, support policies are formulated and improved in order to speed up the development of emerging low-carbon industries.

2.4 *There is a Lack of a System for Low-carbon Evaluation and Assessment Index*

The main obligatory targets, as set out in the work plan of low-carbon pilot regions, include energy consumption reduction index per unit of GDP, carbon emission reduction index per unit of GDP, the ratio of non-fossil energy in primary energy consumption, and forest coverage index, etc. However, the current governmental evaluation system, method and mode do not match the low-carbon developmental mode with a big gap, and the method used to assess carbon intensity reduction remains to be enacted and improved. For local governments, the lack of a specific index measuring the condition and the effect of low-carbon development will result in making it difficult to formulate the local policies.

2.5 *The Foundation of Statistical Data is Weak*

The evaluation of low-carbon development requires accurate and comprehensive greenhouse gas emission data. The authenticity and accuracy of the local GDP and greenhouse gas emission data have been the long-term problem remaining to be resolved by the central government. Though various regions have prepared a greenhouse gas emission list and have set quantitative objectives based on that list, detailed data regarding the condition of emissions of various industries and fields are far from enough. It is mainly because of very little or no initiative on the part of the enterprises in providing such data, and the limited scope of investigation. The investigation of the enterprises below a designated size is incomplete due to objective reasons, causing a lack of data. So the only way is to make a simple estimation based on output value, power consumption and other statistical data, which leaves room for great error.

In addition to all of the above, there is also a lack of fund support, financial and taxation policies and other incentive measures. Subject to the restriction of administrative management duties, the economic means adopted by local governments are limited, so the fundamental solution is to take action at a national level. The construction of a system for the computation of greenhouse gas emission statistics remains to be enhanced. At present, the technology for the preparation of the list is specially designed for province-level greenhouse gas emission calculation, and cannot be adapted to and used for the levels of cities (district and county), parks and enterprises. Moreover, talent cultivation and team construction remain to be enhanced, and local low-carbon development lacks support from research institutions, specialized institutions and social organizations.

3 Promotion of Urban Low-carbon Industry Development and Park Construction

For low-carbon development, the construction of a low-carbon industrial system is one of the important tasks. Strategic emerging industries are critical forces for guiding social and economic development in the future. So, developing these industries is a significant strategy for various countries to carry out in order to be leaders in the new round of economic and technological development. Accelerating the cultivation and development of strategic emerging industries (such as energy-saving environmental protection, new-generation information technology, biology, high-end equipment manufacturing, new energy, new material and new-energy vehicles) is also an important aspect for Chinese low-carbon development. For this, pilot provinces and cities have worked out comprehensive plans involving agriculture, secondary industries (the advanced manufacturing industry and the hi-tech industry), service industries and other major industries in terms of the adjustment of the industrial structure. In different industries, pilot provinces and cities develop themselves based on their own conditions, forming local features. As a whole, the adjustment within the industrial structure and energy structure in pilot provinces and cities presents the following three characteristics.

3.1 *Low-carbon Upgrading and Transformation of Traditional Industries is the Foundation of the Adjustment of the Industrial Structure*

Low-carbon upgrading and transformation of traditional industries includes the elimination of outdated production facilities in the electrical industry, the steel industry, the chemical industry, the cement industry and the dyeing and finishing industry, and the enhancement of energy-saving supervision on high-consuming and high carbon emission industries such as the metallurgy industry, the power industry, the chemical industry and the petrochemical industry. By full use of energy auditing and clean production examination, the energy-saving and cost-reducing work of key energy-consuming enterprises is emphasized so as to boost the low-carbon upgrading and transformation of these industries.

Besides, the advanced manufacturing industry and the hi-tech industry are the focus of industrial adjustment. Pilot provinces and cities are based on the foundation and advantages of local traditional industries for the formulation of various specific plans for developing advanced manufacturing industries, including the equipment industry, the steel industry, the automobile industry, the aviation industry, the power industry, the petrochemical industry and the tobacco industry, with full and accurate contents and clarified targets.

3.2 *The Construction of Low-carbon Parks and Communities is the Testing Field for the Implementation of Low-carbon Development*

Now, pilot regions are making efforts to design low-carbon developmental plans for parks and to construct low-carbon demonstration parks in order to achieve the low-carbon transformation of some industrial parks. For instance, Tianjin Economic and Technological Development Zone organized key enterprises to carry out greenhouse gas emission calculation, constructed low-carbon demonstration buildings to display low-carbon technologies and construction products, and set up low-carbon technology supermarkets for facilitating the promotion and application of low-carbon technologies. Chongqing City has selected Liangjiang New Area, Fuling and Wanzhou for the construction of low-carbon industrial parks.[9]

By an implementation plan of low-carbon community construction and in combination with circular economy demonstration construction, some pilot regions incorporate the ideas of green development, circular development and low-carbon development into their community construction, and conduct pilot construction in communities with good achievements in resource conservation and low-carbon life to encourage their inhabitants to live a green and low-carbon life.

---

9 "List of Chongqing 2012 Major Preliminary Projects," http://www.keyproject.cq.cn/Doc/18%E9%99%84%E4%BB%B63.htm.

3.3 *Modern Service Industries become the Focus of Structural Industrial Adjustment*

Pilot provinces and cities spare no effort in developing modern service industries and view it as an important aspect of low-carbon pilot work. For example, Guiyang suggests that "we will continue to energetically develop the tertiary industries led by the tourism industry, the cultural industry, the exhibition industry, the leisure real estate industry and the modern logistics industry." Shaanxi Province strives to develop the engineering design industry, the information consulting industry, the service outsourcing industry, the cultural and creative industry and the low-carbon consulting service industry to try its best to build itself an international first-class service base and national creativity demonstration base. Tianjin City introduces the objective that the service sector value added will account for 50% of its GDP.[10] Liaoning Province strongly supports the exploitation and utilization of non-fossil energy, pays attention to wind power development, actively explores solar energy and biomass power generation, develops nuclear power in a safe and orderly fashion, and promotes the construction of green energy demonstration counties in an orderly fashion. Heilongjiang Province has been developing strategic emerging industries and hi-tech industries rapidly, and has gained great developmental momentum in various service industries such as cultural creativity, finance, logistics, tourism, cloud computing and the internet of things, thus providing strong support for economic growth.[11] Liaoning, Shaanxi, Yunnan, Hubei, Tianjin, Baoding, Nanchang and Xiamen all have set quantitative development objectives in tertiary industries, and have made great efforts at promoting the development of service industries.

4 Guidance Mechanism and Policy Measures for the Low-carbon Development of Pilot Cities

By comprehensively implementing the Scientific Outlook on Development and centering on the fundamental transformation of the economic development model, pilot cities follow the thought of "long-term based, innovation mode and demonstration first" and the idea of sustainable development, under the motivation of technological, institutional and administrative innovation, aim

10 "Tianjin Focusing on Low-carbon Development and Constructing an Ecological City," Xinhuanet, http://news.xinhuanet.com/house/tj/2013-08-26/c_117091773.htm, 2013-08-26.

11 "The 12th Five-Year Plan for Heilongjiang Strategic Emerging Industries," http://gjss.ndrc,gov.cn/zcgh/dfghzc/gdxggh/t20120328_471631.htm.

at achieving low-carbon development, enhancing urban sustainability and promoting human-nature harmony, focus on the optimization of the energy structure and the increase of the efficiency of energy utilization, emphasize on the reduction of greenhouse gas emissions, a decline in carbon emission intensity per unit of GDP and the construction of low-carbon production mode and lifestyle, and adhere to the principles of governmental promotion, planning first, demonstration driving, public involvement, key promotion and step-by-step progress, to speed up the adjustment of the economic structure and the upgrading of the industrial structure. Besides, pilot cities attach importance to the cultivation of low-carbon industries, to the development of a low-carbon economy, to building a low-carbon society and to improving the supporting system, including the management and policy mechanisms, so as to explore a low-carbon development road suitable to the Chinese situation. This is a basic idea for China's low-carbon provincial and urban pilot work. To be specific, the following four aspects are the focus.

4.1 *Combination of Governmental Promotion and Public Involvement*

The government takes full advantage of planning guidance and policy support means to guide and boost low-carbon development, and at the same time, brings into full play enterprise initiative and creativity to set up the production system and the developmental mode which conform to low-carbon developmental requirements. The government may begin an extensive, intense and consistent educational program to raise the low-carbon awareness of the public and create a cultural atmosphere conducive to harmonious development between mankind and nature. In addition, the government may also call on the whole society to participate, and set up a public involvement mechanism and an opinion supervision mechanism, thus consolidating the public foundation of low-carbon development.

4.2 *Combination of Guidance through Ideas and Institutional Innovation*

The shift of ideas is the source of low-carbon development, and institutional innovation is the consistent momentum for it. By consensus propaganda, training education and typical guidance, traditional developmental thoughts are abandoned, and new low-carbon economic ideas dominated by energy conservation and emission reduction are developed. Actions are taken to promote the establishment of a laws and regulations guarantee system, a policy supporting system, a technological innovation system, a motivation and restriction mechanism, as well as a carbon emission monitoring, administration and trading system, which all correspond to low-carbon development. By relying on innovation institutions, laws and policies, the vitality of the microeconomic unit is regulated and stimulated, and various economic activities are guided, encouraged and normalized, thus promoting low-carbon development.

4.3 *Combination of Independent Innovation and the Introduction of Technology*

Technological innovation is one of the critical aspects of low-carbon development. On the basis of the introduction of technology, pilot provinces and cities manage to raise low-carbon technology competitiveness through introduction, implementation and re-innovation. At the same time, pilot provinces and cities strive to strengthen their innovation capabilities, take full advantage of their own features, enhance cooperation with universities and research institutions, and improve the capabilities for development and innovation of key technologies, in order to be in the commanding position in global low-carbon technology in the future. Furthermore, pilot provinces and cities are intensifying the reform of their scientific and technical system, playing the fundamental and leading role of technology in low-carbon development and promoting independent innovation and technological progress in key fields, so as to facilitate energy conservation and emission reduction, control greenhouse gas emissions and enhance their ability to respond to climate change.

4.4 *Combination of Demonstration Driving and Scientific Promotion*

According to Chinese realities, pilot provinces and cities set stage objectives of low-carbon socioeconomic development, and make a scientific plan and an overall layout to expedite low-carbon development step by step. On the basis of economic development practice, pilot provinces and cities conduct pilot work first in high-consuming and high-polluting industries such as the power industry, the transportation industry, the construction industry, the metallurgy industry, the chemical industry and the petrochemical industry, and at the same time, carry out organizational work in low-carbon developmental regions. Besides, with the principle of "limited objectives and key breakthroughs," pilot provinces and cities make bold innovations, summarize experiences, make full use of the demonstration and guidance effects of typical cities and projects, and make it be quickly spread and applied in associated fields and similar areas.

CHAPTER 6

Viewing China's Green Building Development from a Green Building Label Evaluation

Li Xun and Li Congxiao

Abstract

The total energy consumption of China's buildings accounts for 20.7% of social energy consumption, with great potentials for energy conservation and emission reduction. With great advantages in conservation of energy, water, land and materials, environmental protection and other aspects, green buildings will play an important role in achieving China's sustainable development. This research comprehensively analyzes the overall condition of green buildings in China in terms of the developmental status of the green buildings, related policies, future developmental trend, the impact on the climate and other aspects. It gives a detailed introduction to green building label evaluation, standards and management system and incentive policies of the central and local governments on green buildings; besides, it analyzes the scale development of green buildings—concept, planning and evaluation of green ecological urban areas. Green buildings conform to the trend in natural social development, and are expected to continue developing rapidly and play their due role in urbanization.

Keywords

green building – green building label – green ecological urban area – energy conservation and emission reduction

* Li Xun, Vice President of the China Academy of Urban Planning & Design and Secretary General of the Chinese Society for Urban Studies, was awarded the title of young and middle-aged management expert of science and technology with outstanding contributions by the Ministry of Construction in 1995, whose main research field and direction involve regional planning, urban development strategies, urban master planning, small town planning, urban underground space planning, developmental zone planning, and urban sustainable development, etc.; Li Congxiao, Director of the CSUS Green Building Research Center, Ph.D., and professor level senior engineer.

 DOI 10.1163/9789004280021_007

In China's energy consumption structure, the total energy consumption of buildings takes up about 20.7% of social terminal consumption, with a great potential for energy conservation and emission reduction. Since 2008, with the emergence of green buildings around the world, which reflect the concept of energy conservation and environmental protection, China's green buildings have also experienced the growth process from starting to flourish. Green buildings can most save resources, reduce pollution and improve people's living environment, conductive to people's health and safety. Green buildings are relatively mature technically, with certain economic benefits. In general, the development of China's green buildings is of important significance and profound influence in responding to climate change, saving energy and reducing emissions, improving people's livelihood, developing emerging industries, establishing new urbanization patterns, and promoting the sustainable development of China's economy.

1 The Developmental Status of China's Green Buildings

In the past few years, the quantity of green building evaluation label projects has kept a strong trend towards growth. Up to the first half of 2013, a total of 978 green building evaluation label projects had been evaluated in China (see Figure 6.1), and the total building area exceeded 100 million m^2, including 919 design label projects, accounting for 94% of the total and 59 operation label projects, accounting for 6% of the total. Residential buildings take up 55% and public buildings take up 45%. The building area of the green building label projects is about 100,000 m^2 on average, and the project scale is much larger than foreign green building label projects. In the past 3 years, both the quantity of projects and the building area of green buildings each year have exceed the sum total of the same kind of data in previous years.

Seen from the point of view of the regional distribution of the projects, all provinces, autonomous regions and municipalities, except Gansu, Qinghai, Guizhou and Tibet, have green buildings obtaining the label. The regions with more than 30 label projects, with 10–30 label projects and with fewer than 10 label projects take up 1/3 respectively, where the quantity of green buildings in coastal areas including Jiangsu, Guangdong and Shanghai is far ahead. In 2012, the increase in the number of green building label projects of all regions was accelerated, with an obvious increase in the number of green buildings in Shandong, Hebei, Hubei, Zhejiang, Beijing, Fujian, etc., as shown in Figure 6.2.

In the wave of the development of green buildings, real estate developers have become pioneers. Top application units with the maximum green

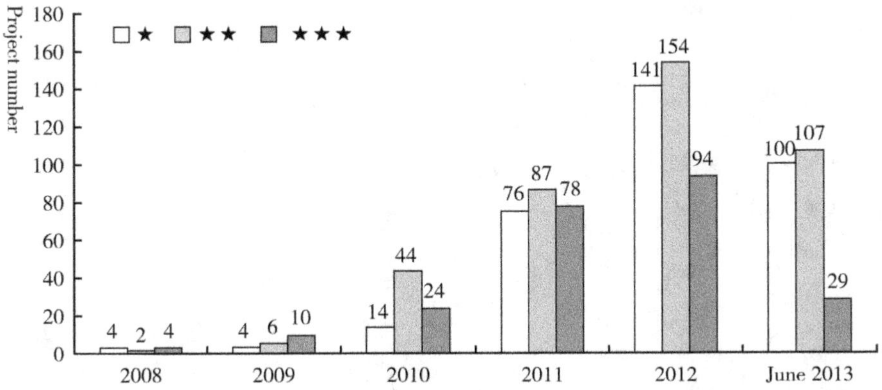

FIGURE 6.1 *Developmental situation of green building evaluation label projects over the years.*

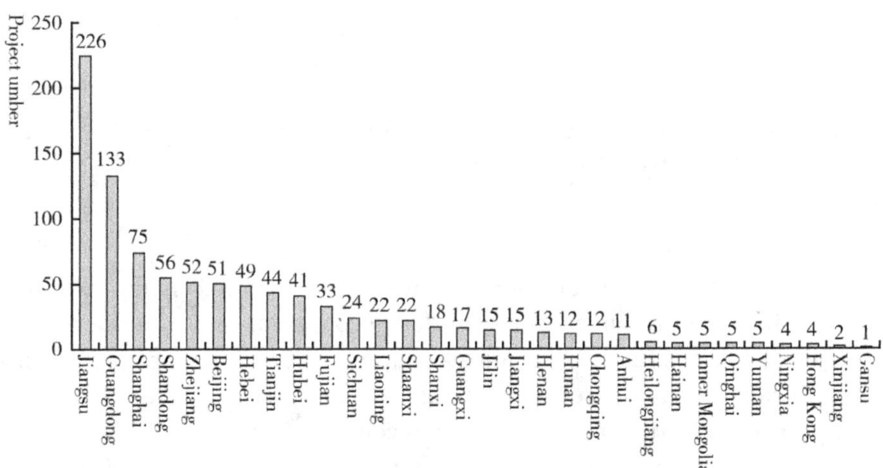

FIGURE 6.2 *Statistics regarding the quantity of green building evaluation label projects in all provinces, autonomous regions and municipalities between 2008 and 2012.*

buildings obtaining the label are respectively Wanda, Vanke, Greenland and other development enterprises (see Figure 6.3), and the quantity of green buildings in the three enterprises accounts for 1/4 of the total, where such groups as Vanke, Greenland, Wanda, and Landsea have the maximum number of application projects for residential green buildings and such groups as Wanda, Greenland, Suzhou Genway, and CMPD have the maximum application projects for public green buildings.

In China, industrial production energy consumption and building energy consumption are two fields with the most social energy consumption. Building

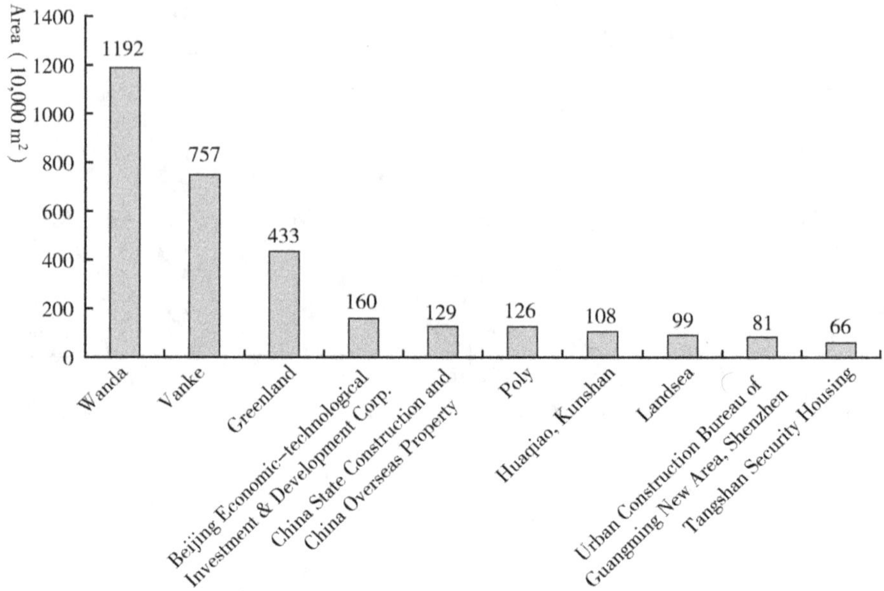

FIGURE 6.3 *The top 10 application units with the maximum area of green building evaluation label projects between 2008 and 2012.*

energy consumption accounts for 24% of the total national energy consumption, while industrial energy consumption accounts for 64%. Considering the characteristics of China's economic development at the present stage, the national economy greatly relies on industrial enterprises, so the energy consumption of industrial buildings should receive great attention. In 2012, China began to review the green industrial buildings and a total of 1,530,000 m^2 of industrial buildings obtained the green industrial building label (see Figure 6.4). Thus, large-scale industrial buildings also join the ranks of green buildings.

Regarding the management of green buildings, in order to carry out the green building label, the Ministry of Housing and Urban-Rural Development has issued the *Management Measures of the Green Building Evaluation Label*, the *Green Building Evaluation Standards* and the *Technical Instructions for Green Building Evaluation*, started green building label evaluation in 2008, and introduced regulations on the application for, review and management of green building labels, gradually forming a management system of green buildings. Currently, except for the fact that the Scientific and Technological Development Promotion Center of the Ministry of Housing and the Urban-Rural Development, and the Chinese Society for Urban Studies have authorized a review of the one-star, two-star and three-star green buildings, all

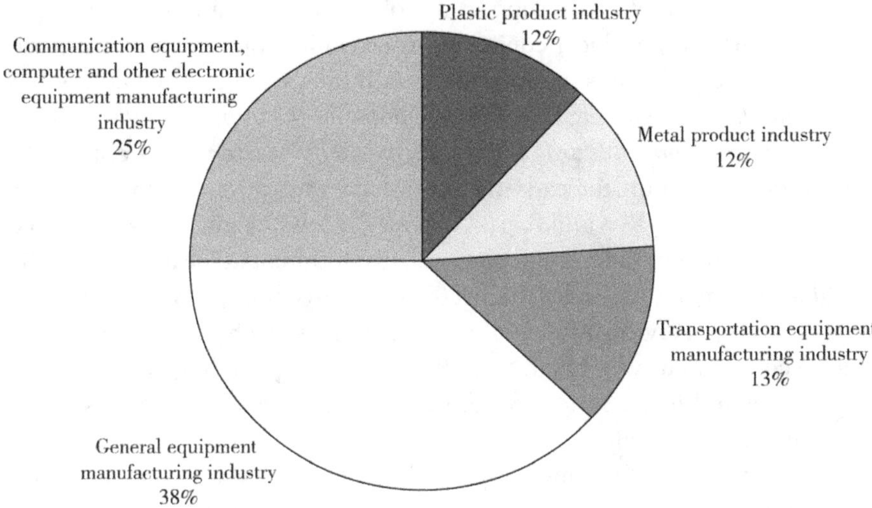

FIGURE 6.4 *Industrial field distribution of green industrial building evaluation label projects in 2012.*

provinces and autonomous regions can organize the review of one-star and two-star green buildings.

Regarding the evaluation of green buildings, China issued the national *Green Building Evaluation Standards* (GB/T50378) in 2006, which is comparable to foreign mainstream evaluation standards such as LEED in the USA and BREEM in Britain, and there is not much difference in content and grade settings. In order to further reflect the connotation of green buildings, China is formulating green building evaluation standards for different kinds of buildings in order to propose targeted energy conservation and environmental protection measures. Based on the current development, it is expected that a comprehensive standard system of green buildings will have basically been formulated in the next 1–2 years; China's green buildings will show a gratifying situation with a great variety and distinctive characteristics, and a large number of green buildings will spring up with the labels of green office, hospital, hotel, shopping mall, railway station, airport, industrial building, etc. China's green buildings, compared with the ones around the world, are at an advanced level in terms of both coverage of green buildings by standard details and specialty of specifications. In addition to new buildings, the existing buildings that take up the most proportion of the total buildings are the focus of energy conservation and environmental protection. At present, the formulation of green renovation evaluation standards for the existing buildings has also been started, which will play an effective supporting role in the evaluation of the existing buildings. Besides, all local construction administration departments, combined with actual local situations and conditions, have formulated the local *Green Building Evaluation Standards*, the *Rules for the Implementation of Green Building Evaluation Standards* and the *Technical Instructions for Green Building Evaluation*, to support the carrying out of local evaluation label work of one-star and two-star green buildings. Currently, a total of 15 Chinese provinces, autonomous regions and municipalities have formulated local green building evaluation standards. Besides the related standard details, related comprehensive evaluation service systems of green buildings, such as the green performance comprehensive evaluation method of existing buildings, the evaluation technology and index systems of green buildings, are also being researched and improved gradually.

Regarding the performance of green buildings, green buildings encourage the use of passive technology, such as natural ventilation and lighting; they control energy with energy-consumption indexes in order to carry out constraint management in aspects such as heating, ventilation, water use and lighting; they encourage the application of renewable energy and reduce the consumption of traditional fossil energy. The remarkable results reflected in "4 conservations and 1 environmental protection" enable the green buildings to have a great potential for economy, the comfort of inhabitants, etc. For instance, in order to achieve the energy conservation of buildings, the improvement of thermal insulation performance of building envelopes is the most efficient way to reduce the energy consumption of buildings, while the envelopes of green buildings have good performance at this point. The energy consumption of outside doors and windows, glass curtain walls, lighting roofs and other transmitting building components is the weakest link in the thermal performance of building envelopes, while high-performance doors and windows, curtain walls and necessary sun-shading measures advocated by green buildings can greatly reduce the necessary energy consumption. Some technologies adopted in the HVAC (heating, ventilation and air conditioning) of green buildings, such as heating system energy-saving technology, air conditioning system energy-saving technology and ventilation energy-saving technology, effectively improve the overall efficiency and reduce the consumption of energy. The GSHP (ground source heat pump) technology, which is the one most widely-adopted in green buildings, saves 50%–60% of the energy compared with other conventional heating and air conditioning technologies, and its operating cost is reduced to 30%–70%. According to the statistical analysis of partial green building label projects, their energy conservation is equal to 15,400 tce per year, an average greening rate of 37.6%, CO_2 emission reduction

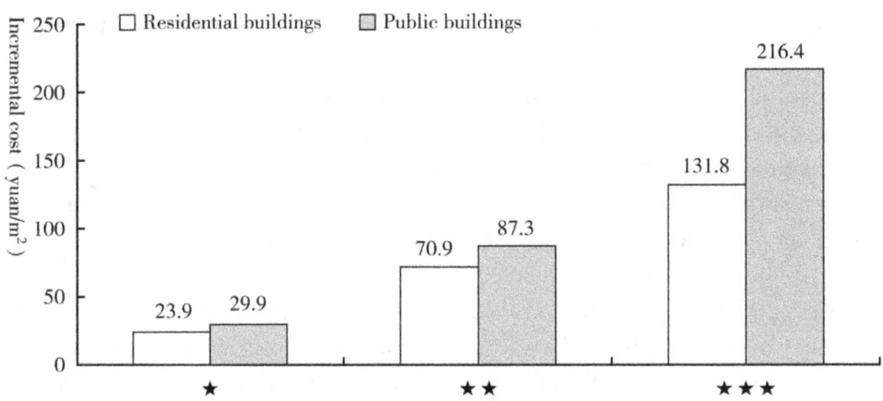

FIGURE 6.5 *Incremental cost of residential buildings and public buildings.*

of 40,400 tons per year, the average utilization rate of nontraditional water sources of 15.2%, the average energy-saving rate of buildings 58.34%, and development and utilization of underground space of 1,511,000 m^2.[1] In spite of such good performance in energy conservation and emission reduction, green buildings are, mistakenly, often considered as high-cost buildings. In fact, according to the incremental cost analysis of green buildings in 148 projects reviewed by the Chinese Society for Urban Studies in 2012, the incremental cost of one-star, two-star and three-star green buildings in residential building projects is respectively 23.9 yuan/m^2, 70.9 yuan/m^2 and 131.8 yuan/m^2, and that of one-star, two-star and three-star green buildings in public building projects is 29.9 yuan/m^2, 87.3 yuan/m^2 and 216.4 yuan/m^2, respectively, as shown in Figure 6.5. Through calculation, for one-star green buildings, the static payback period is basically 1–3 years, while the incremental cost of most green buildings can be recouped within 2–6 years, so in terms of the 70-year design life of the buildings, green buildings can save considerable energy resources within their life cycle.[2]

1 Scientific and Technological Development Promotion Center of the Ministry of Housing and Urban-Rural Development, *Energy Conservation Development Report of China's Buildings (2012)—Renewable Energy Building Application*, China Building Industry Press, 2013, pp. 12–13.

2 Qiu Baoxing, *Situation and Mission of China's Green Building Development and Building Energy Conservation*, Construction Science and Technology, 2012 (10).

2 Incentive Policies of the Chinese Government for Green Buildings

The overall developmental strategy of China's green buildings is, by taking the achievement of the national planning obligatory target as an opportunity, to advance the development of green buildings, greatly develop energy-saving and land-saving housing, and promote China's sustainable development by achieving the "4 conservations and 1 environmental protection" of green buildings. The State Council issued the *Action Plan of Green Buildings* (G.B.F. (2013) No. 1) in January 2013, and the Ministry of Finance and the Ministry of Housing and Urban-Rural Development jointly issued the *Opinions on the Implementation of Accelerating Development of China's Green Buildings* (C.J. (2012) No. 167) in May 2012, both playing an extremely important role in the development of green buildings throughout China. The documents pointed out we should comprehensively promote the green development of buildings in urban and rural areas, comprehensively enforce green building standards by 2014 for public buildings subsidized by the government and security housing in municipalities, cities specifically designated in the state plan and provincial capitals; we should newly build 1 billion m^2 of green buildings during the "12th Five-Year Plan" period, and make 20% of the new buildings in urban areas reach the requirements of green building standards. By 2020, green buildings will account for more than 30% of the new buildings, and the energy consumption level during building construction and use will approximate or reach the current level of developed countries. We should comprehensively execute green building standards since 2014 to the public buildings subsidized by the government, the security housing in municipalities, cities specifically designated in the state plan and provincial capitals, and large public buildings with the individual building area exceeding 20,000 m^2, and encourage real estate developers to construct green residential areas, so as to practically promote construction of green industrial buildings. The documents required regularizing the market subjects' behaviors by means of policies, planning and standards, comprehensively applying such economic means as price, taxation and finance, and playing the fundamental role of the market in resource allocation. In the meantime, the country should strive to develop green building materials, advance building up industrialization, promote resource utilization of building wastes, and actively promote the construction of green rural housing. The documents also required incorporating the completion of the action plan for green buildings and the implementation of measures into the system of the evaluation of responsibility of the energy-saving target of provincial people's governments.

According to the documents, regarding the financial support policies, financial incentives for buildings reaching the two-star level of national green building evaluation standards and above should be provided. The incentive standards of 2012 were as follows: 45 yuan/m^2 for two-star green buildings and 80 yuan/m^2 for three-star green buildings. In the *Planning Outline of China's Middle and Long-term Scientific and Technological Development* (2006–2020), "building energy conservation and green buildings" are considered as preferential projects under the key field "urbanization and urban development." The *Special Planning of the Scientific and Technological Development of Green Buildings during the 12th Five-Year Plan* (2012) issued by the Ministry of Science and Technology required carrying out common key technology researches on green buildings and researches and demonstrations of the promoting technology of the industrialization of green buildings, and establishing technical standards and specifications and comprehensive evaluation service system of green buildings; the *Special Planning of Building Energy Conservation during the 12th Five-Year Plan* (2012) issued by the Ministry of Housing and Urban-Rural Development required striving to promote the development of green buildings and achieve the popularization of green buildings.

The governments at all levels pay great attention to green buildings; they have formulated one after another their own green building development policies, and have introduced or are preparing to introduce a series of incentive policies, achieving positive results. For instance, the People's Government of Beijing Municipality has granted an additional subsidy of 50% for high-star green buildings based on national financial subsidies, announced that as of June 1, 2013, all new buildings in Beijing should adopt green building standards, and encouraged the construction of buildings subsidized by the government and large public buildings with individual building areas exceeding 20,000 m^2 according to the two-star level and above of green buildings standards in the *Action Implementation Plan of Green Buildings in Beijing Municipality*. The People's Government of Shenzhen Municipality has also specified that from 2013, all new buildings should enforce the green building standards to promote the popularization of green buildings on a large scale. The People's Government of Jiangsu Province, with the greatest number of green building labels, proposed in the *Action Implementation Plan of Green Buildings in Jiangsu Province* that, during the "12th Five-Year Plan" period, the total area of the projects reaching the green building standards in the whole province should exceed 100 million m^2, including a newly-increased 15 million m^2 in 2013; in 2015, new urban buildings in the whole province should be comprehensively designed and constructed according to the one-star and above green building standards; in 2020, 50% of the new urban buildings in the whole province should be designed and constructed according to the two-star and above green building standards.

Considering the responses from all of the regions, the developmental trend of China's green buildings goes from individual buildings to regional buildings. New buildings in more and more cities, including Beijing and Shenzhen, will gradually start to comprehensively implement green buildings.

From an overall perspective, green buildings are showing a developmental trend from product to industry, and are gradually maturing. With the development from original single green buildings to large-scale construction and development of green buildings, products, technology and services related to green buildings have also created new industries. A batch of new technology and new products related to such aspects as conservation of energy, water, materials and improvements of construction environments are gradually applied; design, construction, operation management, consulting services, new energy and other industries are also appearing accordingly. The industrial scale is continuously increasing, and products with high energy consumption and high pollution are being eliminated, promoting a virtuous circle between the economy and the environment. In the *Opinions on the Implementation of Accelerating the Development of China's Green Buildings* (C.J. (2012) No. 167), jointly issued in 2012 by the Ministry of Finance and the Ministry of Housing and Urban-Rural Development, in addition to clearly proposing the development of large-scale green ecological urban areas, it also stresses the promotion of science and technology and the industrial development of green buildings, and it strengthens the construction of the comprehensive capacity of green buildings, to achieve a practical sustainable development of green buildings and green ecological urban areas.

With economic development, building areas in China have sharply increased for several decades and are also increasing each year. The completed building area in urban areas reached 1.75 billion m^2 in 2010 and exceeded 2 billion m^2 in 2012. Compared with about 48 billion m^2 of existing buildings in China, green buildings only account for about 0.2%. The construction industry can play an important role in achieving the goal of energy conservation and emission reduction in China.

3 Scale Development Trend of China's Green Buildings—Green Ecological Urban Areas

The construction of ecological cities is becoming a mainstream direction in current urban development around the world, and is also an important goal of China's urban planning and construction. The central government carries out scale development of green buildings by promoting the construction of green ecological urban areas. The Chinese Ministry of Housing and Urban-Rural Development promotes the construction of low-carbon ecological cities from several aspects, including green buildings, green ecological residential areas, ecological reconstruction of existing urban areas, and green ecological (new) urban areas. Particularly, in recent years, centering on various issues in the construction of ecological cities, the Ministry of Housing and Urban-Rural Development has strengthened pilot demonstrations with continuous promotion. Up until April 2012, a total of 280 cities above prefecture level (included) proposed to take such developmental patterns as "low-carbon city" or "ecological city" as their urban development goals, accounting for 97.6% of related cities. These plans have laid a foundation for scale development of green buildings, enabling the expansion of benefits of individual green buildings in combined development with communities and cities, creating multiple effects.[3]

In June 2011, the Ministry of Housing and Urban-Rural Development issued the *Interim Procedures for Application Management of Low-carbon Ecological Pilot Cities (Towns)* (J.G. (2011) No. 78), playing a role in guiding and regulating the construction of local ecological cities. Since 2012, the Chinese Ministry of Housing and Urban-Rural Development and the Chinese Society for Urban Studies have carried out investigations and surveys in nationwide ecological pilot cities, and have proposed six basic access conditions for applying to become a national low-carbon pilot ecological city, including: 1) A compact and mixed land use pattern; 2) Proportion of renewable energy ≥ 20%; 3) Green buildings ≥ 80%; 4) Protection and utilization of biological diversity and natural patches; 5) Green transport: walk, bicycle, public transport ≥ 65%; 6) Reject industrial projects with high energy consumption and high emission. The basic requirements for "green ecological urban areas" are as follows: formulate an overall plan, a regulatory detailed plan and such special plans as construction, municipal administration and energy in line with green, ecological and low-carbon concepts, and establish corresponding index systems.

In September 2012, the Ministry of Finance and the Ministry of Housing and Urban-Rural Development jointly issued the *Opinions on the Implementation of Accelerating the Development of China's Green Buildings* (C.J. (2012) No. 167), pointing out we should comprehensively accelerate the development of China's green buildings and green ecological urban areas by means of the government's financial subsidies, etc. In October 2012, the Ministry of Finance and the Ministry of Housing and Urban-Rural Development jointly issued the *Notice on Applying for Financial Subsidies of Green Ecological Urban Areas in*

3 Qiu Baoxing, *Further Accelerate Development of Green Buildings*, Urban Studies, 2011 (7).

2012 (C.J.B.H. (2012) No. 71), making it clear that urban areas reaching the standards of "national green ecological urban areas" can obtain 50 million RMB from the central government as subsidies.

In April 2013, the Ministry of Housing and Urban-Rural Development issued the *Developmental Plan for Green Buildings and Green Ecological Urban Areas during the 12th Five-Year Plan* (J.K. (2013) No. 53), pointing out that 100 new urban areas will be chosen for planning, construction and operation according to the standards of green ecological urban areas during the "12th Five-Year Plan" period. According to the Plan, related departments of the Ministry of Housing and Urban-Rural Development have further quantified the review requirements for green ecological urban areas: population density ≥ 10,000 people/km^2; gross plot ratio in built-up areas ≥ 1.1; proportion of renewable energy ≥ 20%; utilization rate of nontraditional water sources ≥ 20%; reutilization rate of industrial water ≥ 90%; per capita comprehensive water consumption must be lower than the standard lower limit of similar regions and countries; proportion of green buildings; net loss of ecological conservation areas such as natural wetlands ≤ 10%; local plant index ≥ 0.7; green coverage rate ≥ 30%; green travel ratio ≥ 65%; and reasonable block length of density of road network ≤ 180 m.

Currently, there are several problems and deficiencies in the development of ecological cities, mainly including: high developmental orientation, excessive geomorphic transformation, excessive developmental intensity, high construction costs, insufficient understanding of green buildings; few characteristics of cities and similar index quantification. This means that the construction of green ecological urban areas needs to clearly establish related principles, such as setting up an ecological baseline,[4] protecting local geo-ecological environments and controlling energy consumption and pollution. Therefore, there needs to be careful research and prudent promotion. Since 2007, in order to scientifically guide the construction of low-carbon ecological cities, the Ministry of Housing and Urban-Rural Development has jointly established, one after another, the Sino-Singapore Tianjin Eco-city, the Wuxi Taihu New City, the Shenzhen Guangming New Area, the Shenzhen Pingshan New Area, and the Tangshan Bay (Caofeidian) Eco-city together with Tianjin, Wuxi, Shenzhen and Tangshan, and has carried out research on demonstration areas. Meanwhile, the Shanghai Nanqiao New City, the Hebei "4+1" Demonstration City (Area), the Kunming Chenggong New City, Xiamen, Dezhou, Huainan and other cities have also carried out plans and construction practices. Moreover, foreign

4 Wu Yong, *Developmental Status and Prospects for Green Buildings, China's Green Buildings*, compiled by the CSCEC Construction Co., Ltd., China Building Industry Press, 2001, p. 26.

governments and some planning and construction groups have shown a strong cooperative interest in participating in the promotion of the development of China's green ecological urban areas, e.g. the Sino-Singapore cooperative project in 2007 (the Sino-Singapore Tianjin Eco-city), the Sino-Sweden cooperative project in 2010 (the Wuxi Sino-Sweden Low-carbon Eco-city), the Sino-Germany cooperative project in 2010 (the Qingdao Sino-German Eco-park), the Sino-Japan cooperative project in 2011 (the Tangshan Bay Eco-city Sino-Japan Cooperation Demonstration Area), and the Sino-US cooperative low-carbon eco-city cooperation pilot projects in 2013 (Weifang City and Rizhao City, Shandong Province, Hefei City, Anhui Province, Langfang City, Hebei Province, Jiyuan City and Hebi City, Henan Province). Through international cooperation, China needs to establish long-term stable working mechanisms, refine and intensify the pilot demonstration contents, and carry out research and construction regarding such aspects as urban planning, low-carbon green design construction and operational management of infrastructures, building energy conservation and green buildings, renewable energy application, green transport systems and ecological green space systems.

According to the *Opinions on the Implementation of Accelerating the Development of China's Green Buildings* (C.J. (2012) No. 167), jointly issued by the Ministry of Finance and the Ministry of Housing and Urban-Rural Development of China, for the green ecological urban areas reaching a one-star and above evaluation standard in the *Green Building Evaluation Standards* in the comprehensive implementation of new buildings, the central government will provide 50 million RMB as financial subsidies, which are mainly used for subsidies of the incremental cost of the construction of green buildings and related expenditures of green ecological planning, index system establishment, evaluation label and energy efficiency evaluation of green buildings in urban areas.

In 2012, a total of 26 urban areas in China applied for green ecological urban areas, and finally eight urban areas passed the evaluation and were determined as the first batch of national green ecological urban areas (including the Sino-Singapore Tianjin Eco-city, the Tangshan Bay Eco-city in Tangshan City, Hebei Province, the Wuxi Taihu New City in Jiangsu Province, the Meixi Lake New City in Changsha City, Hunan Province, the Chongqing Yuelai Eco-city, the Zhongtian Weilaifangzhou Eco-city in Guiyang City, Guizhou Province, the Kunming Chenggong New City in Yunnan Province, and the Shenzhen Guangming New Area), as shown in Table 6.1.

Furthermore, green buildings begin to develop from cities to villages and small towns. As a strategic choice to promote the construction of China's resource-saving and environment-friendly society and to change the

TABLE 6.1 *Eco-city details*

No.	Name	Planning area (km^2)	Planning population (10,000 people)	Per capita planning construction land use (m^2)
1	Sino-Singapore Tianjin Eco-city	30	35	57.91
2	Tangshan Bay Eco-city	74.3	80	92.59
3	Wuxi Taihu New City	150	100	100
4	Shenzhen Guangming New Area	156.1	80~100	72.54
5	Shenzhen Pingshan New Area	168	70	93.58
6	Shijiazhuang Zhengding New District	200	140	96.42
7	Qinhuangdao Beidaihe New Area	425.81	20 (in 2020)	109
8	Cangzhou Huanghua New City	110	50 (in 2020)	95.92
9	Zhuozhou Ecological Livable Demonstration Base	43	~50	85
10	Hefei Binhu New Area	196	100	122
11	Shanghai Nanqiao New City	71.39	75–100	61–81.33
12	Shanghai Hongqiao Central Business District	86	53	—
13	Chongqing Yuelai Eco-city	3.44	5.7	43.16
14	Changsha Meixi Lake New City	7.64	17.8	40.44
15	Chizhou Paradise Lake New City	14.53	14.55	85.03
16	Guiyang Zhongtian Weilaifangzhou Eco-city	9.53	17.26	40.74
17	Kunming Chenggong New Area Eco-city	160	150	107

developmental pattern of cities and towns, green buildings will play an increasingly great role in China's new urbanization construction. Compared with the gradually saturated land use area in cities, villages and towns have a great developmental potential and will become another main developmental space for buildings. Besides, combined with the construction of the new socialist countryside, China will construct a batch of energy-saving rural housing facilities and support 400,000 peasant households to renovate their dilapidated housing in rural areas.[5] Related specific incentive policies on green rural housing have been put on the agenda, e.g. Shandong Province will provide 20,000 RMB for each household as subsidies for green rural housing.

4 Effects of Green Buildings on the Climate

With the intense development of urbanization in China, the development of green buildings plays a key role in reducing the energy consumption and greenhouse gas emissions of buildings. By analyzing 79 green building projects which have obtained different-star green building labels, the average energy-saving rate reaches 58%, the average water-saving rate reaches 15.2%, the community greening rate reaches 38% and recyclable materials reach 7.7%. In the whole country, China's overall goal of building up energy conservation is kept within the range of 1 billion tce, and scale development of green buildings is an important method to achieve this goal, which may be kept at no more than 0.8 billion tce. The conservation of water and materials is especially important, so that industrial energy consumption can be reduced.[6]

In the "4 conservations" of green buildings, in addition to direct energy conservation, conservation of land, water and materials also indirectly achieve results of energy conservation. By measuring dozens of green building project samples, if 300 green buildings are built each year, 2.55 billion kWh will be saved during the "12th Five-Year Plan" period, equal to saving 876,000 tce; emissions of CO_2 will be reduced by 2.298 million tons; 90 million tons of water will be saved; and 330 million tons of recyclable materials will be produced. In line with the requirements of the *Action Plan for Green Buildings* of the State Council, China plans to complete an average of 200 million m^2 of green buildings each year during the "12th Five-Year Plan" period, i.e. saving 84.9 billion kWh, equal to saving 29.17 million tce; it also plans to reduce CO_2 emission by 76.52 million tons, save 3 billion tons of water and produce 11 billion tons of recyclable materials with obvious results in energy conservation and emission reduction.

According to the evaluation of some research, the energy consumption in the life cycle of green public buildings is 33% lower than that of ordinary

5 Wang Youwei, *Research on the Developmental Trend of China's Green Buildings*, Construction Technology, 2013 (11).

6 Qiu Baoxing, *Situation and Mission of China's Green Building Development and Building Energy Conservation*, Construction Science and Technology, 2012 (10).

public buildings, and 38% lower than that of large public buildings; the energy consumption in the life cycle of green residential buildings is 13% lower than that of ordinary residential buildings.[7] What is more, measures for roof greening promoted in green building evaluation standards, in addition to storing rainwater and purifying the air and water, can relieve the urban heat island effect by changing the underlying surface reflectivity. The environmental management and maintenance of green buildings and green ecological urban areas during the operation can reduce the production of anthropogenic heat in cities with an energy-saving, pollution-free and comfortable external environment, playing an important role in establishing a virtuous circle of green building development and climatic effect. It is worth pointing out that, in the evaluation of China's industrial green buildings, the incorporation of the industrial technological process into the energy conservation evaluation guarantees the greatest comprehensive high efficiency of energy utilization of industrial buildings. Furthermore, the local rural green buildings with Chinese unique building characteristics combine traditional good energy conservation with modern technology, and provide a comfortable living environment with a minimal impact on the environment, achieving the unity of people and nature.[8]

In short, with the intensification of research and continuous technological innovation and improvement, with good performance in low energy consumption, good thermal insulation performance, a high greening rate and the utilization of waste heat and renewable energy, as well as the optimization design with full life cycle of buildings according to local conditions, green buildings will play a positive role in reducing greenhouse gas emission and relieve the urban heat island effect of urbanization; moreover, they are of great significance in responding to climate change, providing a comfortable living environment for human beings, and achieving the sustainable development with harmonious coexistence between human beings and nature. Green buildings conform to a natural developmental trend of the society, and are expected to continue developing rapidly during the "12th Five-Year Plan" and "13th Five-Year Plan" periods, playing a more obvious and important role in the wave of urbanization.

---

7 Peng Bo, *Case Study on Full Life Cycle Energy Consumption and CO_2 Emission of Green Buildings*, Master's Thesis of Tsinghua University, 2012.

8 Qiu Baoxing, *Action Outline of Green Buildings in Northern China*, City, 2012 (10); Wang Youwei, *Several Trends in the Development of China's Green Buildings*, Construction Technology, 2013 (11).

CHAPTER 7

Review of the Urban Low-carbon Transportation System in the Pilot Regions

Huang Liya

Abstract

The surge in transport energy consumption and carbon emission leads to increasing pressure on energy conservation and emission reduction of the industry. To construct a low-carbon transportation system has been an important aspect in the transformational development of the industry. For this, the Ministry of Transport of the People's Republic of China set phased objectives for low-carbon development, organized and launched pilot work with Chinese cities as the carriers. By centering on the low-carbon transportation system and on the basis of developmental realities, ten pilot cities in the first batch are actively exploring a road towards low-carbon transportation development. As a result, a number of distinctive developmental roads have been created, and considerable carbon reduction results have been accomplished.

Keywords

low-carbon transportation system – pilot city

The transportation industry is a fundamental industry in a national economy, as well as one of the major energy consumption and carbon emission industries. In recent years, with the rapid development of the society and the economy, the demand for transportation and the level of energy consumption have surged. To control carbon emissions from transportation and develop low-carbon transportation has been great concerns of the industry and society, and has been considered as the focus of the national climate change plan of attack. To promote green and low-carbon development, the Ministry of Transport of the People's Republic of China has organized and implemented,

* Huang Liya, senior engineer of the Planning Research Institute of the Ministry of Transport of the People's Republic of China.

 | DOI 10.1163/9789004280021_008

since 2011, a pilot construction of a low-carbon transportation system throughout the country. With city, region and field as the carriers, the idea and action of low-carbon transportation have continuously penetrated into every link of industrial development, and gradually become the basic characteristics of the modern transportation industry.

The key point of the current pilot work is the construction of an urban low-carbon transportation system. As the main production and living location for human beings, the city is an important space for transport agglomeration and activity. Urban transportation contributes to a large proportion of the traditional fossil energy consumption and greenhouse gas emissions in the whole industry. Therefore, urban pilot work is conducive to highlighting an emphasis, grasping the key, integrating resources and conducting systematic promotion, thus accumulating experience and laying a foundation for the whole transportation industry in low-carbon development.

1 Background of Pilot Work

1.1 *Carbon Emission Status Quo of the Chinese Transportation Industry*

Transportation industry has the mission of providing transportation service for all industries through its transportation infrastructures and facilities. It is always accompanied by a great amount of energy consumption, mainly including liquid or gas carbon-based energy. Thus, the transportation industry is a typical high-carbon industry and among the three main sources of carbon emissions. Studies by the International Energy Agency (IEA) show that transport carbon dioxide emission makes up approximately 25% of global carbon dioxide emission.

According to the statistical data of the National Bureau of Statistics, Chinese transportation, warehouses and the postal industry make up the largest part of national oil gross consumption. Currently, the proportion is about 34% (in terms of oil consumption, the proportion of commercial expressway and waterway transport among all means of transport is about 54% and 21%, respectively). Since 1980, the proportion has increased at the annual average rate of more than 10%, far higher than the average rate of increase of national oil consumption (around 5.5%) during the same period. A study predicts that the proportion of oil consumption of the transportation industry in national oil consumption is likely to exceed 50% by 2020. According to *The Second National Communication on Climate Change of the People's Republic of China* issued in 2004, carbon dioxide emissions from transport were nearly

415.74 million tons, accounting for 7.5% of national gross emissions, ranking No. 3, only after the industrial department and the energy production and processing transfer department.

At present, the Chinese transportation industry is mainly divided into expressways, railways, waterways and civil aviation based on the mode of transportation. Regarding the expressways, the main transport vehicle is the motor vehicle, and the energy used includes gasoline, diesel, electric power, natural gas, liquefied petroleum gas and biomass energy. The motor vehicles used gasoline and diesel account for the vast majority of the motor vehicles; the number of the motor vehicles used natural gas has increased rapidly in recent years.

As regards the railways, the main energy-using equipment is the locomotive, which is classified into steam locomotives, diesel locomotives and electric locomotives, with coal, fuel oil (mainly diesel) and electricity as the driving power, respectively. The proportion of both steam locomotives and diesel locomotives has declined, while the proportion of electric locomotives has constantly increased. Of the energy consumption of the whole railway system, since 2000, fuel oil consumption has maintained around 5 million tons, the electric consumption has surged, and the coal consumption has declined.

As regards waterway transport, the main energy-using equipment includes ships for inland transportation, lake transportation, ocean or short-range transportation, and handling facilities in various ports, with the energy used including diesel, heavy oil, gasoline and electricity. The proportion of electricity used is gradually rising, while the proportion of coal and gasoline used has gradually dropped. Energy for ship transportation mainly includes diesel and fuel oil, and since 2000, the unit energy consumption of ship transportation has shown a downward trend.

Regarding civil aviation, the energy-using equipment is mainly all forms of civil aircrafts, with aviation kerosene as fuel. The total energy consumption of civil aviation continues to increase, with an annual average increase of 402,400 tce. By contrast, the intensity of energy consumption presents a downward trend, which dropped from 1.42 kgce/conversion ton-kilometer in 1960 to 0.46 kgce/conversion ton-kilometer in 2009, with an annual average decrease of 19.6 gce/conversion ton-kilometer. The energy consumption of civil aircraft has exceeded that of railway locomotives since 2002. Compared with the motor vehicle, the civil aircraft consumes less energy; the energy consumption of civil aircraft is equal to 4.4% of that of motor vehicles.

Proportion of Chinese transportation sectors in energy consumption and carbon emissions in 2010 are shown in Figure 7.1 and Figure 7.2.

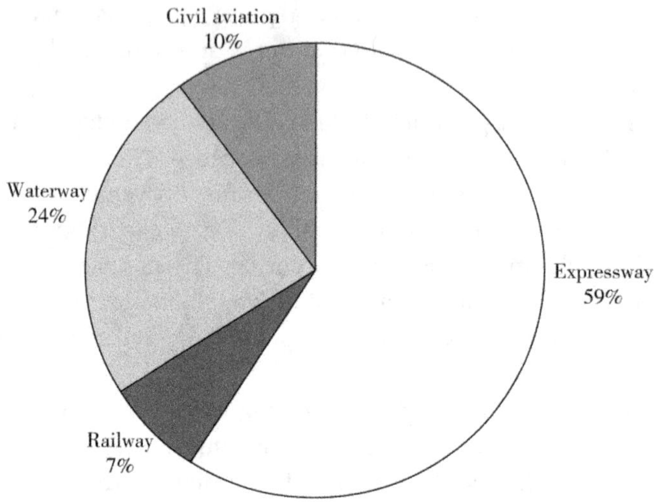

FIGURE 7.1 *Proportion of Chinese transportation sectors in energy consumption in 2010.*

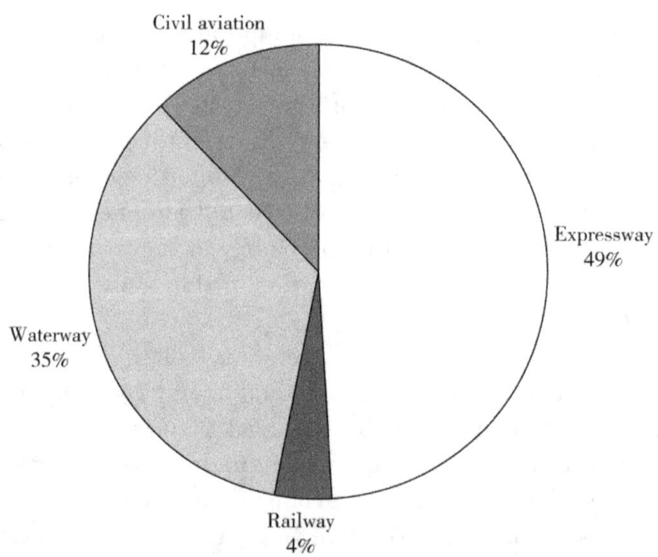

FIGURE 7.2 *Proportion of Chinese transportation sectors in carbon emissions in 2010.*

1.2 *Low-carbon Development Objective of the Chinese Transportation Industry*

Due to the status quo and future developmental trend of carbon emissions in the transportation industry, developing low-carbon transportation has been the focus of Chinese strategic deployment planning in response to climate change from the late period of the "11th Five-Year Plan." The *Resolution of the Standing Committee of the National People's Congress on Making Active Responses to Climate Change* issued in 2009, specially put forward that the construction of a transportation system characterized by low-carbon emissions would be a basic developmental strategy in the transportation industry in the future. In the splitting up and distribution of departmental tasks of the 2010 report on governmental work by the General Office of the State Council, the Ministry of Transport was given the mission of speeding up the construction of a low-carbon transportation system. Since the holding of the national transport conference in 2010, Vice Premier Zhang Dejiang, Minister Li Shenglin, Minister Yang Chuantang and other major leaders have stressed several times that it is absolutely necessary to energetically develop low-carbon transportation and expedite the construction of a low-carbon transportation system in a bid to implement the idea of low-carbon development in all links of transport production, life and travel.

Now, during the "12th Five-Year Plan," the transportation industry is at a crucial stage of transformational development. The Ministry of Transport has determined the strategic thought of "one main line, and five efforts." The "one main line" means the transformation of the developmental mode and the accelerated development of the modern transportation industry, and one of the "five efforts" is to make an effort to construct a "resource-saving and environment-friendly industry," thus constructing a low-carbon transportation system. The construction of a low-carbon transportation system is an important agent and aspect in the construction of a resource-saving and environment-friendly industry, and its level of development is an important symbol for judging the level of a resource-saving and environment-friendly industry. The acceleration of the construction of a low-carbon transportation system is not only the continuation and expansion of energy conservation and emission reduction, but it is also a new start for the transformational development of the transportation industry under a new situation.

From 2011, the Ministry of Transport has issued a series of documents including the *Guidance on Construction of a Low-carbon Transportation System, Energy Conservation and Emission Reduction in Expressway and Waterway Transportation during the 12th Five-Year Plan*, the *Action Plan for Response to Climate Change in the Transportation Industry*, and the *Work Plan for*

Greenhouse Gas Emission Control in the Transportation Industry during the 12th Five-Year Plan, and has set low-carbon developmental objectives for expressway transportation, water transportation and urban transportation.

• • •

The *Guidance on the Construction of a Low-carbon Transportation System* indicates the low-carbon development objectives of the transportation industry as follows:

In the transportation industry, greenhouse gas emission control will achieve a marked effect by 2015. The awareness of energy conservation will be further enhanced, the idea of low-carbon transportation will be more deeply rooted in people's hearts, and carbon emission intensity will gradually drop in all links of transport production, operation and consumption. The comprehensive capacity of tackling climate change will be dramatically strengthened, work on the construction of the low-carbon transportation technology innovation system and the policy & law system will be effectively carried out in an all-round way, and the carbon emission statistics, monitoring and evaluation system will be basically established. Transportation low-carbon emission will be initially accomplished and become the important support for the development of the modern transportation industry. Great efforts will continue to be made to strive to set up a transportation system conforming to the requirements of the national climate change action and featuring low-carbon emission by 2020.

The objectives for energy consumption and carbon dioxide emission intensity in expressway transportation, water transportation and urban transportation are shown below:

(1) Expressway transportation
   - Energy intensity index: by 2015 and 2020, the energy consumption of commercial vehicles in unit transport turnover will decline by 10% and 16% over that in 2005 respectively, of which, the energy consumption of commercial passenger cars in unit transport turnover will decline by 6% and 8%, and the energy consumption of freight cars in unit transport turnover will decline by 12% and 18%, respectively.
   - Carbon dioxide emission intensity index: by 2015 and 2020, carbon dioxide emissions of commercial vehicles in unit transport turnover will decline by 11% and 18% over that in 2005 respectively, of which, the carbon dioxide emissions of commercial passenger cars in unit transport turnover will decline by 7% and 9%, and carbon dioxide emissions of freight cars in unit transport turnover will decline by 13% and 20%, respectively.

(2) Waterway transportation
   - Energy intensity index: by 2015 and 2020, the energy consumption of operating ships in unit transport turnover will decline by 15% and 20% over that in 2005, respectively, of which, the energy consumption of inland ships in unit transport turnover will decline by 14% and 20%, the energy consumption of ocean ships in unit transport turnover will decline by 16% and 20%; besides, the gross energy consumption of port production unit throughput will decline by 8% and 10% over that in 2005, respectively.
   - Carbon dioxide emission intensity index: by 2015 and 2020, the carbon dioxide emissions of operating ships in unit transport turnover will decline by 16% and 22% over that in 2005, respectively, of which, carbon dioxide emissions of inland ships in unit transport turnover will decline by 15% and 23%, and carbon dioxide emissions of ocean ships in unit transport turnover will decline by 17% and 21%; besides, the carbon dioxide emissions of port production unit throughput will decline by 10% and 12% over that in 2005, respectively.

(3) Urban transportation
   - Energy intensity index: by 2015 and 2020, the energy consumption of urban transportation in person-time will decline by 18% and 26% over that in 2005, respectively, of which, the energy consumption of urban buses in person-time will decline by 14% and 22%, the energy consumption of taxis in person-time will decline by 23% and 30%, respectively.
   - Carbon dioxide emission intensity index: by 2015 and 2020, the carbon dioxide emissions of urban transportation in person-time will decline by 20% and 30% over that in 2005, respectively, of which, the carbon dioxide emissions of urban buses in person-time will decline by 17% and 27%, the carbon dioxide emissions of taxis in person-time will decline by 26% and 37%, respectively.

•••

2 Action on the Pilot Construction of the Chinese Low-carbon Transportation System

In order to popularize the idea of low-carbon transportation in the industry and accelerate the construction of a low-carbon transportation system, the Ministry of Transport has carried out work on a pilot construction of a Chinese low-carbon transportation system throughout the country. It is designed to actively explore all viable modes and reasonable paths suitable for the low-carbon development of transportation, sum up the experience of the pilot, and then make a systematic promotion and form an effective promotion mechanism for a transportation system characterized by low-carbon emissions. The pilot work officially started in 2011; with two years of efforts, it can be divided into several modes, such as city pilot, regional pilot and thematic pilot.

The developmental history of pilot work is shown below.

In 2011, the Ministry of Transport released the *Work Plan for Construction of a Pilot Low-carbon Transportation System* (hereinafter referred to as "Work Plan"). Since the city is a transport element agglomeration and a network junction that, normally, has plenty of transport infrastructures, and also is a region with an obvious imbalance between transport supply and demand, the Work Plan confirmed that the pilot work should focus on the city to reach the objective of promoting pilot work in all areas, making a key breakthrough, integrating the resources and accumulating experience.

In the selection of the first batch of pilot cities, relating work on low-carbon pilot provinces and pilot cities of the National Development and Reform Commission and other relevant low-carbon economic pilot provinces (autonomous regions and municipalities) was mainly considered. As a result, 10 cities, namely, Tianjin, Chongqing, Shenzhen, Xiamen, Hangzhou, Nanchang, Guiyang, Baoding, Wuxi and Wuhan, were selected. The pilot work mainly covered the construction of low-carbon transportation infrastructures, the promotion and application of low-carbon transportation facilities, the optimization of a transport organizational mode and operational method, the construction of intelligent traffic engineering, the availability of a low-carbon public transportation information service, and the establishment or improvement of the transport carbon emission management system. The pilot work of various cities went from 2011 until 2013, and was carried out based on three stages, i.e., pilot start, organization implementation and pilot evaluation.

During the implementation of the pilot work, a "four-stage" pilot work mechanism was established, that is, the "competent ministerial-level transportation authority—the competent provincial-level transportation authority—the pilot competent city transportation authority—the pilot project executor." The Ministry of Transport was mainly responsible for developing the overall work plan, reviewing the work implementation plan of the pilot cities, enacting and implementing pilot supporting policies and organizing pilot performance assessment; the competent provincial-level transportation authority was mainly responsible for a first reviewing work implementation plan of pilot cities, submitting it to the competent ministerial-level transportation authority for a second review, organizing the monitoring of the pilot actions, enacting and implementing supporting policies; the pilot competent city transportation authority was mainly responsible for organizing and drawing up a specific implementation plan, submitting it to the competent provincial-level transportation authority for a first review, setting up a system of pilot statistics, monitoring and evaluation and a relevant work system, organizing the specific pilot work, summing up the pilot work and providing cooperation regarding pilot performance evaluation; the pilot project executor was responsible for formulating the pilot project implementation work plan, organizing the construction and management of the pilot project, and reporting the operational data as required.

In 2012, the Ministry of Transport issued the *Notice on the Launch of the Second Batch of Urban Pilot Work in the Construction of a Low-carbon Transportation System* and the *Implementation Rules for Supporting Regional and Thematic Projects by Special Funds of Transportation Energy Conservation and Emission Reduction* (*Trial*) and other documents. While the second batch of urban pilot work was carried out, the pilot mode was stretched to regional and thematic projects.

There were 16 cities (namely, Beijing, Kunming, Xi'an, Ningbo, Guangzhou, Shenyang, Harbin, Huai'an, Yantai, Haikou, Chengdu, Qingdao, Zhuzhou, Bengbu, Shiyan and Jiyuan) selected as the second batch of pilot cities to conduct the pilot work. The pilot duration went from 2012 until 2014. Their pilot work was the same as that of the first batch of pilot cities. But, the second batch of pilot cities was required to further improve the organizational structure of the transportation energy conservation, emission reduction and low-carbon development, set up a statistical and computation system for transportation energy consumption and carbon emissions, and enact policies and measures conducive to transportation energy conservation, emission reduction and low-carbon development by gaining experience from the first batch of pilot cities.

At the same time, the Ministry of Transport organized and performed regional and thematic management pilot work in the transportation energy conservation and emission reduction special fund projects. The regional pilot project referred to any transportation energy conservation and emission reduction project undertaken by the competent transportation authority in a specific administrative region in accordance with the *Guidelines for the Construction of a Low-carbon Transportation System*; the thematic project referred to any transportation energy conservation and emission reduction project undertaken by the competent industrial department of transportation in a specific industrial sector in accordance with the *Guidelines for the Construction of a Low-carbon Transportation System*. The innovation on the two pilot modes was aimed at forming an integrated management mode combining the specific project management, the regional management and thematic management in the utilization of the transportation energy conservation and emission reduction special funds, thus accumulating experience.

After a careful selection within the industry, 2012 regional pilot projects were dedicated to the construction of 10 low-carbon transportation cities, namely, Beijing, Chongqing, Shenzhen, Xiamen, Hangzhou, Nanchang, Guiyang, Baoding, Wuhan and Wuxi; regarding the thematic pilot projects, Tianjin port, Qingdao port, Lianyungang port and Shekou Container Terminals Ltd. were selected to conduct a "low-carbon port" construction, and in addition, Hebei, Jiangsu, Henan, Guangdong, Yunnan and other provinces were selected to launch the construction of the "low-carbon expressway."

3 Achievements in the Pilot Work on a Low-carbon Transportation System

In the promotion of the pilot work in an all-round way, the pilot project undertakers prepared, reviewed and implemented the pilot implementation plan based on pilot commencement and promotion procedures under the organization of the competent local transportation authority. The competent transportation authorities at all levels conducted field investigations on the pilot project undertakers and gave full play to their role in coordinative promotion and specific guidance in order to constantly intensify the pilot work. For this, the Ministry of Transport developed and introduced supporting policies for promoting the construction of low-carbon transportation, strengthened the support for pilot undertakers, and publicized the advanced and mature experience of the pilot project undertakers. When the competent provincial-level and city-level transportation authorities of pilot undertakers carried out the pilot implementation plan approved by the Ministry of Transport, they placed emphasis on the follow-up to the implementation progress of the project, solving problems in a timely manner, suming up experience, and seeking the local governmental support.

Up to now, the first batch of city pilot has completed more than half of the work, and is about to carry out pilot evaluation; while the second batch of city pilot, regional pilot and thematic pilot work has just entered the implementation phase, without obvious achievements. Therefore, this research makes a brief summary of the work progress of the first batch of pilot cities.

To reach the overall objective of a "system construction," the first batch of pilot cities are carrying out diversified pilot work on the basis of the local geological position, natural conditions, economic level, transportation development level, and organization and management level with the principle of "taking actions according to local conditions and highlighting features." Through pilot work, pilot cities are expected to fulfill energy conservation and emission reduction, as shown in Table 7.1.

TABLE 7.1 *Energy conservation and emission reduction expected to be accomplished by the first batch of pilot cities*

City	Amount of energy conservation (10,000 tce/year)	Emission reduction (10,000 tcd/year)
Tianjin	9.1	22.3
Chongqing	29.2	73
Shenzhen	9.1	23.8
Xiamen	8.5	7.9
Hangzhou	9.5	23.2
Nanchang	7	21.8
Guiyang	3.0	11.3
Baoding	10.8	20.5
Wuxi	1.4	5.5
Wuhan	4.4	21.9
Total	92	231.2

NOTE: TCD MEANS TONS OF CARBON DIOXIDE EQUIVALENT.

Since the launch of pilot work, the typical actions and experience of pilot cities are shown in Table 7.2.

3.1 *Tianjin*

As one of the four municipalities of China, Tianjin is situated at the beginning of Eurasia and at the center of the Bohai Economic Rim; it is one of the Chinese modern industrial headstreams and the center of Chinese northern sea transportation and industry. Besides, it has the fourth largest industrial base and the third largest foreign trade port in China. The implementation of the Binhai New Area strategy, the balanced development between port and city, and the construction of the international port city have created new opportunities for Tianjin to carry out the construction of a low-carbon transportation system.

TABLE 7.2 *Key promotion of low-carbon transportation system construction by the first batch of pilot*

Content	Tianjin	Chongqing	Shenzhen	Xiamen
City characteristics	Northern coastal municipality	Southwestern municipality	City specifically designated in the state plan	City specifically designated in the state plan
	Binhai New Area	Southwestern comprehensive traffic hub, and shipping center of the upper reaches of Yangtze River	City in the special economic zone	Tourism city
Key promotion of low-carbon pilot work	1. Low-carbon Tianjin port; 2. Construction of low-carbon expressway, technical application of maintenance materials	1. Construction of a comprehensive transport system; 2. Low-carbon expressway; 3. Technical demonstration of quick charge of buses	1. Road, waterway, railway and aviation traffic management system; 2. Transit metropolis, global leading in application scale of new energy and clean energy bus, taxi and automobile; 3. Construction of low-carbon harbor district	1. Intelligent information public traffic system; 2. Slow transportation system

3.1.1 Construction of the Low-carbon Tianjin Port

Tianjin port has taken a series of measures such as the RTG "oil to electricity," the underground thermal energy utilization, the coal wharf dragon-bone water lift screening and crushing system and ship-to-shore power technology research and application, striving to make it a low-carbon harbor district. At the same time, by focusing on Tianjin port container multimodal transport, Tianjin has genetically promoted the development of multimodal transport to achieve a smooth connection among the various transportation modes, thus better playing its role in shipping energy conservation and emission reduction.

Hangzhou	Guiyang	Nanchang	Baoding	Wuhan	Wuxi
Eastern provincial capital	Western provincial capital	Central-southern provincial capital	Northern prefecture level city	Central provincial capital	Eastern prefecture level city
Tourism city	Ecological city	Poyang Lake Ecological Economic Zone	Optics Valley of China	Comprehensive traffic hub, and shipping center of the middle reaches of the Yangtze River	Economically developed area
1. Five-in-one bus system; 2. Low-carbon water transport engineering	1. Bus "oil to gas"; 2. Promotion and demonstration of alternative fuels such as methyl alcohol and coke-oven gas	1. Reduction of goods energy consumption; 2. Optimization of transportation structure and increase of water transport ratio; 3. Increase of public transport ratio	1. Application of expressway low-carbon technologies; 2. Application of energy-saving and environment-friendly gas ship in Baiyangdian Scenic Spots	1. Construction of a comprehensive transport system; 2. Scale application of clean fuels; 3. Convenient public bicycle system	1. Low-carbon effect of IOT technology; 2. Urban-rural bus integration

3.1.2 Acceleration of the Recycling of Materials and the Application of Energy-saving Construction Technologies in Expressway Construction and Maintenance

Tianjin Municipal Highway Administration has prepared the *Manual of Expressway Construction Low-carbon Technology* to sum up the old pavement material regeneration technology, the warm mixed asphalt technology and the scrap tire rubber modified asphalt technology and to promote the application of such technologies, thus accelerating low-carbon development in expressway construction.

3.2 *Chongqing*

Chongqing is the only comprehensive traffic hub integrating railway, expressway, waterway and civil aviation in Chinese western inland regions. Its construction of a low-carbon transportation system centers on the construction of a western comprehensive traffic hub and shipping center of the upper reaches of the Yangtze River, and advantages of various transportation modes in energy conservation and emission reduction are given full play, with the aim of accomplishing comprehensive energy conservation and emission reduction benefits.

3.2.1 Optimization of the Transport Structure

By relying on the Yangtze River water-land drop and pull transport demonstration project and on the railway, expressway and waterway multimodal transport demonstration project in Guoyuan port, Chongqing has managed to generate comprehensive energy conservation and emission reduction benefits from multiple modes of transportation, and reduce carbon dioxide emissions by the optimization of the transportation structure; in addition, Chongqing port is making great efforts to promoting inland river anchored ships and shore electric systems to strive to accomplish energy conservation and emission reduction in the water transportation industry.

3.2.2 First Construction of a Low-carbon Expressway Pilot Project

Chengdu-Chongqing multiple track (Chongqing section) was selected for its overall low-carbon optimization design. Warm mix asphalt technology and rubber asphalt technology were applied in the 150,000-m^2 expressway pavement, "green illuminating engineering" for tunnels was implemented, with 4,472 sets of LED lamps used in four tunnels. All toll stations in the section are provided with an ETC traffic lane. Service areas were constructed according to green building standards, reserving hybrid gas fuel/electric vehicle filling stations, and installing solar power supply equipment. A carbon sequestration forest system was constructed on both sides of the expressway, with the forest width ranging from 30 to 80m.

3.2.3 Promotion of Hybrid Electric Vehicles and Developing the Application of Pure Electric Buses in the Public Transport Industry

Chongqing is planning to demonstrate, promote and apply 6,000 clean energy vehicles and new energy vehicles. Chongqing actively guides qualified districts and counties to promote the application of clean fuel automobiles such as the CNG and LNG. Chongqing Public Traffic Group is experimenting with the development of an electric bus rapid charging system, with six rapid charge buses put into use. The relevant technology is gradually improving.

3.3 *Shenzhen*

3.3.1 New Energy and Clean Energy Buses

The Shenzhen public transport industry is taking the lead in the scale of application of new-energy and clean-energy vehicles in the world. At present, there are 2,011 new-energy vehicles that have been put into use, including 1,350 hybrid power buses, 20 hybrid power double-decker buses, 253 pure electric buses, 26 pure electric minibuses, 300 pure electric taxis, 60 fuel-cell track bicycles and two fuel-cell buses. As a result, Shenzhen is a city that enjoys the most varieties in the application of new-energy vehicles, the largest scale and quantity in the public transport industry worldwide. In terms of overcoming fuel cell and charging difficulties, Shenzhen has put forward the purchasing, operation and maintenance scheme of new-energy buses, characterized by "financial lease, vehicle-electricity separation and charging-maintenance combination," in which the professional organization is responsible for battery charging and maintenance management to maximize the life cycle of batteries, so as to realize the sustained commercial operation of new-energy vehicles.

3.3.2 Construction of a Typical Green Low-carbon Harbor District

Shekou container terminal in Shenzhen is taking a series of measures such as the RTG "oil to electricity," the application of the anchored ships electricity technology, the usage of LNG trailers, the usage of LED energy-saving lamps, the application of solar photovoltaic technology and the pure electric harbor district cruiser to promote port wharf energy conservation and emission reduction, so as to create a green low-carbon harbor district.

3.4 *Xiamen*

3.4.1 The Intelligent Information Public Transport System

Xiamen Public Transport Group owns 5,569 types of vehicles, 247 regular bus routes, three BRT express lines and 165 long-distance passenger transportation routes. The sharing ratio of public transit of the whole city was 30.87% in 2011. Intelligent information development of public transit is the key to the construction of the Xiamen low-carbon transportation system. On the one hand, vehicle dispatch and utilization are increased efficiently by GPS real-time monitoring, security warning and speed limits, and a convenient public transport network and handheld public transport query system have been opened; on the other hand, a series of measures, such as the normalization of driving habits, a decline in the frequency of accidents, the saving of security costs, the optimization of energy consumption driving under speed control and the reduction of deadhead kilometers under intelligent dispatch, have been taken in order to realize high efficiency and low-carbonization in public transport.

3.4.2 Slow Traffic System

As Xiamen enjoys a unique geographical position and a beautiful natural landscape, a slow traffic system has been constructed along the coast of the whole city to provide leisure sites for citizens, which further illustrates the connotation of low-carbon comfortable transportation.

3.5 *Hangzhou*

3.5.1 Construction of the "Five-in-one" Public Transport System

Located in Jiangnan region (south of the Yangtze River), Hangzhou enjoys a dense water network. It has proposed the construction of a "five-in-one" public transport system that smoothly connects subways, buses, taxis, public bicycles and vaporetti, which better reflects the characteristics of an urban low-carbon transportation system. Hangzhou has taken some important measures, such as the rapid development of public transport, the increase of the urban public transport sharing ratio and the reduction of the number of private car journeys, in order to accomplish urban transportation energy conservation and emission reduction. The year 2009 witnessed a 42.45% sharing ratio of Hangzhou's urban public transport. Now, Hangzhou is trying to vigorously develop public transport, improve the overall carrying capacity and the service level of its public transport system, advance the application of energy-saving and environment-friendly vehicles in its public transport, and actively promote green and low-carbon public bicycles, so as to encourage more travelers to adopt low energy consumption and low-carbon modes of transportation.

3.5.2 The Advancement of Low-carbon Water Transport Engineering

Hangzhou vigorously promotes the development of waterway transport, and has initially formed a waterway channel of "one port, two main lines and two branch lines." Hangzhou's low-carbon waterway construction and reconstruction project mainly focuses on the construction and application of an "integration" management system in the Hangzhou navigation area and the synthetical harnessing project of shipping channels (ship lock + shipping channel) to realize the upgrading of the channels, implement the ship type standardization, apply a comprehensive service system and put waterway low-carbon advantages into full play.

3.6 *Guiyang*

3.6.1 Public Bus "Oil to Gas"

Guiyang Public Transport (Group) Co., Ltd. has tried to refit the diesel engine since 2006, and has successfully reformed the diesel engine into a liquefied natural gas engine. At present, it has built six gas filling stations, and performed the "oil to gas" transformation for 1,216 buses, realizing clean fuel utilization and obtaining direct energy conservation and emission reduction results.

3.6.2 Actively Promoting the Use of Alternative Fuels such as Methyl Alcohol and Coke-oven Gas

Since Guiyang has the basic resources for methyl alcohol products focusing on the coal chemical industry, it has unmatched advantages in promoting methanol vehicles. At present, M100 methanol fuel has been applied to some taxis and large buses; 300 vehicles have already been put into use, with good operating results. Guiyang Public Transport (Group) Co., Ltd. plans to complete the refitting of 936 taxis and 700 large buses by the end of 2012. At that time, there will be 12,800 tons gasoline to be replaced, with a decrease of 12,200 tons of carbon dioxide emissions.

3.7 *Nanchang*

3.7.1 Reduction of Goods Energy Consumption

Expressway freight is the largest source of energy consumption and carbon emission in the transportation industry in Nanchang. With more than 46,000 operating trucks, energy consumption and carbon emissions of such trucks respectively accounted for 68.35% and 68.95% of that of the whole city in 2010. To construct low-carbon transportation, the first thing to do is to reduce freight energy consumption and carbon emissions. Nanchang vigorously promotes drop and pull transport, encourages freight enterprises to develop towards scale, network and profession and transform from a traditional freight industry to a modern logistics industry. In the meanwhile, Nanchang has managed to set up energy consumption and environment-friendly entry and exit mechanisms for operating vehicles and ships, step by step, to promote the renewal of operating vehicles and ships, so as to maintain economic, energy-saving performance of vehicles and ships in terms of oil consumption.

3.7.2 Adjustment of the Transportation Structure and Increase of the Waterway Freight Ratio

With the construction of the Poyang Lake Ecological Economic Zone, Nanchang has implemented lake control engineering without delay, has increased the grade of the Ganjiang River waterway, and has expedited the construction of three inland terminal depots in a bid to promote the development of inland transportation. During the "11th Five-Year Plan," the volume of inland freight rose by 149%, and is expected to reach an annual average increase of 12% during the "12th Five-Year Plan." Container throughput is expected to reach 200,000 TEU, 4 times as large as the current throughput.

3.7.3 Increase of the Public Transport Sharing Ratio

At present, the sharing ratio of Nanchang public transport is only 13%. For this, Nanchang is trying its best to increase it. First, Nanchang is constructing railway, and will open subway lines No. 1 and No. 2 as soon as possible; second, Nanchang is striving to develop its bus rapid transit and open the BRT rapid lines; third, Nanchang is building a special road network for public transport; fourth, Nanchang is trying to increase its supply of public transport vehicles to attain the goal of 15 public transport vehicles per 10,000 people and a 20% public transport sharing ratio by the end of the "12th Five-Year Plan."

3.8 *Baoding*

3.8.1 Promotion of the Application of Expressway Construction Low-carbon Technology

Baoding is speeding up the R&D and promotion of new materials, new processes and regeneration technology, is advancing green corridor engineering and performing energy-saving transformation of lighting facilities. From 2011 to 2013, Baoding applied low-carbon paving technology, such as warm mixed asphalt, in the construction of the Zhangjiakou-Zhuozhou Expressway, the Rongcheng-Wuhai Expressway, the Zhangjiakou-Shijiazhuang Expressway Phase III, the Baoding-Fuping Expressway, the Quyang-Huangyegang Expressway and other expressways. A LED lamp transformation and reconstruction project was implemented in all of its expressway tunnels, service areas and toll stations. After the completion of the four tunnels, 785.53 tce will be saved annually; about 10,732,900 kWh will be saved after the completion of the lighting equipment transformation, and the emission of carbon dioxide will drop by 8,324.45 tons annually. In addition, Baoding applies solar energy in the illumination of the Electricity Valley overpass bridges, outer loops and the Baiyang Lake waterway, with notable achievements in energy conservation and emission reduction.

3.8.2 Promotion of the Usage of Energy-saving and Environment-friendly Gas Ships

Baoding plans to replace all outboard petrol motor ships under its jurisdiction with energy-saving and environment-friendly ships within 10 years. As planned, 1,411 outboard petrol motor ships will be outdated, and each ship owner will receive 35,000 RMB as subsidies, for which the amount of 49,385,000 RMB is needed in total funds. Based on a 14% annual replacement rate, it is necessary to invest about 4.90 million RMB in annual funds. The replacement of the existing outboard petrol motor ships with battery-driven ships or gas ships will lead to a 40% conservation of energy and a pollutant emission decline of 67%, showing that it is conducive to energy conservation, pollution alleviation and water quality maintenance.

3.9 *Wuhan*

As the thoroughfare of nine provinces, Wuhan is located at a key intersection of the national comprehensive transport network of "five vertical and five horizontal highways." It is the traffic hub of the central regions and even of the whole country, as well as the shipping center of the middle reaches of the Yangtze River. It covers multiple modes of transportation, including expressways, railways, waterways and aviation. Besides, it is a core city in the pilot area for the comprehensive reform of the construction of a national energy-saving and environment-friendly society. For this, Wuhan focuses on the construction of a comprehensive low-carbon transportation system.

3.9.1 Optimization of the Integrated Transport System

The implementation of key transport projects, such as Wuhan Tianhe International Airport Phase III, the Fourth Ring Road, Airport No. 2 Tunnel, Hanjiang River Channel Improvement and Wharf, will greatly boost the optimization and low-carbon development of Wuhan transport structures. Besides, as Wuhan enjoys a good condition of water transportation in the middle reaches of the Yangtze River, it will promote the usage of river-sea transportation ships during the pilot period, with a deadweight of 1.05 million tons and an increase of 16 new 10,000-dwt self-propulsion ships. In addition, it actively popularizes the eco-driving method and increase the ratio of water transportation to make itself the shipping center of the middle reaches of the Yangtze River, bringing into full play its strengths in water transportation.

3.9.2 Wide Application of Clean Fuel

Since 2000, Wuhan has started to attempt to perform "oil to gas" for buses and taxis. Until now, it has 457 LPG buses, 1,092 CNG buses, 327 LPG taxis and 10,734 CNG taxis, with an annual total of up to 133,301.74 tons of standard oil having been replaced.

3.9.3 Construction of a "Free + Integrity" Convenient Public Bicycle System

At present, Wuhan has 50,000 bicycles and 1,102 bicycle pick-up points. The duration of free leasing should not exceed two hours each time, and

5 minutes after the drop-off, the bicycle can be leased free of charge for another two hours, for an unlimited number of times. In case the duration of the lease exceeds two hours, the bicycle renter may be fined and receive a poor credit record. The mode of the "free + integrity" convenient public bicycle system has been popularized in Yichang and Xiaogan, Hubei Province, and in Chizhou, Anhui Province, etc.

3.10 *Wuxi*

3.10.1 Giving Full Play to the Strengths in IOT Technology R&D and Application, and Promoting the Low-carbon Development of Transportation

Wuxi is among the domestic cities in the forefront in terms of its level of economic development. In 2010, every 100 people owned 18.49 motor vehicles and 8.27 private cars in the whole city, showing that Wuxi enjoyed rapid development of a modern service industry and distinctive hi-tech industry, suffered high pressure on transport development, and was in urgent need of transformation towards modern low-carbon transportation. Wuxi has given full play to its strengths in technology, such as the Internet of Things (IOT), to study an intelligent monitoring system for IOT-based national or provincial arterial highway bridges, an intelligent public logistics information platform, a handheld public transport system, waterway sensing and port sensing, and to explore the development of "sensing taxi-booking by mobile phone" system. Through the application of IOT technology, Wuxi has realized the optimization for the transportation route, cargo configuration, vehicle configuration and driving conditions, thus improving transportation efficiency and reducing energy conservation and emission reduction per unit of freight volume.

3.10.2 Implementing the Public Transit Priority and the Urban-rural Public Transit Integration Strategy

In Wuxi, every 10,000 people have 16 public transport vehicles, and clean energy vehicles make up 16%. At present, the 2-yuan RMB ticket system is applied to urban-rural public transit throughout the entire city, and a 40% discount is given by card, which will summarize and facilitate public travel. Public transit covers almost every administrative (organizational system) village and town.

Over the two years of pilot work, almost all cities have carried out work based on the pilot implementation plan by looking at the first batch of pilot cities. Pilot cities have initially formed the low-carbon transportation concept and idea. Apart from active promotion, organizational enhancement, timely mobilization and publicity by competent urban transportation authorities in pilot cities, positive collaboration on the pilot work has been obtained from local governments; for example, great support has been gained from local governments in terms of local planning and supporting policies. In particular, among the first batch of 10 pilot cities and national low-carbon pilot cities, eight cities have implemented the construction of a low-carbon transportation system in accordance with low-carbon pilot provinces and cities, taken full advantage of the developmental opportunities afforded by low-carbon pilot work, and mobilized the industry and social forces to boost the low-carbon development of transportation in an all-round way.

By looking at the whole industry, the atmosphere and momentum for low-carbon development have been gradually formed within the industry due to the strong driving force of pilot cities and project undertakers. A great number of low-carbon practical technologies have sprung up and have moved toward maturity during the pilot work. However, the pilot work also brings about a series of problems. It can be said that the development of low-carbon transportation in China still faces a range of difficulties in organizational structures, statistics and monitoring, standards and specifications, policies and laws, incentive mechanisms and systematic promotion, which are in urgent need of being resolved in practice.

CHAPTER 8

Preparation Methods for and Content Framework of the Chinese Urban Greenhouse Gas Inventories

Bai Weiguo

Abstract

Low-carbon urbanization focuses on low-carbon economy and city development. The preparation of the urban greenhouse gas inventory is a fundamental work in low-carbon development. Feasible preparation methods of the greenhouse gas inventory are conducive to enhancing the greenhouse gas inventory normalization. Given the fact that the preparation methods for the Chinese urban greenhouse gas inventory lag behind its low-carbon development, this chapter summarizes the experience based on the *Guide for Province-level Greenhouse Gas Inventory Preparation* (*Trial*) and the *ICLEI Urban Greenhouse Gas Inventory Guide*; it sorts out critical problems that may be encountered, and puts forward ideas for solving critical problems by proper reference to the two guides based on inventory positioning; and then it determines the content framework by taking advantage of solving critical problems, so as to facilitate the study on the methods for preparing the urban greenhouse gas inventory.

Keywords

Chinese city – greenhouse gas inventory – inventory positioning

As China is experiencing rapid urbanization, it faces increasingly severe resource constraints and environmental pollution pressure. Low-carbon urbanization has been the only way leading to the realization of sustainable development. Low-carbon urbanization focuses on low-carbon economy and city development which normally follow the low-carbon idea characterized by low energy consumption, low pollution, low emission, high efficiency and high

* Bai Weiguo, post doctorate of the Institute for the Urban and Environmental Studies of the Chinese Academy of Social Sciences, engaged in studies on the method for preparing the urban greenhouse gas inventory.

 DOI 10.1163/9789004280021_009

output. The preparation of the urban greenhouse gas inventory is an important prerequisite for low-carbon development. The greenhouse gas inventory can help the central and local governments figure out greenhouse gas emission, and formulate practical and appropriate emission reduction objectives, mission and technical paths, thus accurately grasping the developmental direction and mode.

Up to now, there are over a hundred cities that are actively carrying out low-carbon pilot construction such as the "low-carbon city" development demonstration project jointly launched by the Ministry of Housing and Urban-Rural Development of the People's Republic of China and the World Wildlife Fund (WWF) in 2008, and the two batches of pilot cities approved by the National Development and Reform Commission of China, as well as other pilot projects organized by other institutions or spontaneously launched by cities. The two batches of pilot cities approved by the National Development and Reform Commission of China are required to "prepare a low-carbon development planning" and "establish a greenhouse gas emission data statistical and management system." The preparation of the urban greenhouse gas inventory is a fundamental task in low-carbon development. Determining feasible methods for preparing the greenhouse gas inventory is conducive to enhancing the greenhouse gas inventory normalization.

1 Progress of the Preparation Methods of Greenhouse Gas Inventories

With over ten years of experimentation and study, two normative and influential categories of the preparation methods of the greenhouse gas inventory have been formed. The first category refer to the standards and norms in terms of geographical boundaries, such as the *IPCC Guide for a National Greenhouse Gas Inventory*,[1] the *Guide for Province-level Greenhouse Gas Inventory Preparation (Trial)*,[2] the *ICLEI Guide for Urban Greenhouse Gas Inventory*,[3] and the *GRIP*

1 IPCC, 1996 IPCC Guidelines for National Greenhouse Gas Inventories, Intergovernmental Panel on Climate Change, http://www.ipcc-nggip.iges.or.jp/public/gl/invsl.html.

2 Province-level greenhouse gas inventory preparation group: Guide for Province-level Greenhouse Gas Inventory Preparation (Trial), National Development and Reform Commission, 2010.

3 ICLEI, "International Local Government GHG Emissions Analysis Protocol Version 1.0," 2009.

Greenhouse Gas Regional Inventory Protocol[4] (for a brief introduction, please see Table 8.1); the second category refer to the standards and norms in terms of operational boundaries, such as the *Greenhouse Gas Protocol: Enterprise Calculation and Report Standard*,[5] *ISO 14064 Standards*,[6] *PAS 2050 Guide* and *Standard PAS 2060*[7] (for a brief introduction, please see Table 8.2). The greenhouse gas preparation mode is classified into the production mode and the consumption mode. The former means the greenhouse gas emissions from the production of goods and services within the boundary in the inventory; and the latter means greenhouse gas emissions beyond the boundary in the inventory resulted from the consumption of goods and services within the boundary in the inventory.

There are both distinctions and relations between the two categories. The methods for preparing the greenhouse gas inventory in terms of geographical boundaries are mainly developed by the organization with administrative function such as the IPCC, research institutes and governmental organs, with the specific region as the object of study and the preparation mode as the production mode and the production + consumption mode; while the methods for preparing the greenhouse gas inventory in terms of operational boundaries are mainly developed by the organization with an enterprise background, with the enterprise, project, product and other organizations as the object of study. In addition, since the enterprise always carries out procurement, production, processing and sales across the border, the inventory boundary is based on organizational activities, with the preparation mode as production + consumption mode. As for the methods for preparing the greenhouse gas inventory in terms of geographical boundaries, the *IPCC Guide for National Greenhouse Gas Inventory* was first developed and promoted by the United Nations; the *ICLEI Guide for Urban Greenhouse Gas Inventory* is based on urban greenhouse gas emission characteristics, is normalized by drawing experience from the *IPCC Guide for National Greenhouse Gas Inventory* and the *Greenhouse Gas Protocol: Enterprise Calculation and Report Standard*, and is publicized by the platform of ICLEI (International Council for Local Environment Initiatives). In China, the national greenhouse gas inventory for the years 1994 and 2005 was prepared in accordance with the *IPCC Guide for a National Greenhouse Gas Inventory*; the

4 GRIP, "The Greenhouse Gas Regional Inventory Protocol," 2012, http://www.getagriponemissions.com/history.html.

5 WRI/WBCSD, "The Greenhouse Gas Protocol: A Corporate Accounting and Reporting Standard," Revised Edition, WRI/WBCSD, 2009-5-15. www.ghgprotocol.org.

6 ISO 14064, 2006 [EB/OL], [2009-06-26], http://www.iso.org/iso.org/iso/catalogue_detail?csnumber=38381, Sorting based on public data.

7 Sorting based on public data.

TABLE 8.1 *Methods for preparing the greenhouse gas inventory in terms of geographical boundaries*

Name of guide	Issuing authority	Preparation mode	Objective	Scope of application	Application
IPCC Guide for a National Greenhouse Gas Inventory	Intergovernmental Panel on Climate Change	Production	Provide basically the same framework and method for comparison of greenhouse gas inventories of different countries and regions	Country and region	Information notification of most countries in 148 contracting states
Guide for Province-level Greenhouse Gas Inventory Preparation (Trial)	National Development and Reform Commission of China	Production	Figure out provincial greenhouse gas emissions to put into practice greenhouse gas emission medium and long term planning index decomposition	Region (province-level)	Chinese low-carbon pilot provinces
ICLEI Guide for Urban Greenhouse Gas Inventory	International Council for Local Environmental Initiatives	Production + consumption	Figure out the source of main greenhouse gases, make comparisons at home and abroad, define development emphasis and work out low-carbon development paths	City	1,200 cities and regions across the world
GRIP Greenhouse Gas Regional Inventory Protocol	University of Manchester, UK	Production + consumption	Provide a basis for the governments' statistics and monitoring of greenhouse gas emissions, comparison of different cities and years, and tapping emission reduction potential	City	18 cities and regions in Europe

TABLE 8.2 *Methods for preparing the greenhouse gas inventory in terms of organizational boundaries*

Name of guide	Issuing authority	Preparation mode	Objective	Scope of application	Application
Greenhouse Gas Protocol: Enterprise Calculation and Report Standard	World Resources Institute/World Business Council on Sustainable Development	Production + consumption	Inclusion of enterprises' greenhouse gas emissions in financial statements is the basis for greenhouse gas action plan of many industries and enterprises	Enterprise	Enterprise and industry
ISO 14064 Standards	International Organization for Standardization	Production + consumption	It stresses the standard requirements but it only provides general guidance about how to meet such requirements	Enterprise and project	Enterprise and project
PAS 2050 Guide	British Standards Institution	Production + consumption	Figure out greenhouse gas emissions in the whole life cycle of product or service in all links from raw materials to production (or service supply)	Product and service	Carbon mark introduced by the UK and Korea
Standard PAS 2060	British Standards Institution	Production + consumption	By reduction and compensation, perform the provisions to which carbon neutral organizations must conform, so as to guarantee the accuracy and verifiability	Country, community, company and individual	All types of organization

Guide for Province-level Greenhouse Gas Inventory Preparation (Trial) was prepared with reference to the *IPCC Guide for a National Greenhouse Gas Inventory* and by drawing experience from the 2005 national greenhouse gas inventory preparation, and now it is used for guiding low-carbon pilot provinces in preparing their greenhouse gas inventory. At present, the methods available for the preparation of the urban greenhouse gas inventory are insufficient and cannot meet the needs of low-carbon development.

As the methods for preparing the urban greenhouse gas inventory lag behind low-carbon development in China, there is an urgent need to develop methods for the preparation of urban greenhouse gas inventories. It is an important approach to draw experience and effective actions from the methods for preparing key regional greenhouse gas inventories so as to provide some suggestions for the preparation of the Chinese urban greenhouse gas inventory.

2 Experience from the Preparation Methods of Regional Greenhouse Gas Inventories

The methods for preparing the regional greenhouse gas inventories have been an important tool for preparing the greenhouse gas inventories to respond to climate change, setting the greenhouse gas emission goals and formulating the low-carbon action plan. The *Guide for Province-level Greenhouse Gas Inventory Preparation* (*Trial*) and the *ICLEI Guide for Urban Greenhouse Gas Inventory* have made complete and systematic regulations on the scientificity of the preparation methods, data transparency, format consistency and result comparability of greenhouse gas inventories, and have clarified the inventory boundary, scope and sector division, etc. Owing to widespread usage and high practicability, the two guides are important preparation methods of foreign and domestic key region greenhouse gas inventories, and can provide experience on the preparation methods of Chinese urban greenhouse gas inventories.

2.1 *Guide for Province-level Greenhouse Gas Inventory Preparation* (*Trial*)

The *Guide for Province-level Greenhouse Gas Inventory Preparation* (*Trial*) was prepared by experts from universities and research institutes under the organization of the National Development and Reform Commission of China, and it was completed in 2010. As a fundamental work in response to climate change, the province-level greenhouse gas inventory can help to identify the source and sink of the main greenhouse gas emissions, figure out the condition of the greenhouse gas emission in key fields, analyze the future greenhouse gas emission trend and the potential for reduction, make comparisons among different regions, so as to facilitate the formulation of policies and actions with respect to greenhouse gas emission control. The main content of the *Guide for Province-level Greenhouse Gas Inventory Preparation* (*Trial*) is as follows:

calculation of the emission/sink of six kinds of greenhouse gases, namely, CO_2, CH_4, N_2O, HFCs, PFCs and SF_6; determination of the inventory boundary based on the provincial administration precincts; calculation method: greenhouse gas emission/sink = activity data × emission factor; division of sectors into energy activity, industrial production process, agriculture, land utilization change and forestry, waste disposal; the inventory scope includes scope 1 (electricity in scope 2 is listed as information item); data are collected from top to bottom.

2.2 *The ICLEI Guide for Urban Greenhouse Gas Inventory*

The ICLEI developed the *ICLEI Guide for Urban Greenhouse Gas Inventory* by virtue of the City Climate Change Protection (CCP) that started in 1993. The Guide can help local governments grasp the urban energy utilization rate and the structural change correctly, and fully tap the energy saving and the emission reduction space; measure urban strengths and weaknesses in the international and domestic urban low-carbon economy, so as to determine future low-carbon developmental priorities; figure out the condition of the urban greenhouse gas emission and clarify important emission sources and sinks, to make the low-carbon city roadmap scientific and the carbon emission reduction MRV (measurement, report and verification) pertinent and practical. The main content of the *ICLEI Guide for Urban Greenhouse Gas Inventory* is as follows: calculation of the emission/sink of the six greenhouse gases, namely, CO_2, CH_4, N_2O, HFCs, PFCs and SF_6; determination of the inventory boundary based on the administration precincts; calculation method: greenhouse gas emission/sink = activity data × emission factor; division of sectors into two accounts (government and community) to calculate seven key fields, i.e. energy industry, transportation, construction, industrial process, agriculture, land utilization change and forestry, waste disposal; the inventory scope includes scope 1 and scope 2; data are collected from bottom to top.

2.3 *Summary of the Experience*

Both the *Guide for Province-level Greenhouse Gas Inventory Preparation* (*Trial*) and the *ICLEI Guide for Urban Greenhouse Gas Inventory* are designed to provide guidance for the preparation of the regional greenhouse gas inventories. They enjoy a normative content framework, and cover all key points in the preparation of regional greenhouse gas inventories. As the methods of preparing the regional greenhouse gas inventories, there are both distinctions and relations between the guides which are based on the requirements of the preparation of greenhouse gas inventories regarding regional low-carbon development realities. Common points: calculation of the six greenhouse gases, namely, CO_2, CH_4, N_2O, HFCs, PFCs and SF_6; determination of the inventory boundary based on the administration precinct; the method of calculation: greenhouse gas emission/sink = activity data × emission factors. Differences: the former has five sectors, while the latter has two accounts (seven key fields); the inventory scope of the former only includes scope 1, while the latter includes scope 1 and scope 2; the data of the former are collected from top to bottom, while the latter is from bottom to top.

The common points of the two guides show that the regional methods for the preparation of greenhouse gas inventories also have many similarities. At the same time, the differences of the two guides reveal that the *Guide for Province-level Greenhouse Gas Inventory Preparation* (*Trial*) draws on the experience from the national greenhouse gas inventory preparation and its sector division is the same as that of the national greenhouse gas inventory. Besides, it sets the inventory scope based on the production mode, and has complete statistical data, so it is passive of emission reduction. The *ICLEI Guide for Urban Greenhouse Gas Inventory* draws on the experience of the *IPCC Guide for National Greenhouse Gas Inventory* and the *Greenhouse Gas Protocol: Enterprise Calculation and Report Standard*, with the sector division covering all key fields regarding urban greenhouse gas emissions and the scope of the inventory taking a large proportion of urban consumption mode into account. Besides, its data are collected from bottom to top, so that it is more practical. Thus, it is active emission reduction. The differences between the two guides impose challenges on China to develop its own methods for preparing urban greenhouse gas inventories. How to reasonably draw lessons from the two guides has become a difficulty that China is facing.

3 Critical Problems and Solutions

The year 2012 saw a level of 52.57% of urbanization in China. As a whole, the consumption mode and the production mode such as agriculture contributed to a large proportion of the greenhouse gas emissions in most Chinese cities, which was greatly different from the greenhouse gas emissions of province-level and foreign cities. This complicates the preparation of the greenhouse gas inventories. Therefore, it is necessary to first comb through critical problems (mainly the inventory boundary, scope and collection of data) based on the characteristics of greenhouse gas emission in Chinese cities; and then draw experience from the *Guide for Province-level Greenhouse Gas Inventory Preparation* (*Trial*) and the *ICLEI Guide for Urban Greenhouse Gas Inventory* to put forward ideas for solutions. Given the differences between the two guides, the connection between Chinese urban greenhouse gas inventories with those of province-level and foreign cities should be clarified in drawing experience from the two guides, namely, the inventory positioning, with the aim of guaranteeing reasonable reference to both guides.

3.1 *Critical Problems*

3.1.1 Difficulty in Defining the Inventory Boundaries

The city is an administrative division, and more likely a region. It includes not only built-up areas, but also outskirts, villages and forest farms. Quite a number of cities have a complete economic system including industrial production and agroforestry production, so that the emission source or sink of greenhouse gas comes not only from construction, transportation, residential life, but also from industry, agriculture, forestry and waste disposal, which may be divided into multiple sectors and categories. As a result, the sources of greenhouse gas emissions are quite complicated, and agriculture contributes to a great amount of greenhouse gas emission, further resulting in difficulty in defining the greenhouse gas inventory boundaries.

3.1.2 Difficulty in Defining the Scope of the Inventory

Compared with state and province, the city covers smaller area, and at the same time, it is a densely populated and economically-developed area. Thus, the city has complex and innumerable exchanges with outside resources, causing considerable greenhouse gas emission coming from the urban consumption mode. The greenhouse gas emissions from the urban consumption mode involve data collection and repetitive calculations, which results in difficulties in defining the scope of the inventory, such as cross-border imports and exports of electricity and heat, transportation, waste disposal, remote production and usage of raw materials, remote production and consumption of commodities.

3.1.3 Difficulty in Data Collection

The city is weak in statistical capacity. Low-carbon development has appeared only in recent years. Greenhouse gas emission statistics and calculation systems remain to be established and improved. All of these lead to a severe shortage of fundamental data available for preparing the greenhouse gas inventories. Besides, rapid urbanization and constant implementation of energy-saving and emission reduction measures lead to a great variation in key fields of greenhouse gas emission, poor data time sequencing and great data uncertainty; at the same time, the demand for data has increased, and the requirement for data access is higher than before. As a consequence, it is very hard to collect high-quality data.

3.2 *Inventory Positioning*

First of all, the urban greenhouse gas inventories are supposed to keep up with the provincial greenhouse gas inventories based on the characteristics of Chinese administrative management. China has included greenhouse gas emission control in the medium- and long-term planning, and has specifically brought forward the objective in the *12th Five-Year Plan for National Economic and Social Development of the People's Republic of China* that the carbon dioxide emission per unit of GDP is expected to decline by 17% over that for 2010.[8] The *Work Plan for Greenhouse Gas Emission Control during the 12th Five-Year Plan* stipulates that "local governments, at all levels, shall bear sole responsibility for the control of the greenhouse gas emission within their own administrative areas, and the major governmental leaders are the first people responsible" and "all work tasks shall be broken up into smaller chunks and carried out at the grass–roots level," which reveals the administrative management mode of breaking down the greenhouse gas emission control index and implementing the strict target responsibility system.[9]

Second, Chinese actions on greenhouse gas emission control must be integrated into the international community, and the Chinese urban greenhouse gas inventory is comparable to the greenhouse gas inventory of foreign cities. The control of greenhouse gas emission is a global issue. As China is a large emitter of greenhouse gas, the international community is greatly concerned with Chinese greenhouse gas emissions. Now, China is experiencing rapid urbanization, and urban development leads to a surge in greenhouse gas emissions. So, how to find a proper path towards low-carbon development has been an issue that needs to be urgently resolved. Since Chinese cities are the administrator of the actions of low-carbon development to control greenhouse gas emissions, their urban greenhouse gas inventories are supposed to be comparable to those of foreign cities. This facilitates the comparison of the results of the actions of greenhouse gas emission control between domestic and foreign cities, and plays a positive role in drawing on experience, exerting a second-mover advantage and demonstrating the achievements of Chinese green and low-carbon development; and in addition, it provides a reference for various developing countries in the preparation of their urban greenhouse gas inventories.

---

8 State Council, *12th Five-Year Plan for National Economic and Social Development of the People's Republic of China*, http://www.gov.cn/2011h/content_1825838.htm.

9 State Council, *Work Plan for Greenhouse Gas Emission Control during the 12th Five-Year Plan*, http://www.gov.cn/zwgk/2012-01/13/content_2043645.htm.

Therefore, the positioning of the Chinese urban greenhouse gas inventory is to keep up with provincial greenhouse gas inventories and to be comparable to that of foreign cities.

3.3 *Resolving Ideas*

3.3.1 Inventory Boundaries

Nowadays, Chinese low-carbon development is led by the governments at all levels, and the objective of the greenhouse gas emission control is split up and delegated to the cities. Both the *Guide for Province-level Greenhouse Gas Inventory Preparation* (*Trial*) and the *ICLEI Guide for Urban Greenhouse Gas Inventory* define the inventory boundaries based on the administration precincts, and stress explicit responsibility requirements in combination with the geographical boundaries. So it is advised that the urban greenhouse gas inventory boundaries should be based on the administration precincts. The greenhouse gas inventory helps local governments accurately understand the condition of low-carbon development, and assists in the low-carbon policy-making for the adjustment of the industrial structure and the urban-rural coordinative development, so as to reasonably carry out urban low-carbon planning and construction and meet the low-carbon developmental needs of local governments.

3.3.2 Inventory Scope

The inventory scope is classified into scope 1, scope 2 and scope 3. The production mode is scope 1; the consumption mode is divided into scope 2 and scope 3. Scope 2 means the indirect emission from the importation (exportation) of electricity, heat, cool air or steam outside of the city; scope 3 means other emissions from the consumption mode not included in scope 2. The *Guide for Province-level Greenhouse Gas Inventory Preparation* (*Trial*) includes scope 1, and lists electricity in scope 2 as an information item. The *ICLEI Guide for Urban Greenhouse Gas Inventory* includes three scopes; however, only scope 1 and scope 2 are taken into account, in general, due to data access. By reference to the inventory positioning, it is suggested that the inventory scope should include scope 1 and scope 2. Because urban cross-border imports and exports deal mainly with electricity, scope 2 only contains electricity.

3.3.3 Data Collection

A vast majority of Chinese cities have not prepared the annual *Energy Balance Sheet* required by the greenhouse gas inventory. Owing to the diversification of data sources (statistical data, departmental data and investigative data), data collection requires mutual complementation and verification from top to bottom and from bottom to top. Generally, statistical data come first, followed by departmental data and investigative data. The three kinds of data are complemented by each other and checked by each other to look for errors and the causes of errors, with the aim of guaranteeing the accuracy of the data. "Statistical data" concerns the data provided by statistical departments; "departmental data" concerns the data provided by functional departments or industrial associations, and the data provided by the business backbone from functional departments based on business accumulation and experience; "investigative data" concerns the data acquired by statistical investigation due to data shortage or the need for a data check.

4 Content Framework

The content framework of the Chinese method of urban greenhouse gas inventory preparation is formed by taking the advantage of solving critical problems.

4.1 *Kind of Greenhouse Gas*

The *Guide for Province-level Greenhouse Gas Inventory Preparation* (*Trial*) prepared by the National Development and Reform Commission of China requires the calculation of six kinds of greenhouse gases, namely, CO_2, CH_4, N_2O, HFCs, PFCs and SF_6. This is also true for the *ICLEI Guide for Urban Greenhouse Gas Inventory*. Therefore, these six kinds of greenhouse gases are also included in the preparation of urban greenhouse gas inventories. Besides, as the potential and trend of the rise in greenhouse gas temperature, the data in the second evaluation report are adopted according to the suggestions made in the *Guide for Province-level Greenhouse Gas Inventory Preparation* (*Trial*).

4.2 *Inventory Boundary*

The inventory boundary is subject to administration precincts. The responsibilities, duties and benefits of greenhouse gas emission control are explicit, which not only helps local governments to accurately understand the overall condition of local greenhouse gas emissions, but it is also conducive to statistics, monitoring and evaluation of urban greenhouse gas emission control objectives.

4.3 *Method of Calculation*

For the calculation of the greenhouse gas inventory, the method for totalizing the data regarding material balance calculations is mostly used on indirect calculations. In the method for totalizing data, the material balance is calculated based on the total raw materials, primary and auxiliary products, and recovery products in order to calculate the total material losses. Therefore, it is convenient to use this method. This method is applied in the *Guide for Province-level Greenhouse Gas Inventory Preparation* (*Trial*) and the *ICLEI Guide for Urban Greenhouse Gas Inventory*. So, the preparation of the Chinese urban greenhouse gas inventory should also use this method. The calculation formula of this method is as follows: greenhouse gas emission/sink = activity data × emission factor.

4.4 *Key Category*

Greenhouse gas emission involves multiple sectors. Each sector has a set of relevant processes, emission sources and sinks, that is, the key category. By reference to the *Guide for Province-level Greenhouse Gas Inventory Preparation* (*Trial*), the sectors are classified into energy activity, industrial production process, agriculture, land utilization change and forestry, waste disposal. Energy activity mainly includes fossil fuel combustion, biomass fuel combustion, coal mining and post mineralization, petroleum and natural gas escape, and electricity consumption; industrial production process mainly includes the processes of producing cement, lime, steel, calcium carbide, adipic acid, nitric acid, chlorodifluoromethane, aluminum, magnesium, electrical equipment, semiconductors and hydrofluorocarbon; agriculture mainly includes rice fields, farmlands, animal intestinal fermentation and the animal waste management system; land utilization change and forestry mainly includes the change of forest and other woody biomass carbon storage, as well as the transformation of forests into non-forestry land; waste disposal mainly includes disposal of urban solid waste, domestic sewage and industrial sewage.

4.5 *Inventory Scope*

The scope of the inventory includes scope 1 and scope 2 (electricity). Scope 2 only contains electricity, showing that the method of preparing the Chinese urban greenhouse gas inventory takes into account not only the greenhouse gas from the urban consumption mode, but also the calculation requirements for scope 2 as set out in the *Guide for Province-level Greenhouse Gas Inventory Preparation* (*Trial*).

4.6 *Data Collection*

"Data collection" means acquiring the necessary data and generating new data based on key categories according to the requirements of the selected method, in a bid to meet the calculation needs. According to the statistical data of Chinese urban greenhouse gas, "data collection" mainly uses the method of mutual complementation and verification from top to bottom and from bottom to top.

4.7 *Uncertainty Analysis and Quality Management*

"Uncertainty analysis" is a basic part of a complete greenhouse gas inventory. In the course of preparation of the greenhouse gas inventory, the uncertainty should be reduced as much as possible to ensure that the inventory results are as close as possible to the actual urban greenhouse gas emissions. When implementing the measures to reduce the uncertainty, we should base on the priorities of the entire inventory, the analysis of the key categories, and specific category evaluation, and we should also reduce the uncertainty according to the cause of the uncertainty.

"The management of the quality of the inventory" refers to any coordinating activity conducted by quality command and control organizations in the process of the preparation of the inventory, including quality control and quality assurance. Quality control is a regular technical activity, which is used for evaluating and assuring the quality of the greenhouse gas inventory, and is performed by the inventory preparation groups; quality assurance is a set of planned review systems, which is performed by persons who are not directly involved in the preparation of the inventory.

5 Summary

Chinese urban low-carbon development is facing the pressure on its control of greenhouse gas emissions, and at the same time, is driven by the realization of sustainable development. Now, it is in urgent need of a scientific kind of low-carbon development. As the Chinese urban greenhouse gas inventory is an important basis for decision-making, its reflection of the facts will exert a significant influence on low-carbon development. Feasible preparation methods of the greenhouse gas inventory are conducive to enhancing the greenhouse gas inventory normalization. Thus, the formation of the content framework of the greenhouse gas inventory will strongly promote the studies on the methods used in preparing the greenhouse gas inventories, which will help the fact that the Chinese methods of preparing the greenhouse gas inventory lag behind its low-carbon development, and will help cities advance towards scientific and low-carbon development, resulting in aggregated forces in order to promote low-carbon urbanization.

CHAPTER 9

Collaborative Governance of Low-carbon Development and Adaptation: Based on the Perspective of Sustainable Urbanization

Zheng Yan

Abstract

Low-carbon development and adaptation to climate change are important contents of sustainable urbanization. Taking research and policy practices at home and abroad as examples, this chapter analyzes how cities achieve the collaborative governance mechanism of low-carbon development and adaptation, and points out that during the construction of the low-carbon green urbanization of China, urban managers need to change the traditional urban management pattern and governance concept and carry out collaborative management in such aspects as goals, polices and means in order to respond to climate change. The chapter analyzes specific problems and obstacles present in the collaborative governance of response to climate change by cities, and proposes several ideas and implementation suggestions, including reflections on the collaboration of governance goals in urbanization strategies and urban development planning, the collaboration of governance mechanisms in such aspects as the decision-making process, department coordination, public participation, and the collaboration of governance ways in policy design of key departments and fields, and enhancing the guarantee mechanisms of collaborative governance, etc.

* This chapter is supported financially by the key project of the National Natural Science Foundation of China—"Research on Climate Change Adaptation and Management Countermeasures of Urban Agglomeration Areas in the Yangtze River Delta" (No. 70933005) and the youth project of the National Natural Science Foundation of China—"Research on the Climate Change Adaptation Governance Mechanism" (No. 71203231).

** Zheng Yan, Climate Change Economics Laboratory of the Institute for Urban and Environmental Studies of Chinese Academy of Social Sciences, doctor of Economics, whose main research fields are environmental economics, research on policies of adaptation to climate change.

 | DOI 10.1163/9789004280021_010

Keywords

climate change – low-carbon resilient city – adaptation – sustainable urbanization – collaborative governance

1 Background

Combating climate change is becoming a mainstream policy design of all countries. In June 2007, China issued *China's National Program on responding to Climate Change*,[1] and since then, actively responding to climate change has gradually become an important topic in China's fundamental policies. China's "12th Five-Year Plan" stresses "sticking to attaching equal importance to mitigation of and adaptation to climate change, improving the institutional mechanisms and policy systems, and enhancing our ability in responding to climate change."

As two challenges in responding to climate change, mitigation and adaptation are especially prominent in urban areas. The international community has issued urgent requirements for emission reduction and adaptations at the city level, and, promoted by some non-governmental organizations and urban policy-makers, it has established such inter-governmental organizations as the Metropolis Climate Leadership Group (C40 City), the International Council for Local Environmental Initiatives (ICLEI),[2] the World Climate Change Mayors' Council, and the Cities for Climate Protection. Many cities in China have used these international platforms to carry out various governmental and non-governmental exchanges and cooperation activities.

In 2012, China's urbanization rate reached 52.57%. More than half of the population is now living in urban areas, and China's urbanization and industrialization will also continue its rapid improvement in the future. Driven by the population growth and consumption, cities in China have become the main areas of energy consumption and carbon emission. The developmental thinking of China's urbanization is to actively exert the agglomeration effect of central cities in designated areas to drive the local social and economic development. This makes it necessary for many large cities (especially super-large regional central economic cities such as Beijing, Shanghai and Guangzhou)

1 National Development and Reform Commission, "National Program on Combating Climate Change of China," June 2007.

2 ICLEI, "Financing the Resilient City: A Demand Driven Approach to Development, Disaster Risk Reduction and Climate Adaptation-An ICLEI White Paper," ICLEI Global Report, 2011.

to take on multiple pressures of emission reduction, employment, economic growth, and environmental governance. Meanwhile, against the background of climate change, climate risks and the vulnerability of cities caused by extreme climatic events are increasingly prominent. Therefore, it is particularly important to be able to withstand the risks, enhance the overall adaptability of cities, and to carry out the collaborative management countermeasures of adaptation and emission reduction in cities from the point of view of such aspects as urban spatial planning and urban governance mechanisms.

Driven by the national policy goals of energy conservation, emission reduction and low-carbon development, the establishment of low-carbon cities has become common practice from ideas to actions. However, regarding emission reduction, China's strategic goals, action plans and governance mechanisms for its cities to be able to adapt to climate change still lack targeted and operable policy research programs and plans. Currently, many cities in China have compiled the *Responding to Climate Change during the 12th Five-Year Plan*, but they are not yet changing the contents and the structural design with an emphasis on energy conservation and emission reduction due to national constraints on emission reduction tasks. As the environmental issue that gives the most concern in recent years, climate change involves multiple goals and multiple governance fields, including meteorology, disaster prevention and mitigation, water conservation, agriculture, ecology, health, environmental protection, planning and other decision-making managerial departments. Since 2007, China has formed one leading group after another to respond to climate change, save energy and reduce emission at both national and local levels, and it has established a basic framework for climate change governance, but it still needs to carry out research and experiment on how to collaborate and integrate the goals and requirements of adaptation and emission reduction according to the actual conditions based on the existing policies and mechanisms, and on whether the integration and collaboration are practical in terms of theory and practice. It has been reported that the *National Strategy on Responding to Climate Change*, led by the National Development and Reform Commission of China, will be introduced, and that the *National Planning on Responding to Climate Change (2011–2020)* has been actively promoted locally. Compared with the *National Program on Responding to Climate Change* with strategic guidance, national adaptation strategies and planning documents will have a more prospective strategic vision and more distinct operability. Although a mainstream consensus among educational circles and governments has not been reached on collaborating on low-carbon development and adapting to climate change, carrying out demonstrations and tests on low-carbon resilient cities by taking cities as a pointcut is helpful to promote the integration of mitigation and adaptation from ideas to actions. This chapter describes how to collaboratively achieve the goals of low-carbon development and adaptation to climate change in cities, promote urban sustainable development and improve the quality of urbanization.

2 Collaborative Governance of Mitigation and Adaptation: Research Status

For many developing countries, mitigation of global climate change is a long-term and tough challenge, while adaptation to climate change is a realistic and urgent task. Collaboration on mitigation and adaptation is to seek a "win-win" scheme between mitigation and adaptation, and develop policy measures which can not only control greenhouse gas emission but is also conducive to adapting to climate change.

The IPCC proposed to carry out a collaborative management of mitigation and adaptation actions in the third assessment report for the first time. In 2007, the IPCC specially discussed the collaboration problem of mitigation and adaptation actions in the fourth assessment report,[3] and since then educational circles and governments have begun to pay attention to the research field with developmental prospects. The research on collaboration of mitigation and adaptation mainly focuses on three aspects: definition of the collaborative effect, feasibility of collaborative management and optimization of collaborative management.[4]

2.1 *Collaborative Effect*

Climate change affects human society at different levels. Mitigation and adaptation actions may be complementary at some levels, but are conflicting in some cases, so we need to identify the actions with collaboration opportunities and remove the actions with conflicts and contradictions. Mutual effects of mitigation and adaptation actions are of four kinds: effects of mitigation on adaptation; effects of adaptation on mitigation; no effects between mitigation

3 IPCC (Intergovernmental Panel on Climate Change), "Climate Change 2007: Impacts, Adaptation and Vulnerability," in Working Group II Contribution to the Intergovernmental Panel on Climate Change Third Assessment Report (Cambridge: Cambridge University Press, 2007).

4 Wang Wenjun and Zhao Daiqing, *Research on Collaborative Development of Emission Reduction and Adaptation: Taking Guangdong as an Example*, China Population, Resources and Environment, 2011 (6), pp. 89–94.

and adaptation; mutual effects of mitigation and adaptation.[5] Conscious management of mitigation/adaptation actions dispersedly implemented to various departments can generate win-win effects, called collaborative management of climate actions.

2.2 *Feasibility of Collaborative Management*

There are two different views about feasibility of collaborative management of mitigation and adaptation actions: some scholars are optimistic about collaboration, believing that the collaboration between the two exists and may obtain multiplier effects of actions through a system design;[6] some scholars do not deny the feasibility of collaboration between the two, but they doubt its effectiveness.[7] The British Tyndall Center believes that actions of mitigation and adaptation have common driving factors and are likely to exert collaborative effects at regional and departmental levels, but knowledge and tools of other disciplines are necessary.[8] Zheng Yan et al.,[9] taking China's coastal cities' dealing with heat island effects and urban flood risks as examples, analyze the key fields and their policy choices in implementing a collaborative type of management of mitigation and adaptation.

2.3 *Optimization of Collaborative Management*

Collaboration of mitigation and adaptation is actually a win-win decision, and the optimization of collaborative management is a process to follow in order to seek the optimal path. The governance level and the collaborative management ability will directly affect the results of the collaboration actions of mitigation and adaptation. Due to different time and space scales for adaptation

---

5 Zhuang Guiyang and Chen Ying, "International Climate Regimes and China," World Affairs Press, 2005.

6 Biesbroek, G.R., Swart, R.J., and van der Knaap, W.G.M., "The Mitigation-adaptation Dichotomy and the Role of Spatial Planning," Habitat International 33 (3), 2009, pp. 230–237; Smith, P. and J.E. Olesen, "Synergies Between the Mitigation of, and Adaptation to, Climate Change in Agriculture," Journal of Agricultural Science 148 (5), 2010, pp. 543–552, doi: 10.1017/S0021859610000341.

7 Klein, R.J.T., et al., "Inter-relationships between Adaptation and Mitigation, Climate Change 2007: Impacts, Adaptation and Vulnerability," Contribution of Working Group II to the Fourth Assessment Report of the IPCC AR4.

8 Klein, R.J.T. et al., "Integrating Mitigation and Adaptation into Climate and Development Policy: Three Research Questions," Environmental Science & Policy (6), 2005, pp. 579–588.

9 Zheng Yan, Wang Wenjun, and Pan Jiahua, *Low-carbon Resilient City: Idea, Way and Policy Choice*, Urban Studies, 2013 (4).

and mitigation, different implementing departments and fields, different affected objects, costs and benefits, as well as the scientific bases of decisions, governance ability, technical conditions and other complicated factors, it is difficult to determine the optimal collaboration point of the actions of mitigation and adaptation. Feasible programs should design different policy combinations according to different climatic scenarios, or find out the scope of collaborative actions which can promote developmental goals, reduce costs or more efficiently utilize resources,[10] e.g. the construction of shelter forests in coastal zones can exert more collaborative benefits and use fewer costs compared with the construction of embankments and dams. With a gradual intensification of research and practices, recognition of the collaborative management changes from an operational level, such as simple policy design and technological innovation, to an institutional innovation level, such as implementation of the collaborative management of multiple goals and multiple departments. Successful collaboration strategies are not simple trade-offs between mitigation and adaptation actions, but are means to exert the leadership of local decision-makers, specify preferential topics and carry out overall planning according to the actual conditions, so as to incorporate adaptation and emission reduction strategies into the overall developmental goals.[11]

3 Goal, Idea and Connotation of Collaborative Governance of Mitigation and Adaptation

3.1 *Idea of Collaborative Governance*

Collaborative governance of mitigation and adaptation is a typical multi-goal and multi-department decision-making process, and it requires adopting the adaptive management idea and changing the traditional administration-oriented means of governance. Adaptive management is a flexible decision-making mechanism adopted to face various uncertainties. The foundation of its epistemology is the complexities and uncertainties of the human-ecological system, stressing the "learning by doing" mode. In addition to the general background of global environment and climate change, the rise of

10 Wilbanks, T.J., et al., "Integrating Mitigation and Adaptation as Responses to Climate Change: A Synthesis," Mitigation and Adaptation Strategies for Global Change (12), 2007, pp. 919–933.

11 Laukkonen, J., Blanco, P.K., Lenhart, J., and Keiner, M., "Combining Climate Change Adaption and Mitigation Measures at the Local Level," Habitat International (22), 2009, pp. 287–292.

the adaptive management idea is related to the development of the global environmental movement, civil society and governance theory which have emerged since the 1970s. Application of this concept to the environment and resource management fields first appeared in the late 1970s. Since the 1990s, it has begun to enter the social policy and political management fields.[12] Adaptive management is particularly applicable to the complex social-ecological systems filled with high uncertainties, and is widely applicable to such fields as watershed management, water resources management, biodiversity and habitat management, disaster risk management and environmental impact assessment.[13]

Adaptive planning has several key points: (1) Participation of the interested parties; (2) Specific aim and consensus; (3) Designing future policy scenarios based on uncertainties, stressing the variability and flexibility of policies; (4) Paying attention to the importance of monitoring and evaluation processes for policy improvement; (5) Active learning ability during practices, etc. Responding to climate change is an adaptive management process to repeatedly study and continuously improve its adaptive capacity; this process requires the evaluation of or changes to the existing policy and mechanism design, including the improvement of the goal, function and structure of the existing governance system, collaborating the problems of different time and space scales and different policy requirements, and providing the capacity building guarantee and mechanism guarantee with information, technology, and resources.[14]

3.2 *Goal of Collaborative Governance and Sustainable Urbanization*

For developing countries, adaptation to climate change is a very urgent task. According to the requirements of the *United Nations Framework Convention on Climate Change* (UNFCCC), under the existing emission reduction framework agreement, the developing countries do not take on global emission reduction responsibilities, but they can perform voluntary emission reduction actions. Mitigation must not weaken the ability to adapt to climate change. The goal of low-carbon development and adaptation to a collaborative type of manage-

12 Lessard, G., "An Adaptive Approach to Planning and Decision-making," Landscape and Urban Planning (40), 1998, pp. 81–87.

13 Allen, C.R., Fontaine, J.J., Pope, K.L., et al. "Adaptive Management for a Turbulent Future," Journal of Environmental Management (92), 2011, pp. 1339–1345; Williams B.K., "Adaptive Management of Natural Resources: Framework and Issues," Journal of Environmental Management (92), 2011, pp. 1346–1353.

14 IPCC, *Managing the Risks of Extreme Events and Disasters to Advance Climate Change Adaptation* (SREX) (Cambridge: Cambridge University Press, 2012).

ment is to achieve the strategic goal of responding to climate change for the world and countries. The collaborative governance of mitigation and adaptation is helpful to achieve the goals of ecological cities and sustainable urbanization. During this process, we need to fully consider the priority of urban developmental goals, specify principles and bases of policy choices in case of conflicts among different developmental goals, and carry out scientific evaluation. Therefore, China's principle of promoting collaborative governance during the urbanization process is to preferentially ensure requirements for responding to climate change in case of conflicts and contradictions between the goals and means of emission reduction and those of adaptation.

3.3 *Connotation of Collaborative Governance*

Collaborative governance aims at, against the background of responding to climate change, enriching and improving the connotation of ecological cities, being helpful to promote sustainable development of the urban social—economic—ecological environment, and achieving a harmonious coexistence of urban environment and human society. The connotation has two aspects: on the one hand, collaborative governance aims at focusing on issues of urban climate safety and sustainable development from the perspective of governance innovation and development transformation, against the general background of global environment and climate change; on the other hand, the idea of collaborative governance requires collaboratively considering various complex problems faced by cities and the requirements of subjects with different interests during the urban development planning and decision-making management. In terms of the idea of governance, it is necessary to lay particular emphasis on people orientation, and pay attention to the sustainability of social, economic and ecological levels of cities; in terms of governance goals, it is necessary to lay particular emphasis on enhancing cities' ability at adapting to climate change and reducing risks caused by climatic disasters, rather than taking emission reduction of greenhouse gases as the primary task and goal; in terms of governance measures, it is necessary to strengthen the integration of various departments and fields at the levels of management, technology and research, and reflect on flexibility and adaptability.

4 Steps and Practices of Collaborative Governance

4.1 *Steps of Collaborative Governance*

Based on the idea of adaptive planning, policy planning research on urban collaborative governance needs to grasp several principles: first, there are no perfect policies, and we need to continuously explore and sum up experience during practice in order to carry out collaborative management; second, seek a common ground while maintaining differences: fully respect values and interest appeals of the different participants during decision making; third, attach importance to the role of scientific evaluation: there should be clear evaluation principles and inspection standards for specific goals and implementation effects of the collaborative management. Basic steps include the following.

1) Define the main climate risks for cities in the future and the main goals of collaborative management.

   Managers should recognize the impact of climate change on areas, especially the various potential risks, enhance the understanding of social and economic vulnerability, and take this as the basis for determining the goals and decisions of collaborative management.

2) Define the main fields and departments of collaborative management

   First, comb through the emission reduction and adaptation measures of the existing fields to find problems and deficiencies; second, according to the future features of climate risks for cities, specify the key fields and key departments in which to carry out adaptation actions, and based on this, analyze the likely mutual effects of mitigation and adaptation actions, and determine the fields and departments suitable for carrying out the collaborative management of emission reduction and adaptation.

3) Screen policy choices to carry out low-carbon adaptation and collaboration

   In accordance with specific climate risks, screen various possible collaborative countermeasures of different fields and departments, such as engineering measures, technical and institutional measures, and evaluate the correlation, compatibility and feasibility of the collaborative countermeasures with urban development policies.

4) Design policy evaluation principles and carry out policy optimization and sequencing according to the decision-making principles

   Policy optimization consists of centralizing the advantages in order to solve the most important and the most critical problems with limited resources. The principle of policy selection reflects the values of the decision-makers, such as the principle of no regret, the cost benefit principle, the multi-win principle and the urgency principle.

5) Implementation and post-evaluation of policy planning

   Incorporating selected policies into policies and planning, reflects on them as the will of the decision-makers who will use administrative power or laws to implement the policies. During the implementation, carry out supervision, evaluation and feedback, promptly discover new problems, and make adjustments and improvements combined with new scientific knowledge, information and feedback.

4.2 *Policies and Actions of Collaborative Governance of Climate Change for Cities at Home and Abroad*

Cities are a policy laboratory to respond to climate change. Policy integration and collaborative governance of adaptation and emission reduction are of practical significance in order to make climate change policies and planning, and have also been a new topic arousing concerns at home and abroad in recent years.[15,16] Although there are doubts about the collaborative effects of adaptation and emission reduction and their feasibility, some experiments and practical research have been carried out at home and abroad. In particular, as the governance subjects at the local level, cities are performers of emission reduction goals and actions at the national level and they are also objects of climate change risks and impact, so cities have favorable conditions for carrying out collaborative governance of emission reduction and adaptation.

4.2.1 China's Policy Demands and Progress

According to the *China's National Program on Responding to Climate Change*, key fields in mitigating emission of greenhouse gases include: the energy department, the industrial department, agriculture and forestry, and people's living consumption; key fields in adapting to climate change include: agriculture, forestry, water resources management, coastal zones and coastal areas. It can be seen that, agriculture and forestry departments are common key fields in mitigation and adaptation actions, conducive to carrying out collaborative management actions. For urban areas, problems needing consideration while carrying out collaborative management are more complex.

Key fields for cities to carry out collaborative management of emission reduction and adaptation include: the urban ecosystem, the urban construction field, urban energy and power, the urban public transport system, urban water resources management and watershed management, land use,

---

15 Wang Wenjun and Zheng Yan, "Collaborative Effect of Low-carbon Development and Adaptation to Climate Change and Its Policy Implication," in Wang Wenjun and Zheng Yan, eds., *Report on Combating Climate Change (2011)—Dilemma of Durban and China's Strategic Choice*, Social Sciences Academic Press, 2011.

16 Norman, B., "A Low-carbon and Resilient Urban Future: A Discussion Paper on an Integrated Approach to Planning for Climate Change," Australia: Department of Climate Change and Energy Efficiency, 2010, http://www.climatechange.gov.au/~/media/publications/local-govt/low-carbon-resilient-urban-future.pdf.

urban planning, etc.[17] For instance, Shanghai Municipality stresses, in the *Energy Conservation and Emission Reduction and Responding to Climate Change during the 12th Five-Year Plan*, while promoting energy conservation and low-carbon development and reducing greenhouse gas emissions, it is necessary to further strengthen the cities' capacity building in flood prevention and waterlogging prevention, improve the construction of the emergency early warning system for disaster prevention and mitigation, enhance basic research on adaptation to climate change, and gradually improve the ability in adaptation to climate change. Some policy goals can exert collaborative effects of emission reduction and adaptation. For instance, by constructing shelter forests in coastal zones, water conservation forests and anti-pollution isolation forests, the forest coverage rate in 2015 reached 15.5%, improving the carbon sink ability; by developing efficient ecological agriculture and promoting green manuring, the rate of reusing farmland straw and waste reached 95%; developing urban vertical planting and green roof requires that green roof areas of new public buildings should account for more than 95% of those of suitable public buildings, etc. During the "11th Five-Year Plan," Guangzhou City invested first 58 billion RMB and then more than 100 billion RMB in disaster prevention and mitigation and multi-purpose projects of water resources to strengthen the disaster monitoring, early warning and water resource management, and the city has planned protection and development of such ecological systems as marine environment, wetlands, and mangrove forests. Through the overall planning of urban development, Xiamen City has increased the flood control standard from "20-year frequency" to "50-year frequency," strengthened the monitoring management of the rise in the sea level as well as ocean and coastal zone management, has successively implemented the man-made beach engineering, expanded the planting area of mangrove forest, and created a good habitat for rare and endangered species such as the Chinese white dolphins and amphioxi while also preventing ocean disasters.[18]

4.2.2 International Experience and Progress

The EU has always been active in the process of international climate negotiation, and has made national emission reduction goals and adaptation

17 Zheng Yan, Wang Wenjun, and Pan Jiahua, *Low-carbon Resilient City: Idea, Way and Policy Choice*, Urban Studies, 2013 (4).

18 Cao Lige, Su Buda, and Zhai Jianqing, "Practice and Progress of Adaptation to Climate Change of Coastal Cities," in Wang Weiguang and Zheng Guoguang, eds., *Report on Combating Climate Change (2011)—Dilemma of Durban and China's Strategic Choice*, Social Sciences Academic Press, 2011.

strategies. Britain is a pioneer in climate change policy legislation and local action. In 2001, London established the London Climate Change Partnership with the wide participation of 15–20 governmental agencies, enterprises and media, and the city also assigned special officials to be in charge of coordinating and implementing the "London Adaptation Plan," including a new increase of 1,000 hectares of greenbelts by 2012 to reduce urban heat island effects, carrying out ecological remediation of rivers in London, strengthening energy conservation of family and public buildings, and promoting low-carbon buildings, energy-saving and zero-carbon transportation. Besides, London also launched an advanced program to lower the temperature of underground transport tunnels, which is called London Transport's Tunnel Cooling Program.[19]

As representatives of the Umbrella Group, such developed countries as the USA, Canada and Australia are relatively dilatory in terms of national actions. However, with raising the public's awareness of climate change, advocated by business circles and non-governmental institutions, the local governments are becoming the main forces in responding to climate change. Some cities in North America have launched a series of motions and plans for emission reduction and adaptation. At present, mayors in more than 1,000 cities in the USA have signed the "Climate Protection Agreement," where the successful adaptation actions include: adaptation service infrastructure, flood management, urban afforestation, land use planning and policies, etc.; the emission reduction actions include: reduction of greenhouse gas emissions, improvement of energy efficiency and energy conservation, development of renewable energy, emission reduction plans in the field of transportation, etc. In the meantime, urban managers have begun to focus on how to better use the existing experience to promote the integration of adaptation and emission reduction, so as to increase the cost-effectiveness and collaboration advantage of climate actions.[20] In 2010, New York, in the USA, established the New York City Panel on Climate Change (NPCC), and issued the *New York Climate Adaptation Plan* to cope with climate risks such as the rise in the sea level and high temperatures in the cities; this plan included making a 20-year green infrastructure plan, improving the flood management ability and reducing urban heat island effects (this is also conducive to reducing transportation emissions and energy consumption) through block renovation, land use planning and improvement

---

19 Greater London Authority (GLA), "The Draft Climate Change Adaptation Strategy for London," Public Consultation Draft, GLA, London, February 2010, http://www.london.gov.uk/climatechange/sites/climatechange/staticdocs/Climate_change_adaptation.pdf.

20 Zimmerman, R. and Craig Faris, "Climate Change Mitigation and Adaptation in North America Cities," Environmental Sustainability (3), 2011, pp. 1–7.

of water supply and drainage facilities (such as increasing the forest-cover rate of cities, designing flood inundated areas into greenbelts in parks, improving the rainwater collection system, modifying the drainage design standard, and using and strengthening the drainage function of the natural landscape).[21]

5 Methods of Implementing Collaborative Governance

Collaborative governance of responding to climate change requires reflecting on the collaboration of governance goals in urbanization strategies and urban development planning, reflecting on the collaboration of governance mechanisms in such aspects as decision-making process, department coordination, and public participation, and also reflecting on the collaboration of governance methods in policy design of key departments and fields. This chapter proposes several preferential points in implementing the collaborative governance as follows, so as to promote practical activities of collaborative governance based on them.

5.1 *Goal and Mechanism Design of Collaborative Governance: Construct Multi-goal Collaborative Governance Mechanisms to Respond to Climate Change on the Part of the Cities, and Collaboratively Consider the Adaptation and Low-carbon Developmental Requirements in Urban Developmental Planning*

In some large Chinese developed cities, climate change consciousness and environmental governance demands of the public are increasingly improving, so when establishing urban climate change collaborative governance mechanisms, we should learn from advanced international urban governance ideas and experience. For instance, London's "Climate Change Partnership" implies forces of all sectors of society and widely absorbs different interested parties such as the public, experts and enterprises to participate in decision-making processes. Scientific and reasonable urban spatial planning (such as the population and the industrial layout, the traffic network structure and community development) is conducive to protecting the integrity of the urban ecological environment and reducing unnecessary traffic emissions and building emissions. Designing ideas and practices for compact cities at home and abroad are helpful for mitigating the heat island effects and reducing energy consumption. For example, in order to meet the strict urban green rate limit in urban

---

21 The City of New York (2011), "PlaNYC 2030: A Greener, Greater New York," update April 2011, http://www.nyc.goc/html/planyc2030/html/theplan/the-plan.shtml.

spatial planning legislation, Singapore and London, England, are making efforts to improve the transit share ability of different means of transport in the planning of their urban transportation, reflecting efficient utilization of urban land resources.

5.2 *Policy Approach Design of Collaborative Governance: Policy Collaboration of Key Departments and Fields*

As two fields in responding to climate change, adaptation and the reduction of greenhouse gas emissions have great differences in nature, goals, methods, and realization. Compared with emission reduction, adaptation to climate change is more complex and more urgent, requiring the consideration of regional differences and policy integration of multiple departments. In spite of different priorities, both new urban satellite towns and traditional old urban areas need to comprehensively consider the collaborative effects of building, transportation, ecological afforestation, disaster prevention and mitigation, community training and other aspects.

5.3 *Safeguard Mechanism of Collaborative Governance: Funds, Technology and International Cooperation*

Currently, theoretical discussion and practice on how to adapt to climate change for cities in China are still insufficient. In particular, how to actively use the market to raise funds and promote technology R&D and application promotion not only needs government instructions on strategies, but also needs some policy safeguard mechanisms, including increasing R&D input of low-carbon and adaptation of the collaborative technology, providing taxation and credit support for enterprises to carry out related activities, and incorporating key indices of low-carbon resilient cities into urban planning, legislation and government performance appraisal indices.

CHAPTER 10

Effects of Climate Change on China's Typical Urban Agglomerations and Its Potential Risks

Xu Ying, Zhou Botao, Guo Wenli, Mu Haizhen and Xie Xinlu

Abstract

Climate change and the development of urban agglomerations are two factors causing human society to suffer disasters. With global climate change, the frequency and severity of disasters caused by climate change such as windstorms, floods and droughts increase noticeably, and their threats to the society, economy, health, security and other aspects of human life are getting increasingly more intense. With the acceleration of urbanization, cities have become a disaster-affected body with a large scale and they are more likely to suffer from effects of disasters which cause heavy losses. Taking the urban agglomerations in the Beijing-Tianjin-Hebei region of North China and the Yangtze River Delta region of East China as examples, this chapter explains the main effects of climate change on China's typical urban agglomerations and the possible risks for them, mainly including effects of climate change on the developmental design, energy security, human health and ecological security of the urban areas, and proposes corresponding countermeasures and suggestions.

* The article is the result of research entitled the *Regional Climate Change Evaluation Report*, a special project on climate change carried out by the China Meteorological Administration, 2013-CCSF201339 and the "Research on Climate Change Adaptation and Management Countermeasures of Urban Agglomeration Areas in Yangtze River Delta (70933005)," a project carried out by the National Natural Science Foundation of China.

** Xu Ying, a doctor at the National Climate Center, whose research fields are numerical simulations, future estimation and disaster risks of climate change; Zhou Botao, a doctor at the National Climate Center, whose research fields are climate and climate change mechanisms; Guo Wenli, director of the Beijing Meteorological Service Center, whose research fields are climate and climate change; Mu Haizhen, director of the Shanghai Climate Center, whose research fields are climate and climate change; Xie Xinlu, a post-doctor at the Institute for Urban and Environmental Studies of the Chinese Academy of Social Sciences, whose research field is climate change economics.

 | DOI 10.1163/9789004280021_011

Keywords

urban agglomeration – climate change – risk

1 Introduction

Urban development symbolizes human civilization and progress. With economic development and continuous urban expansion, single cities gradually develop into large urban agglomerations. The urban agglomerations are mostly located in coastal areas and along the Yangtze River provided with convenient transportation, and are also areas at high risk due to climate change. According to the *World Urbanization Prospects* issued by the United Nations in April 2012, half of the 7 billion population around the world lives in cities, and another 2.3 billion will be absorbed into the cities by 2050. Since the reform and the opening up, China has experienced a rapid process of urbanization, and one characteristic is the rise of urban agglomerations led by large cities. According to the *2013 China Human Development Report* issued by the United Nations Development Programme (UNDP), as early as 2007, China's urban agglomerations occupied 21% of the national area, with 49% of China's total population.[1] By 2030, the level of China's urbanization will probably have reached 70%, and the total population in urban areas will probably exceed 1 billion, where the urban agglomerations will continue to play an important role in accumulating population and resources and leading regional social and economic development.

The effects of the interlacing of climate change and urban expansion makes urban agglomerations with a high density of population and wealth face continuously more intense disaster risks. After the formation of urban agglomerations, the variation in the method of regional land use has changed the underlying surface; at the same time, the atmospheric stratification and stability in some areas have changed. Accordingly, the distribution of thunderstorms and rainfall areas have changed, thus changing the occurrence, the law of development and the manifestation of disasters, which makes it more difficult to prevent disasters and cope with their risks. Urban floods, high temperatures and heat waves, heat island effects, and air pollution are prominent problems in urban agglomerations, and in particular, strategies for coping with disasters

1 The United Nations Development Programme and the Institute for Urban and Environmental Studies of the Chinese Academy of Social Sciences, *2013 China Human Development Report*, UNDP, 2013.

caused by rapid changes in these events, such as floods caused by short-term heavy rainfall, now need to be further improved. Therefore, the adaptation to climate change and the reduction of disaster risks are important tasks for urban managers. Taking the urban agglomerations in the Beijing-Tianjin-Hebei region of North China and the Yangtze River Delta region of East China as examples, this chapter discusses the effects of climate change on typical urban agglomerations and its potential risks for them.

2 Related Concept of Urban Agglomeration and Climate Risks

2.1 *Proposal for Urban Agglomerations*

Since the Industrial Revolution in the 18th century, with the great development of social productive forces, worldwide urbanization has accelerated in terms of time, and it has also shown urban regionalization and regional urbanization in terms of space. Especially after the Second World War, highly-dense urban agglomerations with close relationships and a certain quantity of them appeared, one after another, in some regions of the developed countries and developing countries. These urban agglomerations are not only an important support for the economic development of their regions, but they are also important nodal regions of the global economic system.

In recent years, with the continuous intensification of the development and practice of urbanization, many concepts related to urban agglomeration have appeared in China successively. There are more than 20 new concepts, including megalopolises, metropolitan areas, urban agglomerations and metropolitan regions.[2] Currently, urban agglomeration is used most frequently.

2.2 *Definition of Urban Agglomeration*

Urban agglomerations are usually derived from large cities, especially megacities and huge cities; they are the highest spatial organizational form of urban development at the mature stage, and are an inevitable result of regional industrial development and urbanization. Geographically, urban agglomerations are areas with dense city distribution in a region, including different developmental stages and manifestations of urban agglomerations. With the continuous development of urbanization, urban agglomerations are given deeper economic connotations, including both a geographical regional concept and a economic regional concept. At present, China has basically formed 10 urban

2 Chen Meiling, *Research and Discussion on Related Concepts of Urban Agglomeration*, Urban Studies, 2011 (18(3)), pp. 5–8.

agglomerations: the urban agglomeration in the Yangtze River Delta region, the urban agglomeration in the Pearl River Delta region, the urban agglomeration in the Beijing-Tianjin-Hebei region, the urban agglomeration in the Shandong Peninsula, the urban agglomeration in the Liaodong Peninsula, the urban agglomeration in the Chengdu-Chongqing region, the urban agglomeration in the Harbin-Dalian-Changchun region, the urban agglomeration along the west coast of the Taiwan Strait, the urban agglomeration on the Central Plain and the urban agglomeration in the Yangtze-Huaihe area.[3] With obvious regional advantages and robust economic strength, the 10 urban agglomerations have become an important engine of national economic development. China's "12th Five-Year Plan" puts forward the pattern of the strategy for the development of urbanization called "two horizontal axes and three vertical axes," which will focus on the artery traffic in coastal areas and along the Yangtze River; it is based on several urban agglomerations and includes other urban areas and cities as important parts of it so as to drive urbanization further ahead. This will lay the basic pattern of China's urbanization in the future, and will lead the rapid development of more regional small and medium-sized urban agglomerations.

2.3 *Climate Change Risks*

Climate change causes the frequency, the intensity and the manifestation of disaster risks to be more complex, and therefore makes them more difficult to predict. More and more evidence shows that the climate change characterized by global warming is inevitable. Climate risks caused by climate change are one of the most typical modern risks of a "risky society".[4] Now, the word "risk" is used widely, involving different fields such as natural science and social science, with different understandings of its definition in different fields. Generally speaking, "climate change risks" refers to the extent of possible damage to nature and to human society's economic system caused by changes in the climate system due to natural and human disturbance (human activities) activities, expressed as: risk = vulnerability × possibility. "Vulnerability" refers to the extent of the effects of adverse climate change, measured by the degree of destructiveness and exposure; while "possibility" refers to different scenarios

3 United Nations Development Programme and Institute for Urban and Environmental Studies of the Chinese Academy of Social Sciences, *2013 China Human Development Report*, UNDP, 2013.

4 Wu Shaohong, Dai Erfu, Ge Quansheng, et al., *Integrated Risk Governance—China's Integrated Climate Change Risk*, Science Press, 2011.

of climate change. Climate change risks are closely related to the effect of climate change, to its vulnerability, its mitigation, and to its ability to adapt.

Compared with traditional disaster risks, climate change risks are longer and more uncertain. As urban agglomerations are a large scope of geographical space, it is more scientific and necessary to establish large-scale and long-term weather, hydrological and climatic predictions. Since urban agglomerations are formed from cities with different sizes and at different administrative levels, decision-making and governance of climate change risks need to be coordinated. Moreover, different urban agglomerations are different in geography, climate, natural resources, population, social economy and other aspects, so proper coping strategies should be made according to the different climate change risks and disaster effects.

3 Effects of Climate Change on China's Main Urban Agglomerations

Bridgit Leonie, spokesman of the United Nations International Strategy for Disaster Reduction (ISDR), points out that climate change and urbanization are two factors causing human beings to suffer disasters. Climate change makes it more likely that human beings will face more threats of disaster. In the past ten years, the frequency and severity of disasters, caused by climate change, such as hailstorms, floods and droughts, have increased noticeably; their threats to the economy, health, security and to other aspects of human life are increasingly more intense, and they are becoming the most prominent environmental problem in the current world. In the meantime, with the acceleration of urbanization, the population and industries are concentrated in cities. Cities are becoming a disaster-affected body on a large scale, and are more likely to suffer disasters, which cause heavy losses.[5] Taking the urban agglomerations in the Beijing-Tianjin-Hebei region of North China and the Yangtze River Delta region of East China as examples, this chapter introduces the effects of climate change on urban agglomerations and its possible risks.

3.1 *Effects of Climate Change on the Urban Agglomeration in the Beijing-Tianjin-Hebei Region of North China and its Risks*

The urban agglomeration in the Beijing-Tianjin-Hebei region, located in the northern part of the North China Plain, is in the important core area of the Bohai Economic Rim (BER) and Northeast Asia, and is the largest urban

5 Dong Suocheng, Tao Shu, et al., *Effects of Climate Change on Urban Agglomerations in China's Coastal Areas*, Advances in Climate Change Research, 2010 (6(4)), pp. 284–289.

agglomeration in North China. With an area of 166,800 km^2 and a population of 67.22 million in 2008, the urban agglomeration's gross domestic product (GDP) is up to 2.27 trillion RMB and its economic strength ranks the third among China's urban agglomerations, belonging to Level I of urban agglomeration.

The acceleration of urbanization in recent years has led to obvious changes in the properties of the underlying surface of the urban agglomeration in the Beijing-Tianjin-Hebei region; in addition, such problems as shortage of fresh water, environmental pollution and ecological degradation in urban areas caused by heat release, atmospheric pollution and water pollution, have become increasingly apparent, and the heat island effects in cities and urban agglomerations have increased. Against the background of climate warming, these problems may become more and more severe. Extreme events caused by climate change are increasing in number and are more and more intense in extent, directly threatening the ecology, the environment, the agriculture, the water supply and transportation in urban agglomerations. Affected by human activities, the local urban climate change will be more complex and will seriously affect the urban living environment, mainly reflected in the design of urban planning, transportation, etc.[6]

3.1.1 Effects of Climate Change on the Drainage Design Standard of the Urban Agglomeration in the Beijing-Tianjin-Hebei Region

In recent years, such disasters as rainstorms and waterloggings in the urban agglomeration in the Beijing-Tianjin-Hebei region have gradually become more and more severe, the main reasons of which are as follows: first, the properties of the underlying surface have changed, and the increase in the areas of impervious surface such as asphalt and cement, and the expansion of urban scale, make the runoff coefficient, with the same rainfall, increase several times; second, the formation of an urban climate leads to a more uneven rainfall distribution, and the local rainstorms within the cities show an ascending trend. In the design of the urban drainage standard, effects of "designed storm" intensity change are very important. The calculation of the "designed storm" can be restricted by the limit and quality of the age of rainstorm observational data, which is the basis of the infrastructure planning of regional flood control and waterlogging as well as the municipal drainage of cities. Against the background of global warming, the intensity and frequency of rainstorms will change, so the threshold value set in the past cannot adapt to the new change in climate. Take Beijing and Tianjin as examples. Compared

6 North China Regional Climate Center et al., *Climate Change Assessment Report in North China*, China Meteorological Press, 2013.

with the rainstorm intensity calculation of 1984, the short-term rainstorm intensity of Beijing in the past several years has increased; the rainstorm calculation obtained by using the latest data from Tianjin is generally greater than the rainstorm design standard calculated according to rainfall records from 1932 to 1981. For instance, for the once-in-a-year rainstorm intensity in one hour, the original calculation was 34.3mm (for the whole Tianjin Municipality), and the new calculation is 40.7mm (the urban district) and 44.1mm (Tanggu). The value of Tanggu District even exceeds the calculation of the original once-every-two-year calculation (43.1mm), obviously greater than that of the urban district of Tianjin. Therefore, when designing the urban drainage standard, we should consider the possible future climate change to reduce the likelihood of the occurrence of disasters.

3.1.2 Effects of Climate Change on the Energy Conservation Design for Buildings of the Urban Agglomeration in the Beijing-Tianjin-Hebei Region

The energy conservation design for buildings should fully consider the climate factors. China's current energy conservation design standard for buildings has not considered the effects of climate change on the meteorological parameters of the energy conservation design for buildings, so of course it has no sustainable applicability. Climate warming has obvious effects on the energy consumption of buildings, accounting for about 30% of the total energy consumption of the whole society in China, which makes new requirements for the energy conservation design for buildings, with examples below.

(1) As regards the heating in winter, the calculated outdoor temperature for heating and winter air conditioning in Beijing, Tianjin, Shijiazhuang and Taiyuan in North China had an obvious increase from 1951 to 2010. Compared with the years 1951–1980, the increase in the calculated outdoor temperature for heating from 1981 to 2010 was 2.1°C (Beijing) – 3.4°C (Shijiazhuang), and that for winter air conditioning was 2.1°C (Tianjin) – 3.1°C (Shijiazhuang).

(2) As regards summer air conditioning, compared with the years 1951–1980, the dry-bulb temperature in Beijing, Tianjin, Shijiazhuang and Taiyuan in North China had an increase with varying degrees from 1981 to 2010, an increase of 1.1°C and 1.0°C for Taiyuan and Tianjin as well as 0.5°C and 0.2°C for Beijing and Shijiazhuang, respectively. Since the increase in the calculated outdoor dry-bulb temperature for summer air conditioning in Beijing and Shijiazhuang was not obvious, it did not much affect the selection of equipment type and the operational safety; while that in Tianjin and Taiyuan exceeded 1.0°C, so it greatly affected the selection of equipment type and the operational safety.

3.1.3 Effects of Climate Change on the Safe Operation of the Urban Agglomeration in the Beijing-Tianjin-Hebei Region

The effects of climate change on the safe operation of the urban agglomeration in the Beijing-Tianjin-Hebei region are mainly shown in urban transportation, water consumption and power consumption.

(1) Effects of climate change on transportation. Such phenomena as rain, snow, fog, high temperatures and low temperatures affect transportation. With the increasingly developed transportation network of the urban agglomeration in the Beijing-Tianjin-Hebei region, the sensibility and vulnerability of urban transportation to the weather and climate have increased. Against the background of global warming, the air temperature of the urban agglomeration in the Beijing-Tianjin-Hebei region has also changed significantly. Since the middle of the 1990s, the number of days with the daily maximum temperature up to 35°C in the 5 large cities of North China has increased remarkably, producing adverse effects on transportation in summer. In winter, the air temperature has also increased obviously; the number of days with the daily minimum temperature between 0°C and –10°C in winter has shown an increasing trend, and that with temperatures between –10°C and –25°C has obviously decreased, with a temperature not lower than –25°C in recent years, which is good for vehicle performance in winter.

The relationship between comfort and the daily average incidence of accidents is obvious: more accidents in uncomfortable conditions (such as high temperatures, rainstorms or low visibility). The number of comfortable days in these cities changed significantly from 1961 to 2010: the number of uncomfortable days in summer in Beijing, Tianjin and Shijiazhuang obviously increased and that of comfortable days decreased; in winter, the number of comfortable days increased while that of uncomfortable days obviously decreased. Thus, from the perspective of comfort change, the effects of climate change in summer are adverse, while warming in winter is relatively favorable.

Rainfall change also produces great effects on transportation. For example, heavy rainfall may cause roads to be water-logged, making it difficult for the vehicles to travel, which will then cause traffic jams. In the past 50 years, the number of days with rainstorms in the five cities of North China has shown a decreasing trend to different degrees, the storm volume has decreased and the storm's contribution to the traffic has obviously diminished. Only from the perspective of climate, do the adverse effects of rainstorms on transportation diminish; with well-developed transportation, crowded roads and lots of vehicles, traffic jams and even disruptions caused by rainstorms in the cities of North China are increasingly prominent at present.

(2) Effects of climate change on urban water consumption. The effects of climate change on urban water consumption are mainly shown in the relationship between climatic factors and domestic water consumption. The urban per capita daily domestic water consumption in Beijing from 1990 to 2009 was significantly related to the annual average temperature: an increase of 1°C in the annual average temperature made the urban per capita daily domestic water consumption of Beijing increase by 26.0 liters, i.e. the per capita annual domestic water consumption increased by about 9.5 tons. From 1980 to 2008, an increase of 1°C for the annual average temperature in Tianjin made the urban per capita daily domestic water consumption of Tianjin increase by 19.2 liters, i.e. the per capita annual domestic water consumption increased by about 6.8 tons.

(3) Effects of climate change on urban power consumption. The scatter diagram of temperatures–daily maximum power load in Beijing and Tianjin shows a similar "U-shaped" structure. With the continuous rise in temperatures, the power load first reduces gradually and then it changes to a gradual increasing trend. When the average temperature is greater than 20°C, an increase of 1°C in the temperature will make the maximum power load in Beijing and Tianjin increase by 292.04MW and 110.63MW, respectively. High temperatures and humidity in the summer cause a rapid increase in the power load. In June in Beijing and August in Tianjin, the effects of the relative humidity on the power load are more obvious compared with other months, i.e. with the same temperature, an increase of 10% in the relative humidity will make the power load increase by 252.8MW and 107.7MW, respectively. In the future, the increase in the number of days with high temperatures in summer will lead to a rapid increase in the power load.

3.1.4 Climate Change will Increase the Urban Safe Operation Risks

In the future, with the continuous increase in the concentrations of greenhouse gas, the climate in North China will continue warming up, extreme high temperature events will increase, the per capita domestic water and power demands of residents of the urban agglomeration in the Beijing-Tianjin-Hebei region will further increase, and the daily water consumption and power consumption at peak time in summer will increase sharply. Besides, the frequency and intensity of extreme events such as extreme rainfall may increase, which will increase rainstorm waterlogging risks in cities and pose a serious threat to the safe operation of urban transportation. Under the scenario of climate change in the future, risks to the safe operation of various urban infrastructures including drainage systems and transportation systems as well as various systems that safeguard to urban residents' lives, including water supply and power supply, will increase significantly.

3.2 *Effects of Climate Change on the Urban Agglomeration in the Yangtze River Delta Region of East China and its Risks*

The east China region is located at 22.5°N–38.4°N and 112.5°E–122.0°E, covering Shanghai Municipality, Jiangsu Province, Zhejiang Province, Anhui Province, Fujian Province, Jiangxi Province and Shandong Province. By virtue of the favorable geographic position of the coastal areas and along the Yangtze River, excellent climatic conditions, good economic base, lots of scientific and technological talents, high educational level, rapid information dissemination at home and abroad and other advantages, the region's economy has developed steadily and quickly.

Since the reform and the opening up, urbanization in the East China region has produced important effects on the observed temperature change, and especially on the urban agglomeration in the region of the Yangtze River Delta. The overall contribution of urbanization in the region of East China to the regional average warming from 1981 to 2007 was about 1/4, where the contribution of heat island effects of megacitys to warming exceeds 44%. Climate warming has produced important effects on many aspects of the Yangtze River Delta region of East China.[7]

3.2.1 Effects of Climate Change on the Coastal Zone of the Urban Agglomeration in the Yangtze River Delta Region and its Risks

Climate change makes regional differences in the sea level of coastal areas in East China from north to south more obvious. From 1978 to 2007, the rise in sea level of the coastal provinces and municipalities in East China was as follows: Shanghai (115mm), Zhejiang (98mm), Shandong (96mm), Jiangsu (78mm) and Fujian (47mm). In the past several decades, the relative speed of the rise in sea level in the southeast coastal areas of the Shandong Peninsula has been about 3.0mm/year, 0.9mm/year for Jiangsu and the northern coastal area of the Yangtze River Estuary, 4.7mm/year for the coastal areas of the Yangtze River Delta region, and 1.9mm/year for the coastal areas of Fujian. The speed of the rise in sea level in the Yangtze River Delta region is the most obvious, much higher than that of other coastal areas in East China.

7 East China Regional Climate Center et al., *Climate Change Assessment Report in East China*, China Meteorological Press, 2013.

Most beaches in the coastal areas of East China are storm surge-prone areas, where the beaches with serious risks include the coastal areas from Bohai Bay to Laizhou Bay, Xiaoyangkou in Jiangsu to Taizhou and Wenzhou in Zhejiang, Shadi in Fujian to the Minjiang River Estuary and other coastal belts. In 2000, a serious storm surge disaster occurred in China's coastal areas, which caused a direct economic loss of up to 11.2 billion RMB.

The East China region suffers serious coastal erosion, accounting for 67% of the total length of China's coastal erosion. The main forms of coastal erosion in the Yangtze River Delta region are channel siltation, coastline recession, and drowned depression.

The future rise in the sea level will increase the salt-water encroachment distance of the estuary. Take the Yangtze River Estuary as an example. When the sea level rises by 30cm, 50cm and 100cm, the salt-water wedge will advance upstream by 3.3km, 5.5km and 12km, respectively. Relative to rise in the sea level, the high-water level in the original design of the seawall will accordingly occur more frequently. For instance, the design level of the flood control wall of Huangpu River in Shanghai was once-in-a-millennium, but it will be once-in-a-century if the sea level rises by 100cm. In case the sea level rises by 20–50cm, the seawall standard of the Yangtze River Delta will be reduced to once-every-50-year from once-in-a-century, which will greatly increase the risks of damage to the flood control wall, and the increase is non-linear.

3.2.2 Effects of Climate Change on Energy Consumption in the Urban Agglomeration in the Yangtze River Delta Region and its Risks

Effects of air temperature on energy consumption in East China are mainly shown in the consumption of energy for heating in winter and cooling in summer in the lives of the people. Under the dual impact of global climate warming and heat island effects caused by rapid regional urbanization, continuous high-temperature weather has frequently occurred in East China in recent years, with an increase in power consumption. According to statistics, in the economically-developed Yangtze River Delta region with hot summers, the air conditioning load accounts for more than 1/3 of the entire power load at peak time in summer. According to the relative analysis of day-by-day power consumption and maximum power load data in Shanghai during the summers of years 2003–2006 and the meteorological factors of the corresponding period, there was an obvious positive correlation among the daily power consumption, the daily maximum power load and the temperature of the air. An increase of 1°C in the daily maximum temperature in Shanghai would increase the maximum load by 377,000kW. What's more, the heat accumulation effects of continuous high temperatures and excessive minimum temperatures higher than 28°C also significantly affect the increase in the power load. In the future, the frequency and the intensity of extremely high temperature events in East China may increase, which will further increase the energy consumption of the region.

3.2.3 Effects of Climate Change on Human Health in the Urban Agglomeration in the Yangtze River Delta Region and its Risks

3.2.3.1 *Direct Effects of Climate Change on Human Health in the Urban Agglomeration in the Yangtze River Delta Region and its Risks*

The most direct effect of global climate warming on human health is that the heating effect caused by extremely high temperatures will be more frequent and more widespread. Increase in intensity and duration of high temperatures and heat waves, in addition to directly leading to an increase in human deaths, it will cause an increase in the fatality of diseases, mainly cardiovascular and cerebrovascular diseases and respiratory diseases. East China is one of the areas in China where high temperatures and heat waves occur frequently. Besides, due to the intensification of urban heat island effects caused by the rapid development of urbanization in the Yangtze River Delta region, the urban centers experience a greater number of days with high temperatures, higher extreme maximum temperatures and longer-lasting high temperatures than the suburbs and outer suburbs; the air pollution in urban areas is more severe because of the large population density, further increasing the damage of high temperatures and heat waves to human health.

Extreme weather and climate events such as heavy rainfall, floods and typhoons can cause an increase in mortality, disability, morbidity of infectious diseases and mental disorders, and damage to public health infrastructures. Global warming may increase the frequency of extreme weather and climate events such as typhoons and heavy rainfall. With a long coastline and a high concentration of population in the urban agglomeration, East China becomes a high-risk area which is vulnerable to extreme weather and climate events with heavy human losses.

3.2.3.2 *Indirect Effects of Climate Change on Human Health in the Urban Agglomeration in the Yangtze River Delta Region and its Potential Risks*

Indirect effects of climate change on health come from the interference of climate change in the complex ecosystems, mainly including changes in the intensity and range of vector-borne diseases, malnutrition and other health effects caused by non-medium infectious diseases and reduction of the agricultural production, and various health problems caused by the rise of the sea level.

One of the main indirect effects of climate change in the Yangtze River Delta on health is the transmission of the vector-borne disease. Changes in air temperature and rainfall can affect the behavior and the geographical distribution of vector-borne insects, thus changing the incidence of vector-borne diseases. A vector-borne disease that is most obviously affected by climate change is malaria, which is only found in areas where the minimum temperature in winter is higher than 16°C. Except Fujian Province, other areas show many imported cases of malaria. Since global warming makes many places experience a warm winter, malaria may spread to the north and middle parts of East China. In recent years, due to such factors as climate warming, an increase in population mobility and the acceleration of the process of urbanization, the incidence of dengue fever has also noticeably increased. The coastal areas in East China, including north of Fujian, Zhejiang, Shanghai, Jiangsu and south of Shandong, have become one of the high-risk areas for dengue fever in China. Under the scenario of climate warming in the future, the limit of the range of distribution of schistosomiasis in the urban agglomeration of the Yangtze River Delta region will move northward, which is particularly evident in Jiangsu and Anhui.

4 Enlightenment and Thinking

Against the background of global warming and the acceleration of urbanization, uncertainty and risks in responding to climate change in the future by China's urban agglomerations may increase. With the process of urbanization, population and economic dealings, urban construction activities, resource and energy consumption in urban agglomerations will continuously increase, posing more challenges to the urban disaster risk management. For example, extreme climate events such as rainstorms and floods will endanger urban functioning and urban lifeline safety; high temperatures and heat waves will further increase the residents' demands for water resources and electricity in large cities, increasing the pressure on urban infrastructures; and smog will deteriorate the environment and cause health problems. So, the improvement of the climate adaptability of urban agglomerations is a systematic engineering, requiring the combination of technical and capital advantages of all cities and the strengthening of collaborative planning and actions in monitoring climate changes, climate protection infrastructures, social and economic structures, the ecosystem, governance capacity and other aspects. Aiming at possible climate change risks in China's urban agglomerations, we should focus on the following aspects.

4.1 *Create Medium- and Long-term Planning and Decision-making and Coordination Mechanisms so that Urban Agglomerations can Adapt to Climate Change*

Planning runs first in urban development, so scientific urban planning is very important for disaster risk management and adaptation to climate change. First, strengthen the climate change risk assessment of urban agglomerations, specify the industries, the key fields and high-risk areas in all cities that could be affected by climate change, and be aware of the condition they are in; second, consider the goals of climate change risk management and adaptation based on regional developmental planning, and specify the main risks, the priority fields and the key measures of adaptation, especially regarding the population and the industrial layout of high-risk areas such as the coastal areas, areas along the Yangtze River and areas with ecological vulnerability; last; third, establish decision-making and coordination mechanisms in response to climate disasters in urban agglomerations, e.g. establishing major joint conference systems, setting up technical support institutions or committees of experts for regional adaptation to climate change, building up disaster risk statistics and monitoring information platforms, establishing emergency response plans of security assurance in water consumption, power consumption and urban transportation in urban agglomerations under the scenario of various extreme weather and climate events like high temperatures and heat waves, low temperatures, rain and snow, rainfall and waterlogging as well as persistent drought.

4.2 *Strengthen the Climate Protection Capability of Key Infrastructures in Urban Agglomerations*

First, include adaptations to climate change and prevent climate disaster risks in major projects and infrastructures of cities, practically improve the capability of combating extreme weather and climate events, and make efforts to guarantee urban residents' safety in domestic water consumption, power consumption as well as operational safety of urban transportation. Second, with the combination of funds and technology in all cities, the construction of infrastructures (such as transportation, electric power, communication, water supply and drainage, and meteorological monitoring warning) in urban agglomerations must take into consideration climate change risks in the future, and strengthen the inter-urban coordination and links. For instance, smog and air pollution are greatly related to meteorological conditions and urban transportation conditions, the inter-regional information communication and joint defense and control, as well as the construction of the warning, forecast and monitor system of extreme health risks of urban agglomerations, should be strengthened. Last, strengthen the demonstration of the climate feasibility in the urban spatial layout major projects and key infrastructures, implement the elimination of risks and hidden dangers, upgrade and reform the urban flood control and drainage systems; effectively control urban heat island effects through the optimization of the urban green land and water layout; establish meteorological standards for the energy conservation design for urban buildings, and guarantee safe operations, energy conservation and consumption reduction of energy conservation facilities of buildings.

4.3 *Strengthen Climate Risk Assessment and Research Support*

Research and assess new problems and new disasters caused by urbanization in order to support the scientific response. For instance, carry out climate change impact, vulnerability and risk assessment of cities and urban agglomerations and set up future climate change disaster risk zoning; improve the monitoring and warning technology of climate change integrated disaster risks in urban agglomerations and strengthen the construction of weather, climate, hydrological monitoring and warning information platforms in urban agglomerations; strengthen research on the urban safety and the population carrying capacity against the climate change background, draw lessons from experience at home and abroad, study and propose adaptive measures in fields such as urban major projects, living environments, water resources, energy and power, transportation and human health, and experiment with collaborative policies, mechanisms and measures for the reduction of urban emissions, ecological protection, disaster prevention and reduction.

4.4 *Strengthen Guarantee Mechanisms of Adaptation to Climate Change in Urban Agglomerations*

First, strengthen the urban funding in disaster prevention and reduction and adaptation to climate change to simultaneously improve urban social and economic development and the ability to adapt to climate change, and promote the disaster risk-sharing mechanism of the government, the family and the market with equal importance through financial and taxation policies; second, enhance popular science education, propaganda, and emergency drilling, and raise preventive awareness and coping capacity of the public to abrupt meteorological disasters; last, enhance policies and legislation on adaptation to the climate change of urban agglomerations, gradually improve supporting facilities such as health care and social security systems, pay attention to vulnerable groups such as low-income groups, migrant worker groups, the young and the old, and encourage community disaster prevention and reduction and public participation.

CHAPTER 11

Climate Change and the Layout of Urban Planning: Taking Beijing as an Example

FANG Xiaoyi

Abstract

Proceeding from climate change and urban layout, this research discusses how to enhance the urban planning considerations regarding meteorological environmental problems against the background of technological advance. Taking Beijing's master planning as an example, the research shows the overall condition of the Beijing meteorological environment in the course of Beijing's planning and development. In addition, the research analyzes the conditions of meteorological dispersion in the planning area, the potential and application of climatic resources, atmospheric water resources and meteorological disaster risks, and it then evaluates the meteorological environment of Beijing's urban developmental planning. As a result, it concludes with the status quo of the meteorological environment in the capital function core area of Beijing, the area of urban functional expansion, the new area of urban development and the area of ecological conservation development, the matters needing attention in the future planning, and regional developmental advantages in these areas.

Keywords

climate change – urban planning – meteorological environment – Beijing

* The research is funded by "Technical Research on Urban Planning Climate Feasibility Demonstration" (GYHY201106032), a special scientific fund in the public welfare industry (meteorological industry).

** Fang Xiaoyi, Ph.D. at the Beijing Climate Center, whose research fields are applied climatology and meteorological environmental problems in urban development. E-mail: fangxybj@163.com.

 | DOI 10.1163/9789004280021_012

1 Introduction

In the future, the world is considered as a world of city. The fast development of Chinese urbanization has aroused a growing concern from the international community. Stiglitz, winner of the Nobel Prize in Economics, once said that, "Chinese urbanization and American hi-tech development are the two themes that will deeply affect human development in the 21st century."[1]

However, with the surge of urban population and the sharp expansion of urban scale, the structure of land use and the underlying surface features in the urban area have been changed, replacing the original vegetation with buildings, asphalt and cement roads. The production and life of human beings have led to a significant change in urban atmospheric thermodynamics and dynamics. A great quantity of smoke dust, aerosol and particulate matters from industrial emissions, as well as automobile exhaust and dust in urban pavements, all make an impact on urban temperature, humidity, visibility, wind and precipitation. Consequently, they bring about a series of urban problems, including urban endemism such as the "urban heat island," the "urban dry island" and the "urban turbid island," all of which affect the factors of the urban living environment in many ways.[2] Therefore, urban sustainable development is a subject of paramount importance which all countries are faced with and which is closely related to the future of human beings while urbanization is sharply speeding up.

Urban planning is the important basis that guides urban development, construction and management, and scientific planning is the key to urban sustainable development. To achieve urban sustainable development, China should develop a scientific urban planning first so as to promote the conservation of resources and environmental protection.

Urban planning and the urban meteorological environment complement each other. That is to say, the urban planning and design need to be grounded on the urban environment, while urban human activities, underlying surface changes and the different complex architectural layouts all will affect the urban meteorological environment, further changing the local urban microclimate.

---

1 Wu Liangyong, "Urbanization and Urban Modernization," *Sunan Sci-Tech Devel.*, 2003(4): 8–9.

2 Wang Yongwei and Jiang Weimei, "Numerical Investigation on the Influence of Land Use Change on the Urban Meteorological Environment," Journal of Nanjing University (Natural Sciences), 2006, 42(6): 562–581; Chen Yan and Jiang Weimei, "Impact of Nanjing Urban Development on the Boundary Layer," Chinese Journal of Geophysics, 2007, 50(1): 67–74.

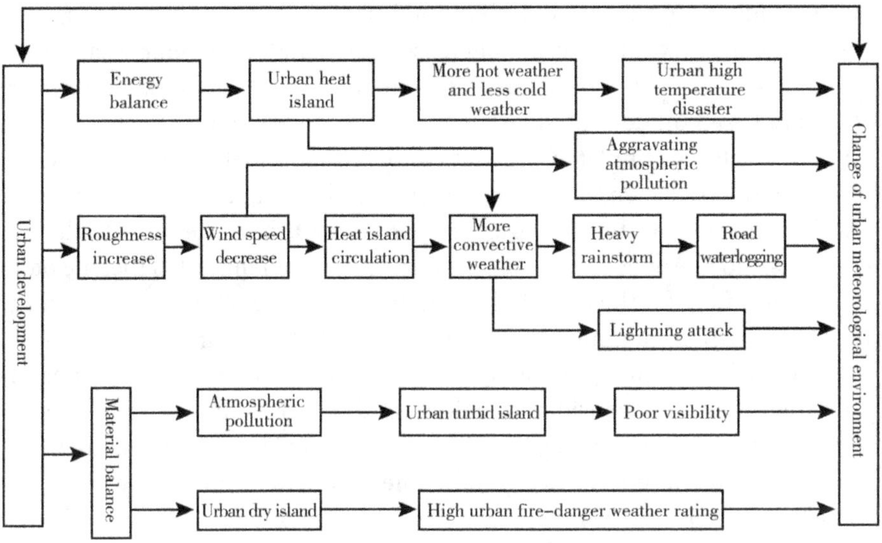

FIGURE 11.1 *Diagram of the influence of urban development on the urban meteorological environment.*

The relationship between urban development and the urban meteorological environment is shown in Figure 11.1.[3]

From urban planning and management, the indifference to the influence of urban development on local meteorological conditions is very likely to cause severe irreversible results. In spite of massive funds in pollution abatement, the improvement effect in atmospheric quality might be offset. On the contrary, the reasonable consideration of meteorological environmental conditions in urban development and a scientific urban planning and layout, will play a critical role in improving urban meteorological environmental conditions and achieving a good livable environment.

2 Meteorological Environmental Assessment Study of Urban Planning

2.1 *Study Overview*

In a broad sense, global change and its possible influence have been an important issue which has highly concerned the international community. In a

3 Li Hui, *Technical Report on the Influence of Beijing's Urban Development on Climatic Resources*, 1995, 23.

city, a change in local meteorological conditions will lead to a rise in temperature, a decline in wind speed and increased frequency of extreme weather, etc. Nowadays, urban planning should take into account not only expanding construction based on development, but also its influence on the water, soil and air. As a result, to completely understand the urban development and the rule of interaction between the urban system and the climatic system from a scientific point of view and to work out appropriate urban planning, traditional concepts and methods are facing unprecedented challenges. On the one hand, one must reflect on the cognition of meteorological environmental issues in traditional urban planning. In the theory of traditional urban planning, the influence of meteorological factors on urban construction and on urban environment has been considered mainly on the basis of a wind-rose diagram and contamination factor. Against the conditions and background at that time, this traditional method plays a guiding role in planning, and is of a certain rationality and objectivity. However, it has limitations in the situation of urban development and technological advance. On the other hand, with the expansion of the urban scale and a change in the urban functional structure, urban environmental problems become more complicated, so traditional concepts and methods can not be very good to deal with them. Now, the core of urban planning has been changed from the arrangement of the space for construction projects to the overall planning of resources, environment, population and the public policy properties of construction, development and environmental protection,[4] which has a new greater requirement on planning, meaning that overall considerations must be made in urban planning to coordinate the living environment and the integrated urban layout.

Against the background of technological advancement in the new period, it is necessary to put the scientific application of meteorological environmental ideas in urban planning and develop the new theories and methods of "atmospheric environmental optimization-based" urban sustainable developmental planning on the agenda. This would help re-recognize and study the relationship between urban planning and meteorological environment from a scientific point of view, and apply the findings of meteorological studies in urban planning. At the same time, the rapid development of computer technology provides a strong support for a numerical simulation of the meteorological environment, and the enormous development of numerical simulation technology makes the integration of meteorology and urban planning possible. Besides, the development of information technology makes digital planning technology

4 Wang Xiaoyun, Wang Guangtao, and Fang Xiaoyi, "Scientific Application of Environmental Ideas in Urban Planning and Construction in the New Period—Study on Atmospheric Environmental Effect in Planning the Olympic Venues," Planners, 2005, 21(10): 84–87.

an important means and driving force for urban planning and management. Under such a work environment, how to improve the relationship between meteorology and urban planning with the aid of emerging scientific technologies against the background of technological advance, conduct studies on atmospheric environmental effects in a scientific, objective and systematic way and make a qualitative and quantitative analysis and assessment on its impact is the hot spot and front field in current research on urban planning.

2.2 *Research Method*

Continuing the process from the relationship between the urban atmospheric environment and current urban planning, we analyze the difference, contradiction and perplexity between the practice of current urban planning construction and the cognition of atmospheric environment and raise questions.[5] On the basis of an observational test of meteorological factors and a wind tunnel test, a multiscale numerical stimulation system is established for Beijing's urban geological information, which focuses on the urban scale (horizontal range from tens of kilometers to hundreds of kilometers, from several hours to tens of hours) and the urban sub-domain scale (horizontal range of kilometers and several hours). Furthermore, evaluation indexes for the influence of Beijing's urban planning construction on atmospheric environment are established on the basis of stimulation results.[6]

An urban scale numerical model has been used in Beijing's Overall Urban Planning. This model adopts the terrain-following coordinates and comprises a 3D Non-Hydrostatic Fine Boundary Layer Model and an Atmospheric Pollutants Transport and Diffusion Model.[7] Fig. 11.2 shows the basic structure

5 Wang Guangtao, Wang Xiaoyun, and Miao Shiguang, "Research and Application of Atmospheric Environment Influence Multiscale Assessment Technology System in Urban Planning," Sciences China Series D, 2005, (35): 145–155; He Xiaofeng, Jiang Weimei, and Guo Wenli, "Numerical Experimentation for Urban Planning Influence on Boundary Layer Structure," Plateau Meteorology, 2007, 26(2): 363–372.

6 Jiang Weimei, Wang Yongwei, and Liu Gang, "Numerical Model System of Multiscale Urban Boundary Layer," Journal of Nanjing University, 2007, 43(3): 221–237; Miao Shiguang, Wang Xiaoyun, and Jiang Weimei, "Evaluation Research for Atmospheric Environment Influence on Urban District Planning," Plateau Meteorology, 2007, 26(1): 92–97.

7 Xu Min, Jiang Weimei, and Ji Chongping, *Numerical Stimulation Test for Beijing Region Meteorological Environment, Journal of Applied Meteorological Science*, 2002, (13): 61–68; Wang Guangtao, Wang Xiaoyun, "One Practice of Modern Urban Planning Theory and Method—Stimulation Analysis for Atmospheric Environment Influence on Foshan Urban Planning," Urban Planning Forum, 2005, 21(10): 84–87; Wang Xiaoyun, Wang Guangtao, and Chen Xianyan, "Research on Pearl River Delta Urban Agglomeration Development Planning and Atmospheric Environment," Urban Planning Forum, 2005, 21(12): 29–32.

of an urban scale numerical model. In order to pay more attention to the parameterization process of urban influence, three special parameterization processes and calculations are incorporated into the model. (1) It factors in the drag force urban buildings experience under wind conditions by replacing this value for the commonly used roughness (Z_0) treatment scheme. This involves including this factor in the formulas for calculating the horizontal wind velocity components u and v and turbulent kinetic energy E, which would allow the impact of sub-grid buildings on wind velocity and turbulent kinetic energy per unit of area to be determined. (2) It takes into account the thermal dynamics unique to cities by including urban anthropogenic heat in formulas used for generating forecasts for ambient temperature and atmospheric heat.

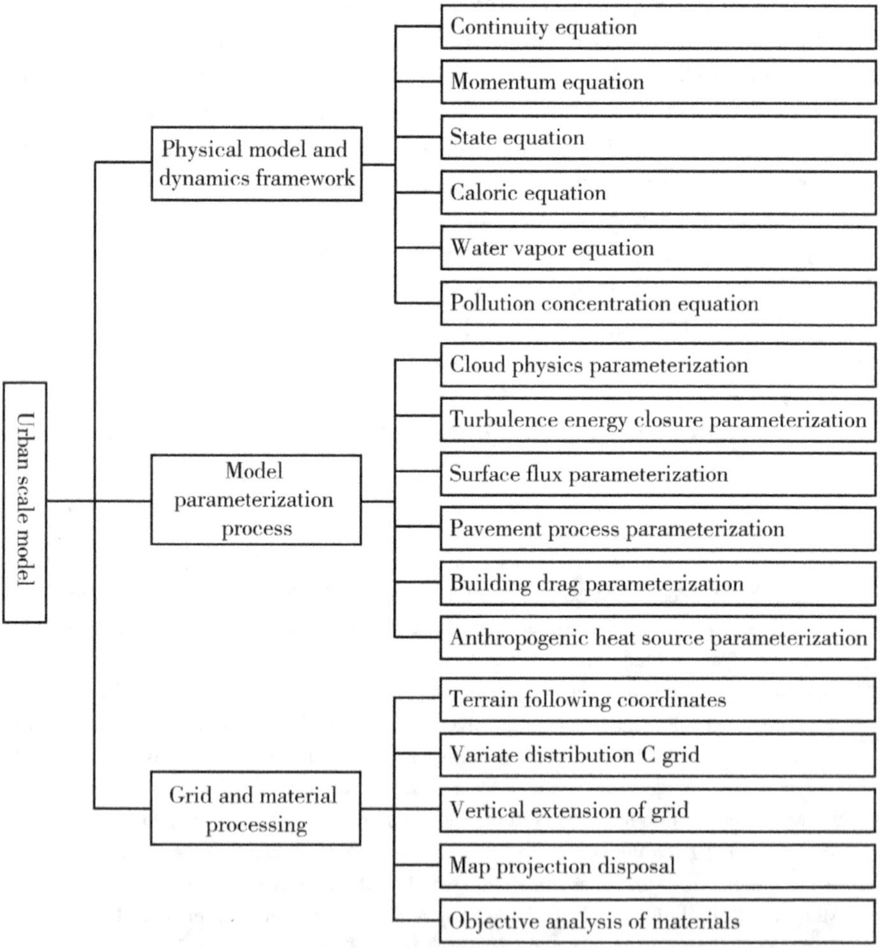

FIGURE 11.2 *Basic composition of the urban scale model.*

(3) It improves the precision and efficiency of the basic data required for simulation by making use of the underlying high-resolution urban land surface parameters obtained from the Geographic Information System (GIS).

3 Beijing Urban Planning Overview and Meteorological Environment Background

3.1 *Beijing Geological Landform and Climatic Features*

Beijing is situated in the northwest of the North China Plain and the mountain-plain transition zone, with mountainous land and plains land accounting for about 62% and 38%, respectively. Surrounded by mountains on the northeast, north and west, its southeast is a flat plain leaning to the southeast, which forms a special kind of terrain facing the sea with the hills for a background, commonly known as the "Beijing Bay."

Beijing's climate is a semi-humid-semiarid monsoon climate in the warm temperate zone. In spring, the temperature of the air rises sharply, with great daily temperature differences; in summer, the weather is hot and rainy; in fall, the weather is suitable, and is mostly sunny with little rain; in winter, the weather is cold and dry. Owing to the hot rainy season, Beijing is suitable for growing a variety of corps and has rich climate resources. However, it also suffers from a wide variation of the annual precipitation rate due to its location in a zone of the convergence of cool and warm air, and so it often experiences drought, rainstorms, gales, hail, cold waves and other climatic disasters.

3.2 *Beijing Urban Planning Overview*

The change of Beijing land use is mainly reflected in the transition to non-agricultural land from agricultural land. To be specific, there are three transition modes,[8] i.e., outside sprawl, axial expansion and suburbanization. Before the founding of new China, Beijing's urban landscape presented a convex pattern, while after the founding of new China, Beijing's urban area presented apparent concentric expansion features due to the large scale of urban construction. From the 1960s to the late 1970s, Beijing made full use of every inch of land for construction, leading to a gradual densification within the 5-km range of the second ring old city, thus forming a dense, concentrated mass layout; since the 1980s, Beijing began to suffer from a housing crisis (incapable of conducting large-scale satellite town construction), so it developed a number of residential

8 Gu Chaolin, "Beijing Land Use/Research on Coverage Change Mechanism," Journal of Natural Resources, 1999, 14(4): 307–312.

communities outside the city. As a result, urban land use has evolved from the inside to the outside, and the speed of the expansion of the urban area increased sharply. The second ring road has been completely opened to the public, and the third and fourth ring roads have been built. All of this has resulted in a massive loss of agricultural land around Beijing.

Since 1993, the *Beijing Urban Master Planning (1991–2010)*, approved by the State Council, has played a vital role in guiding the construction and development of Beijing. However, Beijing has faced a number of new problems from 2004. With a surge in the population and function clustering, traffic has become jammed, water resources and energy are in short supply, environmental pollution is severe, and the pressure on historic city conservation is heavy, causing built-up area sprawl, a large urban-rural division and a great limitation in the monocentric developmental pattern... Therefore, it is in urgent need of summarizing and adjusting the old master planning in order to seek new solutions from a strategic and overall perspective.

And then China set up the division of land space into four main functional areas, i.e., an optimized developmental area, a key developmental area, a limited developmental area and an area where development is prohibited. Based on realities, Beijing has identified the capital functional core area (which includes Dongcheng District and Xicheng District, and presents features such as ancient capital, and functions such as political, cultural center and international exchange center), the urban functional expansion area (which includes Chaoyang District, Haidian District, Fengtai District, Shijingshan District, and is an important area for realizing and expanding capital functions, an export-oriented economic service area facing the whole country and the world, a significant base for promoting production innovation and an area for improving the core competitiveness of capital development), the new area of urban development (which includes Tongzhou, Shunyi, Daxing, Changping and Fangshan, focuses on developing the manufacturing industry and urban modern agriculture, is an important area for evacuating industries and population in the urban center, and is Beijing's economic center for the future) and the ecological conservation developmental area (which includes Mentougou, Pinggu, Huairou, Miyun and Yanqing, is Beijing's ecological shelter zone and water conservation area, and is a vital area for ensuring the sustainable development of Beijing).

To propose suggestions regarding the meteorological environment for Beijing's master planning and the distribution of its main functional area is the issue that concerns planners and must be resolved by us. Proceeding from the fact that Beijing's climate change is closely associated with planning, meteorological factors that affect planning are analyzed in an attempt to answer questions relating to urban climate and meteorological disaster before or during Beijing's urban planning, and include them in the specific urban plan. As a result, suggestions regarding the meteorological environment are made available for Beijing to identify the main scientific and reasonable functional areas of the "capital functional core area, the urban functional expansion area, the new area of urban development and the ecological conservation developmental area."

3.3 *Beijing's Meteorological Environment under Climate Change*

Observational data from the Beijing Observatory from 1871 to 2007 show that Beijing has witnessed an average temperature rise of 1.05°C over the last 136 years, a significant rise (as shown in Figure 11.3); and the increasing rate of the average temperature from 1951 to 2007 is 0.39°C/10 yr, higher than that of the whole country (0.26°C/10 yr) and North China (0.33°C/10 yr). Such an obvious temperature rise is the result of global warming and of the impact of Beijing's fast urbanization on temperature. After analysis, it can be concluded that the warming effect brought by Beijing's urbanization is one of the two factors resulting in local warming, which contributes to 48.4% of the total warming, while the climate system change factor contributes to the remaining 51.6%. Irrespective of the impact of urbanization, Beijing's temperature rise is expected to be 0.2°C/10 yr due to climate change.

Figure 11.4 shows Beijing's surface radiation brightness temperature in the same season in different years. From the figure, it can be seen that the shape of the heat island is almost identical to the scale of urban construction. In 1989, Beijing's heat island intensity was more than 2.0°C, the shape was oval-like, and it mainly covered the central urban area, the central and south area of the Haidian District, the north part of the Fengtai District and part of the Shijingshan District. Besides, the small heat island formed in the Tongzhou

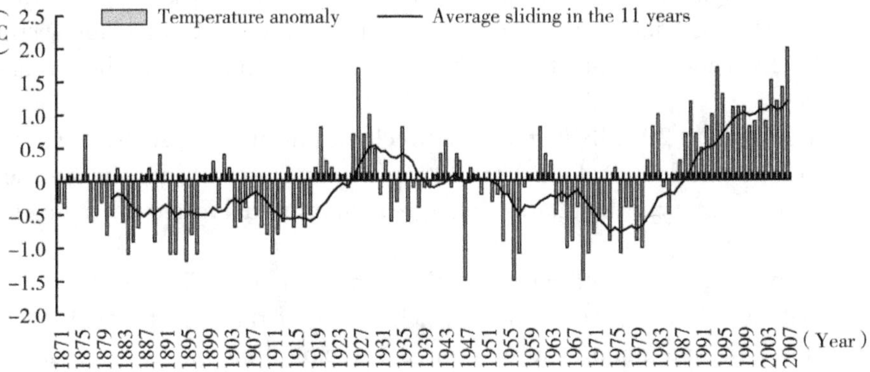

FIGURE 11.3 *Mean temperature anomaly in the recent 136 years and the sliding evolution in the 11 years, from the Beijing Observatory.*

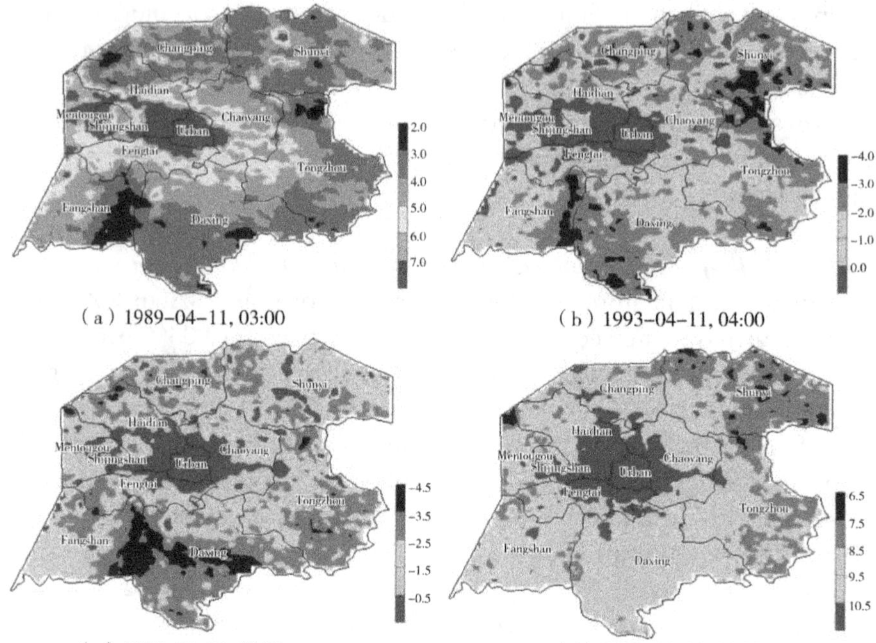

(a) 1989-04-11, 03:00 (b) 1993-04-11, 04:00

(c) 1998-04-02, 03:00 (d) 2005-04-18, 03:00

FIGURE 11.4 *Beijing's surface radiation brightness temperature in the same season in different years.*

District also can be seen. With the expansion of the urban scale, the heat island stretched to the north, south and east with the urban area as the center, of which the eastward extension is most obvious. Since April 1993, a distinct "linear" island between a point due east of the central urban area and the Tongzhou District and across the Chaoyang District has formed, which is still obvious as shown in the distribution diagram. This means that Beijing's urban construction has focused on the Chaoyang District and the Tongzhou District in the 21st century, which brings about the most obvious "urban heat island" effect.

From 1901 to 2011, Beijing annual precipitation presents slight downward trend, with trend value of –95.1mm/100 yr (Beijing Observatory is the representative observatory). Precipitation has a great inter-annual variation, as shown in Figure 11.5.

In the past one hundred years, the Beijing area has had a great inter-annual variation in precipitation, which is one of the climate factors causing drought and flood. The amount of precipitation ranges from 242 to 1,406mm over the previous years, with a variability of up to 0.58, reflecting

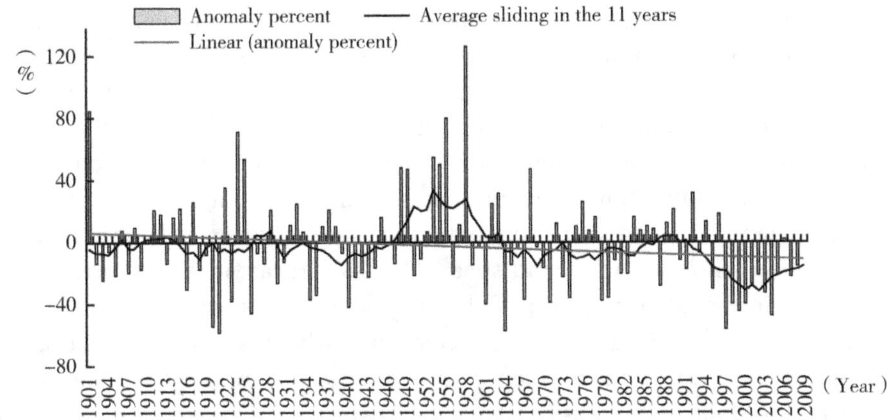

FIGURE 11.5 *Change in Beijing's annual precipitation anomaly percent from 1901 to 2011.*

the fact that every year, Beijing's precipitation tends to suffer from significant deviation, further resulting in unpredictable droughts and floods. The urban rain island manifests that Beijing's leeward area always experiences lots of heavy precipitation in summer and the maximum downpour intensity center of the urban suburbs lies in the northern suburbs. The urban dry island in Beijing presents the features of low relative humidity in the urban areas and a gradual expansion towards the suburbs. The urban wind speed is lower than that of the outskirts, as the urban area houses a number of buildings and air flow is hampered, further resulting in a decline in the wind speed. The wind speed in the urban areas is normally 20% to 30% lower than that in the outskirts. The urban turbid island indicates a higher concentration of air pollutants in the urban areas than in the outskirts, showing foul air and low visibility. The distribution of the visibility in Beijing's suburbs presents a closed oval shape, revealing that the urban areas have a lower visibility than the outskirts.

4 Meteorological Environmental Assessment for Beijing's Urban Developmental Planning

4.1 *Analysis of the Atmospheric Diffusion Conditions in the Area Involved in the Planning*

The urban scale model is used for the simulation of Beijing's city territory, with a scope of simulation that covers an area of 187.5km×181.5km and a grid resolution of 1.5km×1.5km. From the perspective of wind speed, flow field, self-cleaning capacity, temperature and human body comfort,[9] Beijing's atmospheric diffusion conditions are analyzed to provide scientific meteorological support for the layout of the functional areas. Figure 11.6 gives the summary of the results of wind speed, flow field, self-cleaning capacity, temperature and human body comfort. From the figure, it can be seen that Foyeding, Nankou of Changping County, Shangdianzi of Miyun County and other strategic passes have a high wind speed, while the southern flat area and southwestern mountainous area have a lower wind speed. The eastern area has a shorter self-cleaning time, as shown in the self-cleaning capacity diagram. So, it is suitable to locate an industrial zone in the eastern area which is exactly the location of the new area for urban development. From the air flow diagram, Fangshan District and Mentougou District have an air stream convergence, so it is unsuitable for the two districts to house large polluting enterprises. The southern part of the Shunyi District, the eastern part of the Chaoyang District, the eastern part of Daxing District, and the Tongzhou District are larger than other flat areas. In terms of self-cleaning time distribution, the Tongzhou District has a strong diffusion capacity; the Shunyi District, the eastern part of the Chaoyang District, the Daxing District, etc., have relative strong diffusion capacities.

With regard to the general layout of urban buildings, studies on layout planning of buildings within Beijing's sixth ring have been carried out, so as to make a rational planning for the height and density of urban buildings, expand the sky oxyoptex and increase the surface long-wave outgoing radiation. Simulation calculation and analysis results reveal that Beijing's downtown buildings should avoid the layout of being higher in the center and lower around the center, and the space for a "building space corridor" should be reserved. It would be suitable for the northeastern and eastern areas instead of the western and southwestern areas to house high-rise buildings; however, dense high-rise buildings should be avoided.

The effect of the assessment scheme was analyzed by comparison with meteorological environmental difference brought about by the urban territory land use status quo and the urban territory land use planning. In simulation calculation, the input of mean meteorological conditions should be the same, and in addition, the diffusion condition of pollutants from highways and industrial areas should be simulated. In the simulation, the property of land use in the greening belts is grassland covered by trees.

9 *Urban Planning and Atmospheric Environment* research group, *Research on the Relationship among Beijing's Urban Planning Construction, Meteorological Conditions and Atmospheric Pollution*, China Meteorological Press, 2004.

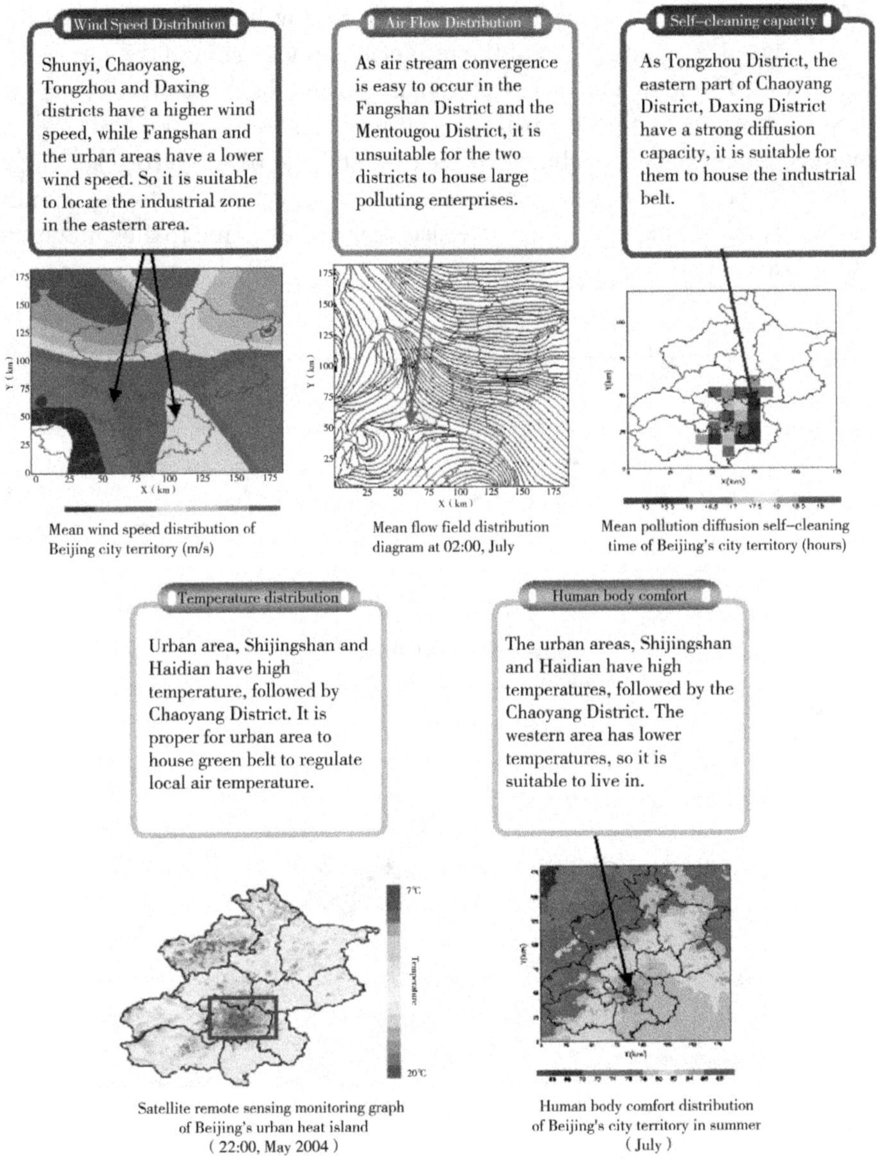

FIGURE 11.6 *Distribution of meteorological factors in Beijing.*

Figure 11.7 shows the distribution of the variation of mean annual surface air temperatures under the two schemes. Compared with the status quo, the temperature variation is great in the planning scheme owing to the partial area of urbanization and the large area of planned green space. The mean annual temperature variation ranges from −0.3°C to 0.6°C. Besides, the distribution difference in wind speed in winter and summer between the two schemes was also analyzed. By comparison of the results of the analysis and the status quo, the planning scheme has increased roughness and decreased wind speed by 0.1 to 0.3m/s owing to the partial area of urbanization and the large area of planned green space; the change in the thermodynamic property leads to the change in wind speed in the mountainous area. Moreover, highway pollutant diffusion and self-cleaning time were also analyzed to compare the differences of the two schemes.

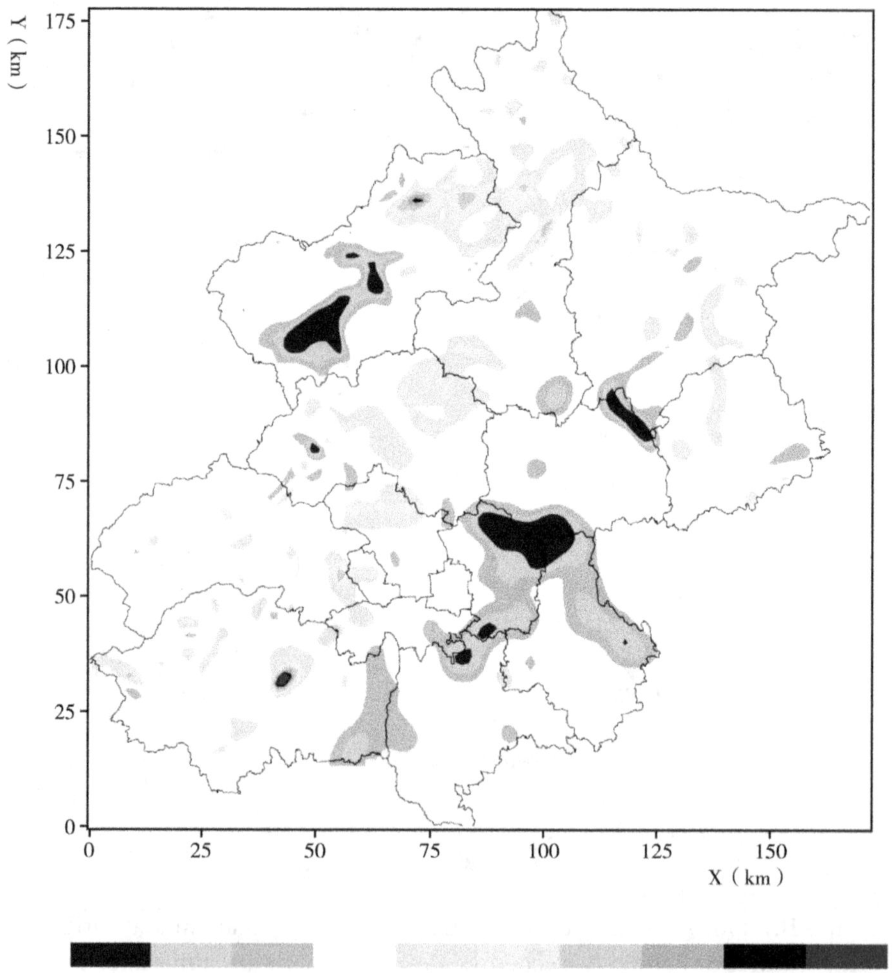

FIGURE 11.7 *Distribution of mean annual ground air temperature variation (planning—status quo, in °C).*

4.2 *Climate Resources Potential and Application Analysis in the Areas Involved in the Planning*

4.2.1 Wind Power Generation Industry—Location Advantages of the Ecological Conservation Developmental Area

The project of *Detailed Survey and Assessment on Beijing's Wind Power Generation Resources*, completed in 2011, collects and sorts out meteorological data from Beijing's 20 meteorological observatories (observation stations) in the previous years, makes information processing for the data of hourly wind speed and direction in representative years, and finally obtains the distribution situation of Beijing's wind energy resources by reference to the wind tower observation data and the results of Beijing's wind energy resources numerical simulation. The areas circled in Figure 11.8 below are rich in wind energy resources (100 to 150 W/m^2), that is, the surrounding area of Foyeding Mountain in the Yanshan Mountain Chain in the northwest, the surrounding area of Guanting Reservoir in Yanqing County, the Kangzhuang area, the Baihebao area, Lingshan Mountain, Huangcaoliang and Miaofeng Mountain areas in the Mentougou District. The first wind power plant was built in Guanting Reservoir, with phase II under construction. Therefore, the new clean energy industry under planning should be housed in an area that is rich in wind resources in an attempt to reduce costs and develop energy efficiently.

4.2.2 Solar Energy Utilization—Location Advantages Shared by All Functional Areas

The Beijing area enjoys favorable natural conditions for the development and utilization of solar energy. In the division of national solar energy resource areas, the Beijing area falls in class II, having around 2,600 hours of annual sunshine (see Figure 11.9). Lhasa, known as the "city of sunlight," only has 3,005 hours. Beijing's annual gross solar radiation amount reaches 5,227.14MJ/m^2. As one of the regions that is developing first and utilizing solar energy, Beijing has had a history of nearly 30 years in the development of the solar energy industry, such that it has formed a complete industrial chain and enjoyed unique advantages in the R&D of products and technology, in the application of energy products, and in the industrial chains such as engineering designs to support the applications. In the preparation for and the holding of the "Green Olympics," Beijing made some attempts in promotion and popularization of the architecture of solar energy integration and solar energy central hot water supply engineering, etc. The valuable experience acquired should be fully utilized. Therefore, Beijing has good and upbeat prospects for the development of solar energy. As direct solar radiation has little regional discrepancy in flat areas, radiation resources are regarded as no variance in consideration of solar

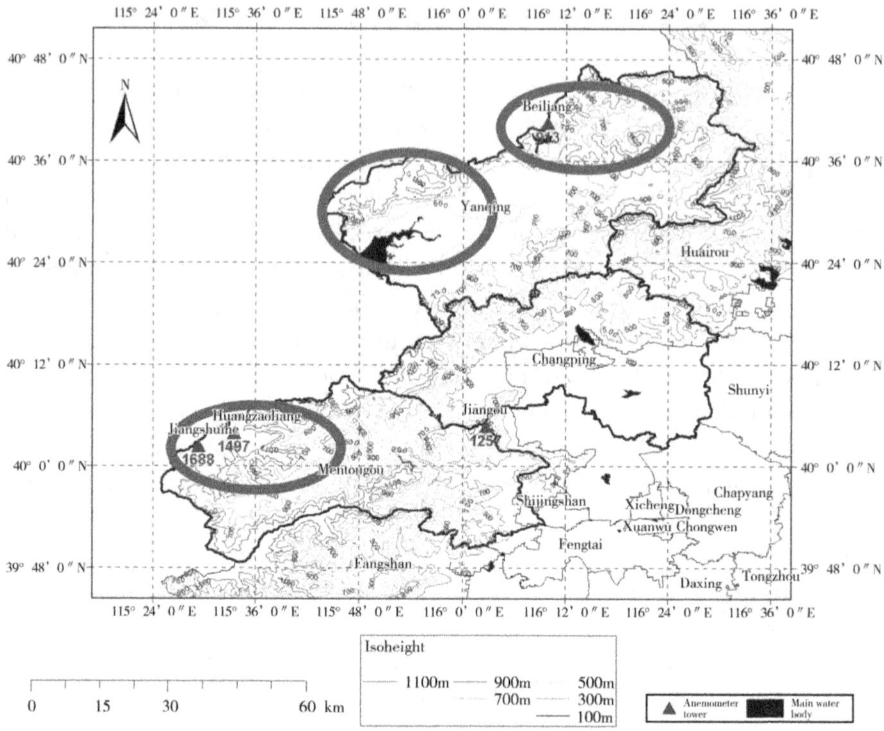

FIGURE 11.8 *Beijing's areas that is rich in wind energy resources.*

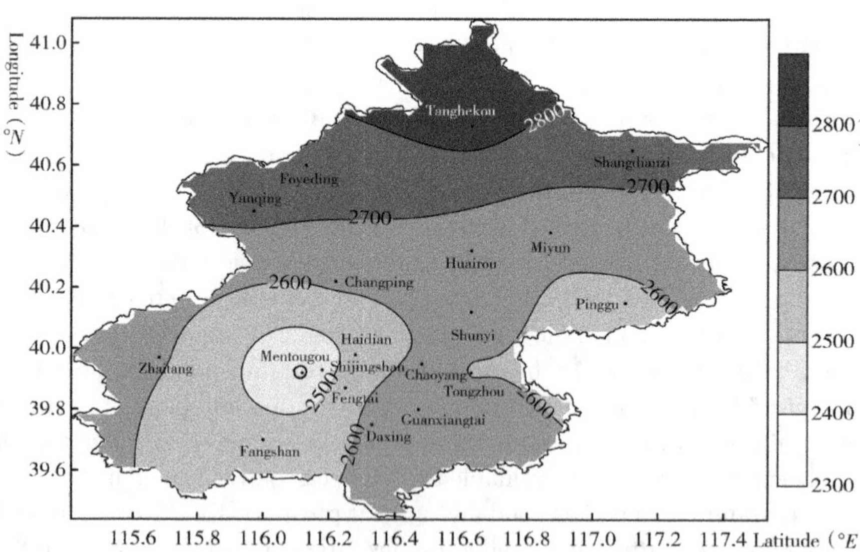

FIGURE 11.9 *Spatial distribution of the duration of Beijing's mean annual sunshine (from 1978 to 2007).*

energy-related industries. For mountainous areas, however, as regional discrepancy is great due to the influence of the terrain, which causes strong direct radiation on sunny slopes (including the mountaintops) and weak direct radiation on shady slopes, less solar radiation areas in the mountains should be excluded for the layout of related industries.

4.2.3 The Geothermal Energy Industry—Location Advantages of the Ecological Conservation Developmental Area

Beijing is abundant in geothermal resources. An initial investigation shows that geothermal distribution area is up to 2,372km^2, and the total heat from the thermal reservoir is equivalent to the calorific value of 11 billion tce; gross geothermal reserves are 18.12 billion m^3, and the annual exploitation capacity is 110 million m^3, equal to 1/10 of Beijing's annual exploitation capacity, revealing that there is still a great potential for its development and utilization. At present, Beijing has four geothermal belts in all, i.e., Huailai—Yanqing Geothermal Belt, Wenquan—Shahe—Xiaotangshan Geothermal Belt, Liangxiang—the urban area—Tianzhu—Lisui Geothermal Belt and Yufa—Fengheying Geothermal Belt.

4.3 *Analysis of the Atmospheric Hydrothermal Resources in the Areas Involved in the Planning*

Beijing's functional area planning observes the principles of "optimization of urban areas, and reinforcement of the outskirts." The ecological conservation area with a large coverage should not only play its function as the ecological protective screen for Beijing's urban area, but also be grounded on its advantages for seeking further development.

Ecological conservation developmental area: location advantages of Beijing's modern suburban agriculture; New area for urban development: location advantages of urban eco-agriculture.

The expert classifier is used for the calculation of Beijing's crop climate regionalization.[10] The calculation result reveals that the west and northwest of Beijing are favorable for the development of Beijing's characteristic suburban agriculture (such as peaches in Pinggu, mopan persimmons in Fangshan, off-seasonal vegetables in Yanqing and American ginseng in Huairou). This means that the development of Beijing's suburban agriculture is accompanied by the realization of ecological conservation. Therefore, the new areas of urban

10 Quan Weijun, Zhao Xinping, and Guo Wenli, "Application of Expert Classifier in Beijing's White Pear Climatic Regionalization," Meteorological Science and Technology, 2007, 35(6): 849–852.

development, such as Daxing, Changping and Fangshan, can all strive to develop urban eco-agriculture.

Specifically, as the Beijing mountainous front area has a warm zone, it is suitable for the cultivation of safety-certified vegetables and fruits, and for developing picking, flower appreciation, sightseeing and other industries, such that the role of ecological protective screen is played while Beijing's characteristic suburban agriculture is developed. Besides, as the southwestern mountainous area mostly has a karst landform and enjoys diversified microclimate conditions, it is a resource for agriculture and tourism. In view of short water resources and poor water pouring conditions, it is difficult to cultivate fruit; however, it has advantages in the development of dried fruit. Chinese medicinal herbs are also among the distinctive varieties in the mountainous areas of Mentougou and Fangshan. Yanqing has the largest basin in Beijing's mountainous area, and is located at the special zone that is in both ends of 2,234m Haituo Mountain Chain and Yanshan Mountain Chain and in the north of the Taihang Mountain Chain. With such a special geological terrain and climate conditions, Yanqing is superior in the development of high-quality fruit, such as grapes, and in the dairy industry. Besides, as Yanqing is cool in summer, it is also favorable for developing holiday villages, as well as oxygen-bar tours in the high mountains and dense forests and tours of historical sites; moreover, ice engraving and skiing in winter can be developed. The northern and northeastern areas, including Changping, Huairou, Miyun and Pinggu, has a flat terrain, thick soil layer, meta-acid soil and frequent rainfall, so it is suitable for the development of castanea mollissima and northern peaches. What is more, it has great greening which facilitates water conservation in case of rain and has springs in case of no rain, so it is also appropriate for the development of forest tourism.

4.4 *Meteorological Disaster Risk Analysis in the Areas Involved in the Planning*

Meteorological disaster is a significant factor affecting public security, which greatly restricts economic construction and social development. The failure to take full consideration of meteorological disasters in urban planning preparation and implementation in the early days has resulted in massive loss of life and property caused by meteorological disasters. Therefore, it is necessary to apply the leading consciousness of urban disaster mitigation in the planning of Beijing's main functional areas, face the fact of disasters during planning, and implement the disaster prevention awareness and action in the specific planning. It is urgent and viable to carry out studies on urban disaster prevention and mitigation measures and policies in an attempt to effectively protect

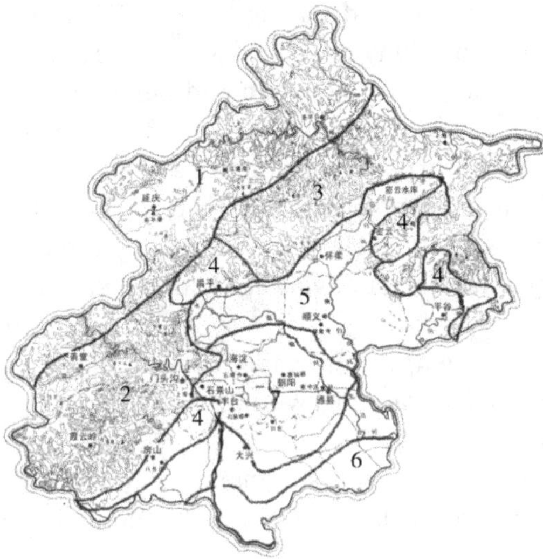

1. The northwestern mountainous area is the area that is prone to drought, wind damage, injuries due to freezing and hail.
2. The western mountainous area is the area prone to storms, floods and mud–rock flows.
3. The northern mountainous area is the area that is prone to storms, floods and mud–rock flows.
4. The high mountain front slope is the area that is prone to drought.
5. Part of the plains area is the area that is prone to droughts and floods.
6. The southeastern outskirts is the area that is prone to droughts.
7. The urban suburbs, Daxing, and part of Tongzhou are the area that is prone to heat damage, lightning damage, floods and pollution.

FIGURE 11.10 *Spatial distribution of Beijing's meteorological disasters.*

people's lives, properties and developmental achievements in the main functional areas.

Data show that Beijing is one of the domestic meteorological disaster-prone areas, as well as one of the metropolises suffering from frequent and severe meteorological disasters in the world. Beijing's meteorological disasters mainly include droughts, storms, wind damage, hail damage, thunder, heat injuries, injuries due to freezing temperatures, fog pollution, snow damage and frost, with distribution as shown in Figure 11.10.

5 Assessment Conclusions and Suggestions

5.1 *Assessment Conclusions*

By summary of the results of the research hereinabove, we can draw the following conclusions:

(1) The area with a high level of urbanization tends to have high temperatures; water and afforestation can reduce temperatures dramatically. Beijing's urban heat island is gradually heightened with development, and presents a trend towards extension to the north, south and east.
(2) Beijing's wind speed is lower than that of the suburbs, and presents a downward trend with each passing year.

(3) Beijing's Shunyi, Chaoyang, Tongzhou and Daxing have a higher wind speed and better transport and diffusion conditions, so they are suitable as industrial developmental zones, while Fangshan and Beijing's urban area have a lower wind speed and poorer transport and diffusion conditions, so they are unsuitable to house polluting enterprises.
(4) Beijing's Fangshan, Mentougou and Shijingshan are easy to encounter air stream convergence, causing poor atmospheric diffusion conditions, so such areas are suitable for greening land or living and tourism instead of polluting enterprises.
(5) The planning scheme has great temperature variations due to partial urbanization and wide green space planning, with a mean annual temperature variation from –0.3°C to 0.6°C.
(6) Due to partial urbanization and wide green space planning, roughness is increased, wind speed is decreased by 0.1 to 0.3m/s, and wind direction has no obvious change.
(7) The decrease in wind speed leads to a decrease in the range of diffusion of pollution and a slight increase in the concentration of pollution. The concentration of pollution in the nearby areas of and in the downward direction close to the pollutant sources is slightly increased, while that in the remote downwind direction is slightly decreased.
(8) The diffusion of traffic pollution in the planning scheme is 3% (0.57 hour) higher than that of current traffic pollution self-cleaning time. For diffusion of industrial pollution self-cleaning time, there is no great change.

Such conclusions provide support for "two-axis, two-belt and multi-center" aspect in Beijing's master planning (from 2004 to 2020).

5.2 *Suggestions*

By summarizing the new energy analysis, atmospheric diffusion analysis, ecological conservation, agriculture and meteorological disasters affecting the areas involved in the planning, the overall meteorological background of every functional area is generalized as follows, with some policy suggestions.

5.2.1 Capital Functional Core Area

The area has severe pollution, is an intense heat island and a dense population, so it is forbidden to open polluting enterprises in the area. The area should give its strengths to full play of its capacity as a center for living, politics, culture and education. Besides, the rich solar energy resources allows the area to energetically promote solar energy buildings, solar energy central heating supplies and to the application of solar energy in public buildings. It should be noted that because of internal waterlogging and high temperatures in summer, attention should be paid to the increase in its capacity for urban water drainage and flood control, the strengthening of greening, and the improvement of the governance of rivers and lakes.

5.2.2 Urban Functional Expansion Area

Currently, the area has a common pollution diffusion capacity, an apparent heat island effect and a rapid development. Thus, the area prefers to have the development of the industries with light pollution and high added value in an attempt to undertake the functional transfer of capital core area. Besides, the rich solar energy resources allow the area to promote solar energy buildings, solar energy central heating supplies and the application of solar energy in public buildings. It should be noted that because of internal waterlogging, high temperatures in summer, thick fog and thunder affecting traffic, attention should be paid to the increase in its capacity for urban water drainage and flood control, the strengthening of its greening, the improvement of the governance of rivers and lakes, and the strengthening of the installation and the testing of the lightning protection facility.

5.2.3 The New Area for Urban Development

The area has a strong diffusion capacity, an insignificant heat island effect and a huge developmental space, so it is superior in the development of new industries, which would enable the area to strive to develop new industries. Besides, the abundant solar energy resources allow the area to promote solar energy buildings, develop solar energy industries and carry out solar energy facility agriculture. It should be noted that because of some storm, drought and dust, it is necessary for the area to strengthen its water drainage and flood control capacity, construction of greening and protective forests.

5.2.4 Ecological Conservation Developmental Zone

The area has high human body comfort, a poor diffusion capacity and complex terrain. As Beijing's urban area ecological barrier, ecological conservation should be integrated into its agricultural development to give play to a strength in tourism and vocations. Besides, the abundant wind energy resources, solar energy and geothermal resources permit the area to develop the wind power industry, to launch a solar energy agricultural facility and carry out geothermal development. It should be noted that because of some storms, hail, droughts and mud-rock flows, it is necessary to enhance the construction of flood control and water conservation, perform integrated governance of mountains, water, forests and fields, to improve its natural forest protection project and carry out water source protection.

CHAPTER 12

Analysis and Countermeasures of China's Fog and Smog Trend

Zhang Ying, Gao Rong and Chao Qingchen

Abstract

This chapter systematically introduces the definition and cause of fog and smog, and reveals the characteristics of the temporal and spatial variation of fog and smog in China, according to historical data. It further points out that the reason for the increase of this kind of weather in China in recent years is the result of a combination of artificial influential factors caused by humans' social and economic activities and climatic factors caused by climate change. Fog and smog can have an obvious negative impact on the climate, the environment, health, the economy and other aspects, so it is urgent for all parties to pay attention to it and take effective measures to control it. In order to control fog and smog, some cities or regions in other countries, such as London in Britain, California in the USA and Mexico City in Mexico, have acquired some successful management experience. Therefore, to cope with air pollution, we should strengthen comprehensive treatment, and seek both temporary and permanent solutions, with regional coordination and department links, focusing on controlling for a long period of time and on its prevention for a short period of time.

Keywords

fog and smog – PM2.5 – control countermeasures – pollution source

* Zhang Ying, an associate researcher at the Institute for Urban and Environmental Studies of the Chinese Academy of Social Sciences, whose research fields are environmental economics, energy economics and quantitative model building of related fields; Gao Rong, a senior engineer at the National Climate Center, whose research field is regional climate simulation; Chao Qingchen, deputy director of the National Climate Center, whose main research fields are climate change science and policy research, and climate diagnostic analysis.

 | DOI 10.1163/9789004280021_013

1 Introduction

Fog and smog are caused by the interaction of fine particles and vapor, where the fine particles are formed by pollutants (gases and particulate matters, such as CO, SO_2, NO_X, NH_3, VOCS, and PM) produced in various emission sources through a series of physical and chemical processes under certain atmospheric conditions. They cause a reduction in the amount of social radiation that reaches the surface of the earth. Particulate matters involved in air pollution generally refer to particles with a diameter between 0.01 μm and 100 μm, wherein the atmospheric particulates (aerosol) whose aerodynamic diameter is equal to or less than 10 μm are collectively known as PM10, while those whose aerodynamic diameter is equal to or less than 2.5 μm are collectively known as PM2.5. With complicated compositions, PM2.5 almost include all elements of the periodic table of elements, covering more than 30,000 organic compounds and inorganic compounds (including sulfate, nitrate, ammonia salt, organic, carbon black, heavy metal, etc.)

Fog and smog is a large-area weather phenomenon with low visibility, generally formed in the ground layer under good vapor conditions, stable atmospheric stratification and low winds. When pollutants in the air accumulate continuously to a certain content, there are many condensation nuclei in the air; if atmospheric stratification is stable, the pollutants cannot diffuse quickly, which will form fog and smog. The cause of fog and smog in China has obvious seasonal variations. In winter, heating in the northern China and stable structure of the atmospheric boundary layer as well as low height of where these layers mix easily cause fog and smog. In summer, high temperatures and high-radiation climate and weather with a high percentage of humidity provokes a photochemical reaction of pollutants and an increase in the hygroscopic property, which will easily cause fog. In autumn, acute radiation and high temperatures in the daytime can accelerate the photochemical gas-particle conversion, while low temperatures in the morning and in the evening together with a high percentage of humidity can easily lead to hazy pollution events in a stable weather system. In spring, remote dust transportation, local dust, acid gas in cities and abundant moisture in the air can also form smog. Therefore, fog and smog occurs in all months with almost the same frequency, slightly more from October to January of the next year; the frequency of smog in wintry half of the year is significantly higher than that in the summer half year; the number of smoggy days in winter accounts for 42.3% of the whole year.

The cause of fog and smog includes natural formation and human activities. The former accounts for about 10% of fog and smog emission, mainly from SO_2 sprayed from volcanic eruption as well as various particulate matters and smoke in wild mountain areas. People focus more on fog and smog produced by humans' social and economic activities, the root cause of which is the continuous accumulation of pollutants (mainly caused by coal combustion and automobile exhaust emission) in the air. Against the background of global warming, the meteorological conditions that form fog and smog have changed significantly, becoming the direct fuse in the formation of fog and smog. In the fog, smog and air pollution in urban areas, there are not only primary particles produced directly from emission sources, but also secondary particles formed through the physical and chemical processes of the one-off gases from industrial and civil coal combustion and from automobile exhaust sources. Fog and smog widely occurs in large industrial cities; the case is particularly severe in cities with quite a bit of sunshine, a warm and dry climate, and lots of automobiles. Generally speaking, fog and smog belongs to local air pollution, so it will mainly affect the areas of emission, but will also affect the surrounding areas under certain climatic conditions, especially in downwind areas.

2 Characteristics of the Temporal and Spatial Variation of Fog and Smog in China

Research shows that the fog and smog in China in the past 50 years has generally shown an increasing trend. The number of foggy days has obviously decreased, while the number of smoggy days has obviously increased, and continuous smog process has increased significantly. Observed from the point of view of spatial distribution, the number of smoggy days has increased in the east and decreased in the west and has obviously increased in large cities more than that in small towns, with such characteristics as long duration, wide range, great impact, and heavy pollution.

2.1 *Fog and Smog Showing a Trend of an Increase in the Number of Smoggy Days and a Decrease in the Number of Foggy Days, and an Increase in the East and a Decrease in the West*

From 1961 to 2012, the annual average number of foggy and smoggy days generally showed an increasing trend in the central and eastern areas of China (east of 100°E), with an obvious inter-decadal variation. Before the 1980s, the number of annual foggy and smoggy days showed an increasing trend, but showed a decreasing trend from the 1980s to the 1990s. Since the 21st century, the number of annual foggy and smoggy days has shown an increasing trend again (see Figure 12.1). In the past 52 years, the most number of annual foggy and smoggy days occurred in 1980, with 35.8 days. On the whole, the number of foggy days

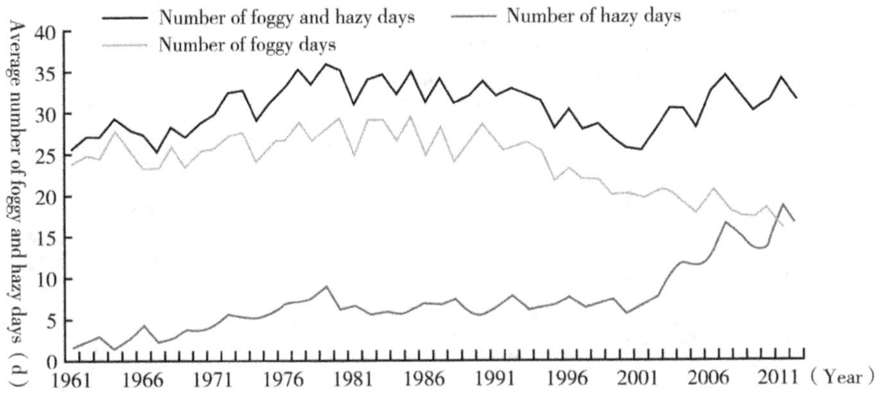

FIGURE 12.1 *Change in the average number of foggy and smoggy days in the central and eastern areas of China over the years (1961–2012).*
DATA SOURCE: NATIONAL CLIMATE CENTER.

is obviously greater than that of smoggy days. Before the 1980s, the annual average number of foggy days in the central and eastern areas of China was basically more than three times that of smoggy days; since the 1980s, the number of foggy days has shown a decreasing trend, while the number of smoggy days has shown an increasing trend, and the ratio between the number of foggy days and the number of smoggy days has gradually diminished. In particular, the number of smoggy days in 2011 and 2012 exceeded that of foggy days.

The change in the number of annual foggy and smoggy days in China showed an increasing trend in the east and a decreasing trend in the west. The number of foggy and smoggy days in the northeastern, northwestern and southwestern areas decreased by 0–0.5 days each year. Except northern Xinjiang, the number of annual foggy and smoggy days in western areas was less than five days, basically; the number of annual foggy and smoggy days in the areas of northern China, the middle and lower reaches of the Yangtze River, and southern China showed an increasing trend, with the fastest increase in the Pearl River Delta region and the Yangtze River Delta region. The annual average number of foggy and smoggy days in Shenzhen, Guangdong and Nanjing, Jiangsu increased by 4.1 and 3.9 days. The number of annual foggy and smoggy days in most central and middle areas of China is 25–100 days, and that in some areas exceeded 100 days.

2.2 *The Continuous Smog Process Significantly Increased*

The number of smog observation stations in the central and middle areas of China observed smog process for three consecutive days or above slightly

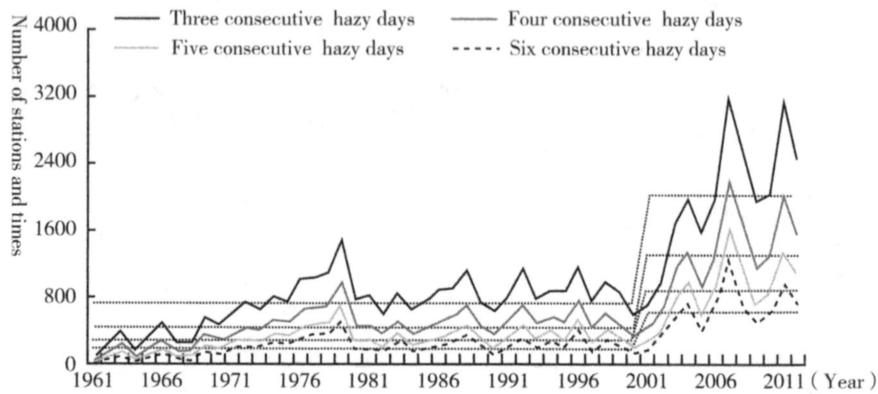

FIGURE 12.2 *Change in the number of smog observation stations observed continuous smog process in the central and eastern areas of China over the years (1961–2012).* DATA SOURCE: NATIONAL CLIMATE CENTER.

increased in the 20th century, but it showed little change on the whole. Since the 21st century, the number of smog observation stations observed continuous smog process has significantly increased (see Figure 12.2). The number of smog observation stations observed smog process for three consecutive days has increased from 725.5 to 2,010, that for four consecutive days has increased from 444.4 to 1,292.1, that for five consecutive days has increased from 291.8 to 881.1, and that for 6 consecutive days has increased from 200.3 to 628.2, respectively 2.8, 2.9, 3.0 and 3.1 times the average of the 20th century. The number of smog observation stations observed longer continuous smog process has increased more.

2.3 *The Increase in the Number of Smoggy Days in Large Cities is More Obvious than that in Small Towns*

Six typical large cities (Beijing, Shijiazhuang, Zhengzhou, Nanjing, Hangzhou and Guangzhou) and six nearby small towns (Zunhua, Raoyang, Xihua, Gaoyou, Cixi and Zengcheng) in the central and middle areas of China have been chosen to compare and analysis China's smog trend. According to the analysis of the change in the number of annual smoggy days, there was little difference in the number of annual smoggy days between large cities and small towns before the middle of the 1970s; however, since the end of the 1970s, the number of smoggy days in large cities has been significantly greater than that in small towns, with an increase of more than 50 days in most years. For different cities, the number of smoggy days in small towns began to increase significantly from the beginning of the 21st century; in the large cities, the number of smoggy days in Hangzhou and Guangzhou has increased obviously since the beginning of the 21st century, but that number in Shijiazhuang since the 21st century has been greatly smaller than it was in the 1980s and the 1990s; the number of smoggy days in Beijing since the 21st century has also significantly increased and the greatest number of smoggy days occurred in 1980, with 135 days; the number of smoggy days in Zhengzhou and Nanjing since the 1980s has shown a fluctuating increase.

3 Reasons for the Increase in Fog and Smog in China in Recent Years

In recent years, the number of foggy and smoggy days throughout China has shown an obvious increasing trend, which is the result of the combination of artificial influencing factors caused by human beings' social and economic activities and climatic influencing factors caused by climate change.

3.1 *Air Pollutant Emissions Produced by Human Activities Continuously Increase*

From the aspect of international experience, the formation of fog and smog is mainly the result of human activities, and the primary reason for the increase in fog and smog in China is the year-by-year increase of air pollutants caused by the increase in social energy consumption. The pollution mainly comes from ground dust, automobile exhaust, industrial emissions and winter heating. China is an energy consumption power, and about 70% of energy consumption is coal. Lots of particulate matters due to coal combustion are emitted into the air, in particular the emission of atmospheric particulates during the heating period (October–March of the next year) is greater than it is in summer obviously. In the past two years, the total greenhouse gas emission in China was the most in the world. The continuous rapid economic development also makes industrial emissions increase greatly. In the meantime, the number of family cars in large cities grew quickly. Car ownership in Beijing in 2012 was 2.6 times of what it was in 2001, causing the emission level of automobile exhaust to increase significantly. Besides, the volatile organic compounds caused by photochemical products, local cooking and automobile exhaust produced in human activities are converted into secondary organic aerosols, also leading to frequent fog and smog weather.

According to the air quality report by the China National Environmental Monitoring Centre for 74 cities in the first half of 2013, the average PM2.5 concentration in those cities in the first half of the year was 76 $\mu g/m^3$, wherein the PM2.5 concentration of only four cities, Zhoushan, Huizhou, Haikou and

Lhasa, was equal to or superior to the air quality of the annual secondary standard, and that of the other 70 cities was substandard. The average PM2.5 concentration in the Beijing-Tianjin-Hebei region was 115μg/m^3, and all cities in the region did not reach PM2.5 average annual secondary standard; the average number of days with air quality reaching the standard only accounted for 31%, and the number of days with severe pollution was up to 26.9%.

What's more, increasingly severe problems of fog and smog are also affected by the surrounding sand dust areas and the general atmospheric circulation. Take Beijing as an example. 30%–40% of the PM2.5 comes from original emissions, 20%–30% from the photochemical conversion in the air, and 30%–40% from regional transport. Dust in areas surrounding the Beijing-Tianjin-Hebei region is transported to the local area and mixes with the continental aerosols of the local area, aggravating the pollution.

3.2 *Meteorological Conditions Favorable to Fog and Smog Occur More Frequently*

Weather causes mainly include static weather and such climatic conditions as high humidity, thin mixing layer and reduction in the number of rainy days.

3.2.1 Reduction in the Number of Rainy Days

The reduction in the number of rainy days weakens the washing effect of rainwater on pollutants in the air. The pollutant particles can remain in the air for a longer time, thus forming fog and smog more easily. In the past 50 years, the number of rainy days in most areas of China has shown a decreasing trend, wherein the number of rainy days in the eastern part of Northwest China, the Yangtze-Huaihe region, North China, and the western part of Northeast China has decreased by 0.5–1.0 days each year, that in Southwest China, South China and the eastern part of Northeast China has decreased by more than 1.0 days each year, and that in Jiangcheng, Yunnan has decreased by up to 2.8 days each year.

3.2.2 Decrease in Wind Speed

Wind conditions can reflect changes in atmospheric stability. Through transportation, the wind delivers the local pollutants to other regions or heights to fully mix with air and finally dilute the pollutants, improving the air quality. The average wind speed in most areas of China shows a decreasing trend, with an average decrease of 0.01–0.05 m/s each year, wherein the average wind speed in some areas decreases by more than 0.05m/s each year. If the daily maximum wind speed is greater than or equal to 10m/s, it is called a strong-wind day; if the daily maximum wind speed is equal to or less than 3m/s, it is called a calm-wind day. The number of strong-wind days and the number of calm-wind days in most areas of China show opposite changes: the number of strong-wind days decreases basically, while the number of calm-wind days increases. The decrease in the number of strong-wind days in the northern area of China occurs at a faster rate than it does in the southern area of China: the number of strong-wind days in the northern area of China decreases by more than 1.5 days each year, while that in the southern area of China decreases by 0–0.5 days; the increase in the number of calm-wind days in the Yangtze River Basin is obvious, with an increase of more than 1.5 days each year, and the number of calm-wind days in Northeast China, North China and South China increases by 0–1.5 days each year.

3.2.3 Increase in Stable Weather

The stability of the atmospheric stratification is favorable to the formation of fog and smog. In the past 10 years, although the occurrence of stable atmospheric conditions in China in winter has shown fluctuation changes from year to year, it presents an increasing trend on the whole, particularly more obvious in North China and East China with more fog and smog. The frequency of stable atmospheric stratification in North China (Beijing) in January increased from 59.7% in 2004 to 64.5% in 2013, and that in East China (Nanjing) in January increased from 33.9% in 2004 to 56.5% in 2013. The increase in stable atmospheric stratification showed meteorological conditions that were unfavorable to an increase in atmospheric diffusion, thus leading to an increase in fog and smog.

Moreover, the rise of the temperature of the stratosphere of the Northern Hemisphere caused by atmospheric zonal circulation and climate change makes pollutants gather in the mid-latitude region of the Northern Hemisphere, aggravating air pollution in some areas.

4 Effects of Fog and Smog

With changes in climatic conditions and continuous deterioration of the quality of the air, the occurrence of fog and smog continuously increases, which gives rise to significant negative effects on the climate, the environment, health, the economy and other aspects of life.

4.1 *Climatic Effect*

The increase in the concentration of fine atmospheric particles in fog and smog may cause extreme weather events to increase. The particulate matters in the air affect the balance of atmospheric radiation, which will make the ground become cooler and cooler and the air hotter and hotter, seriously affecting regional and global climate change, aggravating the heating effects of the regional atmosphere and increasing extreme climate events.

4.2 *Environmental Effect*

Air pollution is one of the most important environmental problems in the world. Since 1975, except Europe, the concentration of PM2.5 in the air has risen noticeably on a worldwide scale. The increment of concentration will give rise to acid rain and atmospheric photochemical smog in cities, causing the reduction of atmospheric visibility, and affecting air, water and road traffic.

4.3 *Health Effect*

The high percentage of pollutants in fog and smog can directly infect bacteria and viruses, and PM2.5, one of its essential components, which is also called respirable particles, can directly enter the human alveolar and even the blood system, directly resulting in cardiovascular disease and other diseases. According to the *Integrated Science Assessment for Particulate Matters* issued by the US Environmental Protection Agency in 2009,[1] sufficient scientific research results show that the fine atmospheric particles absorb a large number of harmful carcinogenic substances and genotoxicity mutagen matters, which will produce significant negative effects on human health, including an increase in mortality, aggravating chronic diseases, deteriorating diseases of the respiratory system and the cardiovascular system, changing the lung function and structure, affecting the reproductive capacity, and changing the humans' immune structure. A recent research by the Chinese Academy of Sciences also basically proves that air pollution has a positive correlation to mortality in respiratory diseases. The morbidity of lung cancer in Beijing has significantly increased in recent years, and 104 patients suffering from lung cancer were confirmed diagnoses on average in 2012. Research on the relationship between the lethality of lung cancer and dust smog in Guangzhou has shown that, with the consideration of a 7-year lag phase, the correlation coefficient between lethality of lung cancer and aerosol extinction coefficient is up to 0.97.[2]

1 EPA, "Integrated Science Assessment for Particulate Matters," 2009, http://cfpub.epa.gov/ncea/cfm/recordisplay.cfm?deid=216546.

2 Tie X. et al., "Lung Cancer Mortality and Exposure to Atmospheric Aerosol Particles in Guangzhou, China," Atmospheric Environment, 2009, 43, 2375–2377.

4.4 *Economic Effect*

General research focuses on the effects of air pollution on health, but we should not ignore the huge economic effects behind the health loss. Some research by the United States National Research Council applied economic models to estimate the effects of coal-based power plants in the US on the overall economy in 2005, and the result showed that the air pollutants discharged from the coal-based power plants caused losses of up to 127 million USD (converted to USD value in 2010) to health and to the environment.[3] Researches on China also came to the same conclusion. In 2004, the *Research Report of China's Green National Economy Accounting 2004* jointly issued by the State Environmental Protection Administration at that time (changed to Ministry of Environmental Protection of the People's Republic of China since 2008) and the National Bureau of Statistics pointed out that the China's economic losses caused by the environment were 51.8 billion RMB in 2004. That was the first time that China had considered the effects of environmental pollution on the economic aggregate. At the beginning of 2013, the *Toward a Future Sustainable Environment—the National Environmental Analysis of the People's Republic of China* issued by the Asian Development Bank also pointed out that the economic loss caused by air pollution in China each year was equal to 1.2% of the GDP, if estimated according to the cost of illness, and up to 3.8% if estimated based on the willingness to pay.

5 Experience of Other Countries in Controlling Fog and Smog and Air Pollution

It should be said that, in terms of the history of human development, the changes in the quality of the air have been affected by the process of the wealth accumulation of post-industrialization on the whole. With the intensification of the process of industrialization, fog and smog occurs in some regions and countries and affects the residents' life, which is not rare. Many countries have eliminated poverty, but their success has been achieved at the cost of environmental damages to a great extent. Research shows that air pollution severely affects the residents' long-term health, so more and more countries and regions have begun to pay attention to and take actions towards controlling and preventing air pollution, and have achieved results.

3 Committee on Health, Environmental, and Other External Costs and Benefits of Energy Production and Consumption, *Hidden Costs of Energy: Unpriced Consequences of Energy Production and Use*, US National Research Council, 2010.

5.1 *London, Britain*

As a country entering the stage of industrialization early, Britain is also one of the countries which began to suffer air pollution, fog and smog early. Although fog and smog in London has a certain relation to its climatic conditions, it comes more from the large-scale use of coal and other fossil fuels during the development of industrialization. In December 1952, London was shrouded in a dense fog for five days, and the visible distance in the daytime in the city was only about 1m. Effects of the extreme fog and smog on London were unprecedented: 12,000 people were killed due to the dense fog and another 8,000 people died later. After the dense fog, the air pollution in London aroused wide concerns from all sectors of the society. The government refused to take actions at the beginning, but later it developed strict measures to control the continuous worsening of fog and smog. The government issued some specific acts, prohibited the use of fuels producing smoke, improved the minimum height of industrial chimneys, and moved the power plants to places far away from cities. These measures made the level of air pollution emissions in urban areas drop sharply.

In 1956, Britain passed the first *Clean Air Act*, which was aimed at reducing the amount of smoke emissions, and was helpful in reducing the emission level of SO_2. It is worth mentioning that the Act clearly specified the regions where only smokeless fuels could be burned and that such coal smoke pollution units as power plants and heavy industries should be moved to suburbs. The Act is considered a milestone in the environmental protection of Britain, because it effectively controls the air pollution problems caused by the residential coal combustion and industrial processes. Although the *Clean Air Act* in 1968 admitted that it was impossible to completely remove SO_2 in the air, it required using tall chimneys in industrial coal combustion, fluid combustion or gas fuel combustion, so as to reduce the dispersal of air pollutants to ground, thus improving the quality of urban air. The *Air Pollution Control Act* in 1974 specified the upper limit of containing sulfur in industrial fuels. The *Clean Air Act* in 1993 consolidated the previous acts and the related regulations, and although it did not establish fixed rigid indexes, it entitled the Minister of State to set up the maximum emission rates of domestic and industrial furnaces. The acts changed London's industrial layout and residents' lifestyle, gradually realizing the natural gasification of residents' life. These measures have greatly reduced the smoke and SO_2 pollution caused by coal combustion.

Although these regulations improved the coal-based pollution emission in London, since the 1980s, the air pollution sources in London have gradually changed from coal combustion to automobile exhaust. For these changes, Britain has introduced new regulations and laws, e.g. in 1993, it started to require all new cars sold in Britain to have a catalyst to reduce the emission of nitrogen oxides. In 1995, Britain passed the *Environmental Law*, which required to make a national strategy regarding pollution treatment and specify the quantitative objectives of the related pollution control. Besides, Britain controls the traffic flow of its urban centers by collecting traffic jam fees and other economic means, and has set up the first low-emission zone in the world almost covering the whole of London in order to restrict entrance into the city of vehicles with exhaust exceeding the bid. Lastly, in order to solve the problems of air pollution caused by dust, in November 2006, the British government and the London City Council jointly formulated the *Guideline on Controlling Construction Site Dust and Pollution Gas Emissions*, specifying that all construction sites in London must take general measures to control air pollution.

These measures have substantially reduced the level of air pollution emissions in London, but the level of air pollution in Britain is still one of the most serious in the EU, so Britain still shoulders heavy responsibilities in air pollution control.

5.2 *California, USA*

In the USA, Los Angeles and the San Joaquin Valley region in the State of California are severely affected by fog and smog. In the 1950s, the air pollution caused by photochemical smoke continuously affected Los Angeles and several other coastal cities in California, causing many residents to suffer from diseases and even death. With the worsening of the problems, more and more people started to pay more attention to the problems of air pollution. Driven by the forces of all parties, all over the United States, represented by California, have spent nearly 60 years to basically achieve air pollution treatment and the control of the sources of air pollution.

Several hundred people died during several photochemical smoke events in Los Angeles in 1952 and 1955, and the United States Federal Government issued many legislative acts and amendments because of this. In 1955, the USA introduced the *Air Pollution Control Act*, the first law for air pollution control; the *Air Pollution Control Act of Motor Vehicles*, introduced in 1965 proposed regulations on automobile exhausts, an important air pollution source. The *Clean Air Act* passed by the United States Congress in 1963 became the most important air pollution control act, pointing out, for the first time, that the air pollution is a trans-regional, national problem. In 1967, the US passed the *Air Quality Act*. Later, the US supplemented, expanded and revised the *Clean Air Act* in 1970, 1977 and 1990, and has continuously strengthened and improved the articles of the Act. According to the acts, the USA issued the National Ambient Air Quality Standards. An amendment to the *Clean Air Act*, implemented since

1990, also stipulated stricter motor vehicle exhaust emission standards, and formulated new control standards of 189 toxic pollutants.

In the original National Ambient Air Quality Standards, the main control objectives focused on total suspended particulates and PM10, but considering the particulate matters (PM), especially those inhaled into lungs, which can produce direct and huge effects on the health of the public, since 1997, the United States Environmental Protection Agency has subdivided the particulate matters into fine particles and coarse particles for monitoring, and has formulated a set of new standards specifically for the particulate matters in the air whose diameter is less than or equal to 2.5 μm (i.e. PM2.5).

The *Clean Air Act* specifies that the United States Environmental Protection Agency should regularly review the air quality testing standards, and it requires all states to regularly submit a detailed *State Implementation Plan* guaranteeing the quality of the air to reach standards, according to the standards and policies formulated by the United States Environmental Protection Agency. What's more, the United States Environmental Protection Agency also publishes, in real time, the national air quality indexes on its official website, intuitively reflecting the air pollution status of all areas by different colors.

Through experience and experimentation over nearly 60 years, the USA has made lots of quantifiable achievements in regional air pollution control. From the point of view of the national level, the GDP of the USA from 1980 to 2010 increased by 127%, the mileage of motor vehicles increased by 96%, the total energy consumption increased by 25%, and the total amount of pollution increased by 36%, but the emissions of 6 air pollutants in the corresponding period diminished by 67%.[4]

5.3 *Mexico City, Mexico*

During the rapid development of urbanization, since the 1980s, Mexico City, the capital of Mexico, has started being troubled by air pollution. Since Mexico City is surrounded, on three sides, by mountains, its geographical position does not allow the dispersal of air pollutants, so the city has become one of the cities with the most severe air pollution in the world. The serious environmental situation has urged the national government and the municipal government to focus on the improvement of the quality of air, and they have successively formulated the *Ecological Balance and Environmental Protection Law*, the *Overall Planning of Antipollution* and other regulations in order to intensify the control of air pollution and environmental protection so as to reach a strategic national height. Mexico City has established air pollution warning systems, made

4 Shen Xinyi, *"Killer" of Air Pollution Control of the US*, World Environment.

two-stage air emergency treatment plans, increased related scientific research input in environmental protection and new energy development, and expanded the green area of the city.

Since air pollution in Mexico City comes from vehicle exhaust emissions to a great extent, regarding policy-making, since the end of the 1980s, Mexico City has started adopting strict motor vehicle standards and transportation measures to solve the problem of the quality of air, and has also started carrying out real-time monitoring of the quality of urban air. Specific actions include the use of unleaded gasoline, the installation of a catalytic converter on motor vehicles, the reduction of sulfur emission of diesel fuels, the strengthening of vehicle maintenance and inspection, and the implementation of stricter driving limitations, such as carrying out "no driving today" activities, putting a high tax on gasoline, moving oil refineries and other plants, etc.

Mexico City has now made obvious achievements through its efforts. From 1990 to 2010, the lead content in the air in Mexico City was reduced by 90%; the suspended particulates causing asthma, emphysema and even cancer were reduced by 70%; the emissions of CO and other pollutants were substantially reduced; the ozone level was reduced by about 75% compared with 1992.

6 Policy Suggestions for Controlling China's Fog, Smog and Air Pollution in the Future

Around the world, with the accelerated development of the process of urbanization, some large cities in some countries have encountered air pollution problems such as fog and smog due to industrial development, energy structure, transportation and lifestyle. Now, many cities in China, especially large cities, are also beginning to face serious air pollution problems. Smog pollution has become an obstacle to China's development, particularly some developed eastern coastal cities. In the meantime, it is expected that the globe will continue to warm up in the 21st century, the number of rainy days and the wind speed will decrease and the frequency of static weather will increase; all these meteorological conditions will provide the formation of heavy smog. Since air pollution has a direct and huge impact on economic and social development and on human health, during China's critical period of economic development and transformation, in order to improve the people's living standard, we need to pay great attention to relevant problems, and learn from the experience of other countries, to better control the effects of fog, smog and air pollution on people's lives. In the past several decades, although other countries, especially the developed countries in America and Europe, have accumulated lots of successful experience in controlling air pollution, fog and smog and other aspects, because of the differences between China and the other countries in geographical and climatic conditions, we need to make further targeted arrangements according to China's actual conditions when learning from their experience. Therefore, to cope with air pollution, we should strengthen comprehensive actions, and seek both temporary and permanent solutions through regional coordination and department links, focusing on control over a long period of time and on prevention for a short period of time. Specific suggestions are as follows.

6.1 *Establish and Strengthen the Mechanism of Multi-area and Multi-department Joint Governing and Controling Air Pollution, and Reduce the Quantity of Condensation Nuclei in the Air*

Air pollution in China shows a trend of regional pollution huddling, so we must implement joint regional prevention and control to effectively solve the problem. For instance, the heavy smog pollution treatment of Beijing needs not only to consider the sources of pollution within the city, but also to consider the sources of pollution in the Tianjin and Hebei area, which requires breaking through management systems of all the administrative units. The European Community collaboratively treats acid rain and air pollution, and Southern California, in the USA, curbs the photochemical air pollution through joint prevention and control, both of which are successful examples. Regarding the joint regional prevention and control of air pollution, China has not established a substantial coping mechanism, but we can consider the institutional innovation in management systems, such as building trans-regional air pollution committees for joint prevention and control. In joint regional prevention and control, we should specify the responsibilities of different areas, and fully consider the environmental capacity, the carrying capacity, cost and other factors of all the areas.

In addition to joint prevention and control among areas and cities, we also need to establish collaborative control mechanisms between the departments and industries, to enable collaboration of all national, provincial and municipal departments, including the Ministry of Environmental Protection, the Ministry of Transport, the Meteorological Bureau, the National Development and Reform Commission, the Ministry of Health, the Ministry of Science and Technology, the Ministry of Industry and Information Technology, and large state-owned energy companies. We also need to achieve data sharing, joint formulation of measures, and the collaboration and cooperation of law enforcement bodies.

6.2 *Improve the Construction of Meteorological Observation Systems, and Establish Forecast and Warning Systems of Air Pollution and Fog and Smog*

The occurrence of fog and smog is closely related to changes in meteorological conditions, so multiple departments should jointly construct a national ambient air monitoring network with a unified national layout, national coverage, and encryption of key areas, relying on the current network of observation stations, and in line with the principles of mutual supplementation, unified standards, and data sharing. Meanwhile, we should actively promote trans-department sharing platforms of regional air pollution observation data, develop technical methods of forecast and warning of meteorological conditions of air pollution, improve the accuracy rate and the refinement level of forecasting and warning, improve the ability at monitoring, forecasting and warning about fog and smog, promptly provide effective forecasting and warning information for the government and the public, take some necessary emergency measures, and reduce damages due to fog and smog on public health and transportation, etc. However, we should notice that the warning systems and measures are temporary and can only reduce the emission level temporarily, so the treatment of fog and smog pollution still needs to start with the fundamental treatment of emission sources.

6.3 *Prevention and Control of Fog, Smog and Air Pollution Needs to Strengthen the Rule of Law, and also Take Full Use of Economic Means*

According to the control experience of other countries, perfect laws and regulating systems are of vital importance to practically guarantee the reduction of emissions and results in pollution control, so we should gradually formulate related laws and regulations in all fields, and ensure specific the formulation of various standards and regulations. In addition to administrative means, we should also use economic means, such as carrying out reform in the fields of electricity and refined oil product prices, and providing sufficient economic incentives for thermal power plants and oil refineries to induce them to carry out environmental treatment voluntarily; we should learn from the experience of other countries, such as collecting "traffic jam fees" from private cars to control the travel of private cars through the leverage of prices, and the income from taxes can be used to improve and construct public transport projects.

6.4 *Enhance the Ability to Control Pollution in Urban Areas*

The areas which are most severely affected by fog and smog and other air pollutants are generally the urban areas with high population, so we should enhance the urban control ability to better control fog and smog. We need to promote the use of clean urban energy, optimize the urban energy structure, change the living and cooking habits of the urban residents, implement the change from coal to gas and from coal to electricity, and strengthen pollution control in areas where coal is used. We should formulate targeted urban traffic control measures, realize the reduction of urban traffic emissions, promote the implementation of stricter emission standards, and strictly control the number of motor vehicles in cities. We need to control the entrance of high-pollution and high-emission vehicles to urban areas with large populations, develop travel modes with relatively less pollution, such as the use of natural gas and electric vehicles, and greatly encourage public transport. According to the climatic laws of various areas, especially the carrying capacity and self-cleaning capacity of air pollutants, we should scientifically determine the urban function of the urban area, effectively control its scale, reasonably plan its layout, force the construction of urban "airways," and avoid the construction of high-emission enterprises in the upstream direction of prevailing winds from the cities, so as to build the urban area well by drawing on advantages and avoiding disadvantages. Furthermore, we should also pay great attention to energy use and pollution emission in rural areas near the urban areas as well as near medium and small towns.

6.5 *Collaboratively Control the Emission of Air Pollutants and Greenhouse Gases, and Formulate a Package Program Covering the Reduction of the Emissions of Multiple Pollutants*

The formation of fog and smog is closely related to climatic changes, and the emission of air pollutants and greenhouse gases has an equivalent homology. So, when controlling the inundation of fog and smog and air pollution emissions, we should pay attention to the synergistic effects of energy conservation and emission reduction. To control the emission of air pollutants, we should focus not only on certain kinds of pollutants, but we should also achieve a reduction of the emissions of all pollution gases and particulates. Air pollution control should give priority to the reduction of the emission of fine particles and the joint control of multiple pollutants such as the emission of PM2.5, MP10, CO, SO_2, NO_x, VOCs and NH_3 at the source. When controlling the primary pollution sources, we should give different focuses according to pollution characteristics, e.g. the SO_2 control should focus on industry, heating and coal-fired power plants; the NO_x control should focus on motor vehicles, coal-fired power plants and industry; the CO control should focus on motor vehicles and coal-fired power plants; the control of the VOCs should focus on motor vehicles, unorganized emissions (kitchen, dry cleaner, etc.) and industrial emissions. Moreover, the emission of some pollution sources, such as CO, SO_2, NO_X and PM, shows obvious seasonal characteristics in some areas (e.g. northern area), where the emission level is high since large-scale amounts of coals need to be used for heating in winter, so we should strengthen the control of these pollutants in winter; for other pollution sources such as the VOCs, we need to carry out corresponding control throughout the year.

6.6 *Gradually Realize the Disclosure of Information, and Form a New Control Pattern with the Participation of All People*

Throughout the control experience of other countries, the participation of the public is of great importance to achieve ideal emission reduction results. Air pollution control is not an isolated problem; it requires long-term and systematic efforts to achieve it. During this process, the participation of the government, industries, enterprises, communities, and the public is of vital importance. The government departments should strengthen publicity and education, realize the improvement of the overall quality of all citizens, and consider the reduction of pollution emission as much as possible beginning from catering, travel and garbage disposal to career choices and property purchases. Related departments should also increase the disclosure of information, and comprehensively accept the supervision and accountability of all sectors of the society.

CHAPTER 13

Countermeasure Analysis for Urban Climate Vulnerability and Adaptability: Taking the Beijing "July 21" Extraordinary Rainstorm as an Example

XIE Xinlu, YE Dianxiu and ZHANG Cunjie

Abstract

In recent years, Chinese cities have witnessed many rainstorms, resulting in urban waterlogging, property loss and personal casualties. Urban vulnerability in extreme climate events such as rainstorms has been thoroughly exposed, which sounds an alarm to urban administrative departments. Therefore, the issue of urban climate adaptability is especially urgent. The event of the Beijing "July 21" extraordinary rainstorm caused massive property losses, personal casualties, and exerted a negative impact on social economy and the ecological environment. It was not only the result of the great rainfall, high intensity, wide range, prominent extremeness and other natural properties of the rainstorm, but it was also closely associated with the imbalanced development and absence of adaptability caused by the destruction of the natural ecology, neglect of climatic risks and the lagging behind of the construction of climate protective engineering in the course of rapid urbanization. Enhancing climate adaptability is an important path leading to the decline of urban climate vulnerability. The governance of climate adaptability should take into account the multi-approach and multi-level build-up of the adaptive capacity in ecology, engineering, technology and institutions, with an aim to promoting sustainable urban development.

Keywords

climatic risk – urbanization – climate vulnerability – climate adaptability

* Xie Xinlu, a postdoctor at the Institute for Urban and Environmental Studies Chinese Academy of Social Sciences, engaging in studies on the economics of climate change; Ye Dianxiu and Zhang Cunjie are senior research engineers at the National Climate Center, engaging in studies on climate and the evaluation of the impact of climate change.

 | DOI 10.1163/9789004280021_014

Climate change has taken place and the mitigation of climate change needs long-term concerted efforts of the international community; however, climate adaptability may get results on a small scale and in the short term. In recent years, some cities in China have suffered from rainstorms frequently, resulting in a number of secondary disasters such as urban waterlogging, flash floods and mud-rock flows in some districts and counties, which affected normal urban operations and caused tremendous property losses and personal casualties. Cities are so vulnerable to the extreme climatic event of a rainstorm that the sustainable urban development is challenged. Therefore, there is an urgent need to address the issues of urban climate adaptability.

1 Frequent Heavy Rainfall and Increasing Severe Urban Waterlogging in Chinese Cities in Recent Years

1.1 *The Increasing Torrential Rainy Days and Maximum Daily Precipitation in Metropolises Since 2007*

From 1951, 31 provincial capitals have presented an upward trend in mean annual torrential rainy days and maximum daily precipitation, and a great inter-decadal variation, as shown in Figure 13.1 and Figure 13.2 below. The 1950s, 1960s and 1990s witnessed many torrential rainy days and high daily precipitation, and the 1970s and the 1980s fewer (lower). The early years of the 21st century have seen a small number of torrential rainy days and low daily precipitation; however, torrential rainy days and daily precipitation have shown an upward trend from 2007. There was a great difference in the variation trend among the 31 provincial capitals throughout the country. Cities in South China, East China and the Huang-Huai area, including Haikou, Fuzhou, Nanning, Nanjing and Jinan, presented an upward trend in the maximum daily precipitation. In spite of the overall downward trend, Beijing's maximum daily precipitation has shown an upward trend since the beginning of the 21st century.

1.2 *Frequent Heavy Rainfall in Various Cities in Recent Years*

Recently, some cities are encountering heavy rainfall more often, as shown in Table 13.1. The heavy rainfall is beyond the urban drainage capacity, causing urban waterlogging, traffic tie-ups, flight delays and personal casualties, which affect the normal urban operations and brings a number of challenges to urban climatic risk administration.

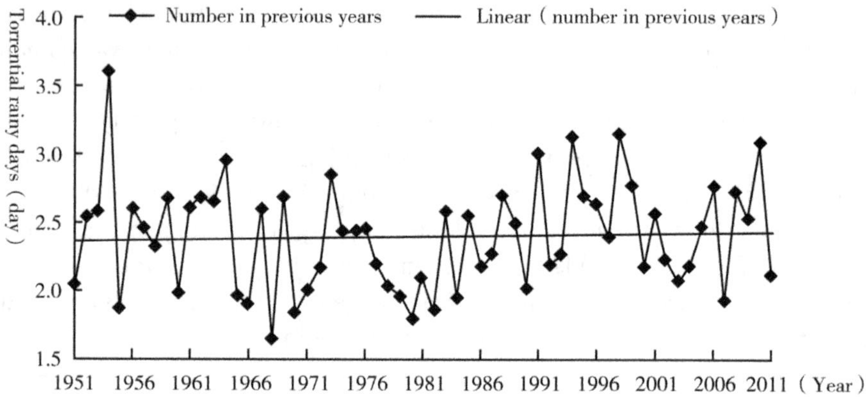

FIGURE 13.1 *Variation in mean torrential rainy days of 31 provincial capitals over the years (from 1951 to 2011).*

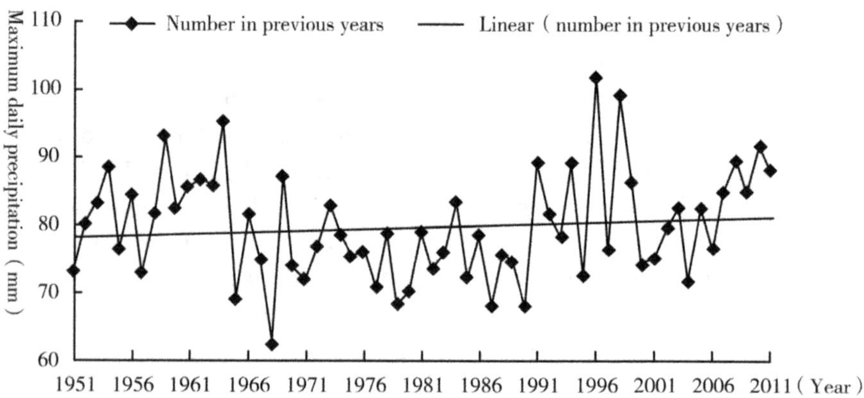

FIGURE 13.2 *Variation in maximum daily precipitation of 31 provincial capitals over the years (from 1951 to 2011).*

TABLE 13.1 *Heavy rainfalls in Chinese large and medium-sized cities and their impacts from 2004 to 2013*

Time	City	Precipitation	Impact
July 19, 2013	Xiamen	Precipitation is over 100mm in most areas, and over 200mm in some areas	Serious waterlogging on multiple main roads, landslides in some tunnels and road sections, severe traffic

Time	City	Precipitation	Impact
			jams in urban areas, and serious inundations in low-lying areas in Xiamen University
July 7–11, 2013	Chengdu, Dujiangyan, Mianyang, Deyang	Rare extraordinary rainstorms occur in the west of the Sichuan Basin, with precipitation ranging from 200 to 400mm in common, from 400 to 800mm in Deyang and Dujiangyan, and from 800 to 1,151mm in some areas. The regional rainstorm is the largest rainfall in the meteorological records since 1954.	Waterlogging in 8 underneath tunnels and 40 road sections in Chengdu central urban areas; 43 deaths and 118 missing people caused by Dujiangyan "July 10" extreme high landslides.
July 21, 2012	Beijing	Mean precipitation is 231.0mm in urban areas, and precipitation is up to 460.0mm in Hebei Town, Fangshan District. There are 18 observation stations (including automatic stations) with hourly precipitation more than 80mm and maximum hourly precipitation up to 100.3mm.	Severe urban waterlogging, flash floods in some areas, severe impact on traffic, heavy casualties and serious economic losses.
July 18, 2011	Nanjing	Heavy rainfall occurred on July 18, with 24-hour precipitation of 225.8mm in Xuanwu Lake, and a maximum hourly precipitation of 94.8mm in urban areas	Urban waterlogging, traffic jams on the main roads, closing or speed limitation measures taken in expressways.
July 2–3, 2011	Chengdu	Rainstorm and heavy rainstorm occurred in some areas in Chengdu from 8 p.m. on July 2 to 9 p.m. on July 3, with a maximum hourly precipitation up to 215.9mm in urban areas	Waterlogging on some roads and low-lying areas, inundations in some areas, traffic blocking in 6 tunnels

TABLE 13.1 *Heavy rainfalls in Chinese large and medium-sized cities* (cont.)

Time	City	Precipitation	Impact
June 23, 2011	Beijing	Beijing was hit by an intense storm in the afternoon and at night on June 23, with a maximum hourly precipitation of up to 128.9mm in Moshikou, Shijingshan District, hitting the highest record since the foundation of the station	Heavily affecting urban traffic, causing 2 deaths.
June 18, 2011	Wuhan	Wuhan daily precipitation was 193mm from 01:00 a.m. to 7:00 p.m. on June 18, setting a new record in daily precipitation in the past 13 years since 1998.	Inundation of many central urban areas, and traffic paralysis.
May 7, 2010	Guangzhou	The maximum amount of one-hour and three-hour precipitation was up to 99.1mm and 199.5mm, respectively	Serious waterlogging in the urban area
July 13, 2009	Beijing	Short-time heavy rain occurred, with mean precipitation of 15mm in urban areas, 74mm in Xinfadi, and 70mm in Yuquanying from 2:00 p.m. to 6:00 p.m. on July 13.	Severe waterlogging on seven bridges including Fengyi Bridge, partial traffic interruption, delay of hundreds of flights in capital airport
August 10–11, 2008	Beijing	Heavy to torrential rain fell over a wide area	An amount of water in some urban road sections, interruption or delay of outdoor Olympic games
August 25, 2008	Shanghai	Short-term downpour occurred in Shanghai, with a maximum one-hour precipitation of up to 117.5mm in Xuhui District, hitting the meteorological records in Xujiahui since 1872.	Severe waterlogging on more than 150 roads, traffic jams, delays of more than 400 long-distance buses and 138 flights in Hongqiao Airport

Time	City	Precipitation	Impact
June 12, 2008	Guilin	A widespread rainstorm occurred in Guilin, with precipitation of up to 180mm	Severe waterlogging in urban areas
August 6, 2007	Beijing	Rainstorm hit suddenly, with one-hour precipitation of up to 82mm	The maximum depth of waterlogging under Anhua Bridge was 1.7m, resulting in traffic stoppage on the main road
July 18, 2007	Ji'nan	The maximum one-hour precipitation was up to 151 mm, hitting the highest record since 1958	Severe waterlogging in Jinan, traffic tie-ups on most road sections, and 25 deaths
July 17, 2007	Urumchi	Daily precipitation was up to 57.4mm, breaking a historical record	Severe waterlogging on some road sections, causing traffic jams
August 5, 2007	Shanghai	Precipitation was more than 100mm in Yangpu, Hongkou, Chongming, Jiading, Nanhui, etc.	A wide area of waterlogging, with maximum depth up to 30–40cm; delays of many flights.
July 10, 2004	Beijing	A rare rainstorm took place in part of the urban area, with maximum one-hour precipitation of up to 52mm in Fengtai District. At 08:00 a.m., July 10 and 08:00 a.m., July 11, the precipitation in Fengtai, Tiananmen and Tiantan all exceeded 90mm, and Tiantan precipitation was up to 109mm.	41 places of waterlogging in urban areas, with maximum depth of more than 1.5m on the Fuxingmen and Lianhua Bridges; severe traffic paralysis.

2 Features of the Beijing "July 21" Extraordinary Rainstorm Disaster

On July 21 and 22, 2012, Beijing encountered a rare torrential rain, the largest citywide rainfall since 1951. Compared with the seven extraordinary rainstorms of before (July 20–24, 1952; Aug. 15–17, 1955; July 29–Aug. 6, 1956; July 11–15, 1958; Aug. 3–7, 1959; Aug. 3–9, 1963; Aug. 9–16, 1979), the torrential rain is characterized by great rainfall, strong intensity, wide scope and prominent extremeness. Citywide mean daily precipitation has occurred over hundreds of years.

2.1 *Great Rainfall*

On July 21 and 22, 2012, citywide mean (20 national stations) precipitation was up to 190.3mm, the mean precipitation for the urban area was 231.0mm, maximum precipitation 337.5mm (Xiayunling), and the precipitation in Hebei Town, Fangshan District was 460.0mm (hydrometric station), as shown in Figure 13.3. The torrential rain on "July 21, 2012" is ranked 6th in history, only after the 1956 rainfall (389.0mm), the 1963 rainfall (281.8mm), the 1958 rainfall (226.8mm), the 1952 rainfall (210.1mm) and the 1959 rainfall (194.5mm).

2.2 *Strong Intensity*

There were 18 observation stations (including automatic stations) with an hourly precipitation in excess of 80mm, and a maximum hourly precipitation

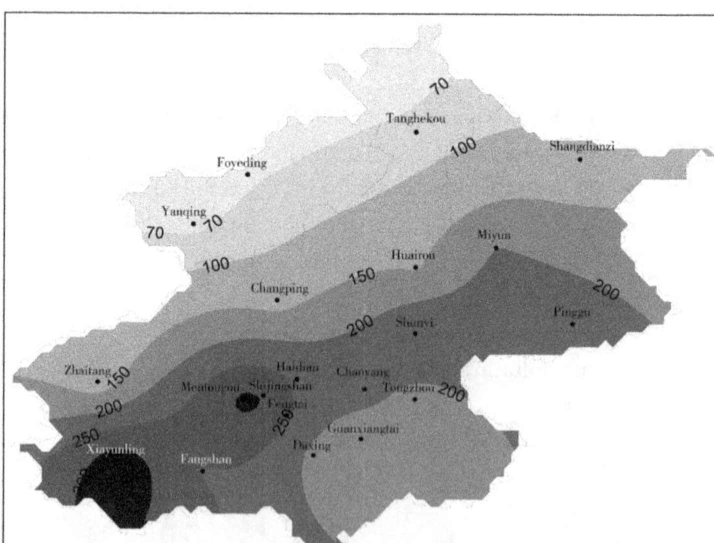

FIGURE 13.3 *Precipitation in Beijing from July 21 to 22, 2012 (unit: mm).*
DATA SOURCE: NATIONAL CLIMATE CENTER.

of up to 100.3mm, which were very rare in the history of Beijing. Besides, in spite of the short-term precipitation process (from 10:00 a.m., July 21 to 06:00 a.m., July 22), the citywide mean rain intensity was 95.2mm/day, far more than that of any other rainfall in history, ranking 1st.

2.3 *Wide Scope*

The heavy rainfall covered a wide area, with a downpour (over 100mm) area accounting for over 86% of the total area.

2.4 *Prominent Extremeness*

Citywide mean daily precipitation has occurred over one hundred of years. There are 11 stations with precipitation of up to the historical extreme since 1951. The mean daily precipitation in the southwest mountainous area (212.6mm) was 72.5mm higher than the historical highest record (140.1mm).

3 Impact of the Beijing "July 21" Extraordinary Rainstorm

The Beijing "July 21" extraordinary rainstorm generated serious adverse effects, resulting in quite an amount of damage and a number of deaths among the population. Besides, it caused a massive direct economic loss of 11.64 billion RMB in the transportation industry, tourism industry and agriculture,[1] more than three times the sum of the direct economic losses caused by meteorological disasters in Beijing in the previous five years, and accounting for 0.72% of Beijing's regional GDP in 2011. The extraordinary rainstorm showed that Beijing was vulnerable to extreme climatic events in terms of society, economy and infrastructures.

3.1 *A Great Amount of Damage and a Number of Deaths among the Population*

The Beijing "July 21" extraordinary rainstorm led to the urgent transfer of 97,000 people, the people involved were 1,602,000 and there were 79 deaths; people mainly died of drowning, being struck by lightning, electric shock and house collapse, including 50 drowning deaths which accounted for 63% of the total deaths (11 driving drowning deaths, which accounted for 14% of the total deaths). Most of the people who were killed were villagers and migrant

1 The data regarding the losses incurred because of the Beijing "July 21" extraordinary rainstorm come from *Beijing News* and special reports by Sohu and other media.

workers, showing that such groups are vulnerable to rainstorms and floods, especially the villagers living near courses of water, pedestrians and basement inhabitants.

3.2 *Heavy Impact on Traffic*

The extraordinary rainstorm of July 21 caused the cancellation of 571 flights, the delay of 701 flights, the shutdown of the airport route from Dongzhimen to T3 airport terminal, and nearly 80,000 stranded passengers. It is really rare to have such a high number of stranded passengers, cancelled flights and delayed flights. Besides, serious waterlogging occurred in 63 places in the whole city, and 95 roads in the urban area were closed due to waterlogging. Nine subway lines accumulated water, and eight railway lines, including the Beijing-Yuanping Railway, were temporarily out of operation. Severe waterlogging occurred under the Nangangwa Railway Bridge in Beijing—the Hong Kong—Macao Expressway, with a flooded road length of about 900m, the average water depth of 4m, maximum water depth of 6m, 127 inundated vehicles, and road closure of more than 40,000 times. The water accumulated under the Guangqumen Bridge was beyond the capacity of the pump station, causing a river water recharge. The depth of accumulated water under the overpass was up to 4m, causing a traffic stop. The extraordinary rainstorm aggravated the imbalance between supply and demand of the urban public transportation service. The traffic system is very vulnerable to extreme heavy precipitation, especially the roads and bridges in low-lying areas.

3.3 *Heavy Impact on Tourism*

The Beijing "July 21" extraordinary rainstorm and torrential flood destroyed lots of tourism facilities, with affected scenic spots including Shidu and Shangfang Mountain Forest Park. Infrastructures, roads, rural tourism houses and fishponds were devastated, causing tremendous losses. As the extraordinary rainstorm occurred during the peak tourist season, it resulted in massive property losses and income losses to operators, so that the livelihoods of operators engaging in tourism were affected to differing extents. Through an investigation in April 2013, the author found out that some stricken scenic spots in Fangshan District were still closed; rural tourism operators in the Shidu Scenic Spot said that the tourist flow had dramatically declined compared with previous years, which had some relationship to the "July 21" extraordinary rainstorm.

3.4 *Impact on Agriculture*

The Beijing "July 21" extraordinary rainstorm and torrential flood caused a crop-stricken area of 769,000 mu (1 mu = 0.0667 hectare), a crop disaster area of

479,000 mu, and a no-harvest area of 122,000 mu; an area of destroyed breeding houses of 178,300m^2, 103 destroyed facilities in animal husbandry breeding farms in the whole city, and a stricken area of the fish industry of appropriately 5,900 mu.

Besides, the scouring effect by the extraordinary rainstorm on urban roads and roofs caused the inflow of various pollutants into rivers and lakes, further resulting in non-point source pollution; in addition, the extraordinary rainstorm may have also caused soil erosion and sediment accumulation, further resulting in the pollution of the quality of the water and some destruction of the ecological environment. Ecological environmental pollution and destruction caused by rainstorms are not seen in the short term, and are not easy to be recovered in the long term. Thus, the urban administrative departments should pay attention to this and enhance their studies on the impact of the rainstorm secondary disasters.

4 Analysis of the Causes of Urban Climate Vulnerability

4.1 *Imbalanced Urban Development*

By taking Beijing as an example, Beijing's built-up urban area was 780 km^2 in 2001 and expanded to 1,231.3 km^2 in 2011, with a growth of 57.9% during the ten years; Beijing's permanent population was 13,851,000 in 2001 and increased to 20,186,000 in 2011, with a growth of 6,335,000 people during the ten years; Beijing's urban road area was 60,620,000 m^2 in 2001 and 91,640,000 m^2 in 2011, with a growth of 51.2% during the ten years; Beijing had 891 bridges and 160 overpasses in 2001, and 1,885 bridges and 418 overpasses in 2011, with a growth of 112% and 161%, respectively.[2]

Rapid urbanization leads to the constant expansion of the built-up area and aggravates the irrational utilization of land. First, urban watercourses and low-lying areas were encroached upon and filled in, and urban high-rise buildings, roads and bridges increased, which increased pavement hardening, decreased urban ponds, lakes and green spaces, and declined the flood storage capacity of natural ecology. Second, the number of urban construction projects in low-lying areas increased, the same is true for the number of highway bridges, overpasses, underpasses and subways in low-lying areas, which increased the number of vulnerable climate sites. Third, the increase in urban road area

2 Data source: *China City Statistical Yearbook 2002, China City Statistical Yearbook 2012,* and *Beijing Statistical Yearbook 2012.*

leads to an increase in surface traffic flow; however, the urban drainage capacity failed to be improved. Beijing drainage systems are mostly designed for rains happening once in 1–3 years, showing a low drainage standard and severe pipeline aging. Urban construction attaches importance to surface engineering, without sufficient attention to the protective engineering for the underground climate and the overall consideration of the relationship between the surface construction and the underground infrastructure. Consequently, the development of the protective engineering for the underground climate lags far behind the urban overall expansion capacity, and the urban pipe network construction cannot adapt to the features and trends of meteorological disasters in the new period.

With a complex landform and a small population and property density, Beijing suburbs and mountainous areas are the areas that are prone to rainstorms and floods, easy to suffer from floods, mud-rock flows and other disasters. However, as the social economy of suburbs and mountainous areas develops and the permanent population spreads to the suburbs, the demand for land increases, causing the occupation of watercourses, the irrational development, etc. By centering on watercourses, facilities and industries including fishponds, breeding farms and rural tourism are exploited. Thus, higher climatic disaster risks are posted while promoting economic development.

4.2 *Lack of Climate Adaptability*

The Beijing "July 21" extraordinary rainstorm exposed a series of problems, such as the absence of climatic risk evaluation in urban planning and construction, poor coordination in disaster emergency linkage mechanisms, a lack of the construction of a climate disaster legal system and incomplete public participation mechanisms.

4.2.1 Absence of Climatic Risk Evaluation

During the Beijing "July 21" extraordinary rainstorm, serious waterlogging and river water back flows occurred in the Nangangwa section of the Beijing—Hong Kong—Macao Expressway and on the Guangqumen Bridge, resulting in personal casualties and vehicle inundations. These roads and bridges were built in low-lying areas, or across the waterway, or they use a concave-down design, which poses potential hazards of waterlogging, river water back flows and other disasters. This shows the neglect of climatic risk evaluation in urban planning and construction and the absence of climate vulnerability in urban construction engineering, causing incomplete formulation of climate adaptive measures. Climatic risk evaluation is aimed at getting to know the frequency, intensity, environment and impact of climatic events, so it is the prerequisite for disaster prevention, mitigation and climatic adaptation.[3]

4.2.2 Poor Organization and Coordination in Disaster Management Linkage Mechanisms

Since urban climatic disasters are characterized by systematization, complexity and linkage, the linkage mechanisms for climatic disaster management is increasingly significant, therefore, integration and coordination are the keys to climate adaptability.[4] The linkage mechanisms involve not only the administration departments of meteorology, municipal administration, civil administration, transport, firefighting, hygiene, agriculture, state land, but also stakeholders such as the media, enterprises, communities and residents. The Beijing "July 21" extraordinary rainstorm revealed a series of problems in the linkage mechanisms. For instance, upon receipt of the early warning of a meteorological disaster, the traffic administration authorities failed to promptly set danger marks; calls for help failed to get through; relevant departments were poor in the organization for rescuing people and property due to a lack of sufficient information; charging enterprises still collected fees in the event of a great amount of water accumulated on the expressway, which showed the absence of a linkage mechanisms between management departments and operating units, and the poor enforceability of emergency plans; some taxi drivers demanded high prices, and traffic coordinators wrote out tickets, revealing the absence of emergency plans for meteorological disasters by relevant departments and enterprises, and undefined responsibilities of staff in the event of an emergency. All of these reflect a series of problems, such as Chinese cities' weakness in disaster management linkage mechanisms, unclear responsibilities of stakeholders before, during and after the disaster, poor systemization of disaster management, and poor organizational coordination.

---

3 United Nation Inter-Agency Secretariat of the International Strategy for Disaster Reduction (UN/ISDR), *Living with Risk: A Global Review of Disaster Reduction Initiatives, United Nations, Geneva, 2004.*

4 World Resources Institute (WRI) in Collaboration with United Nations Development Programme, United Nations Environment Programme, and World Bank, 2011, *World Resource Report 2010–2011: Decision Making in a Changing Climate—Adaption Challenges and Choices*, Washington, D.C.: WRI.

4.2.3 Weak Awareness of Routine Disaster Risk Management

From reports on the "July 21" extraordinary rainstorm, relevant authorities failed to pay sufficient attention to, and put in place, the routine disaster management such as the inspection of hidden disaster dangers and disaster prevention facilities, decontamination of drainage pipelines, and demolition of illegal buildings. Taking basements as an example. Basements were equipped with a switching room, a heating pipeline and a gas pipeline, but the underground space was under a disorderly management, had no or underqualified water and electric facilities, and lacked routine maintenance inspection and hidden danger identification. Such management vulnerability in electrical and water leakage brings hidden dangers to climate security of the basement. Lack of disaster risk awareness in relevant departments and inter-department joint emergency practice will lead to inefficient emergency measures. Thus, it is necessary for relevant authorities to pay attention to it.

4.2.4 Lack of Public Participation Mechanisms

Chinese disaster management mechanisms mainly involve governmental administration departments, organizations and their duties. Early warning, emergency and rescuing information are in a unidirectional transmission from top to down within the government, without public participation. Insufficient public participation causes a certain defect in the risk management structure.[5] The American emergency management authority is composed of the Federal Emergency Management Agency (FEMA), federal districts, states, counties and cities, voluntary organizations, non-compulsory organizations private enterprises, etc., and the American government requires the cooperation of the management authority in the federal response plan; the municipal government of Tokyo, Japan pays special attention to its partnership with the residents of the community.[6] During the Beijing "July 21" extraordinary rainstorm, some private car owners, migrant workers and other citizens organized rescue teams spontaneously, which played a positive role in disaster rescues. Public participation should not only be manifested in rescue processes, but also in the monitoring and management process of unlawful acts such as watercourse encroachments, so as to prevent flood disasters and enhance the urban climate adaptability.

---

5 Qi Ye and Ma Li, "Moving Toward More Positive Climate Change Policies and Management, Chinese Population," Resource and Environment, 2007, (2): 8–12.

6 Zuo Xiong (ed.), *Research and Practice of Abrupt Meteorological Disasters Emergency Management*, China Meteorological Press, 2011, p. 333.

4.2.5 Post-disaster Insurance and Guarantee Mechanisms Needing to be Improved

Insurance is an important means for risk sharing and transfer, as well as an important financial support for post-disaster reconstruction. According to statistics by the China Insurance Regulatory Commission Beijing Bureau, the insurance estimated loss was approximately 790 million RMB at midnight of July 26, accounting for 6.8% of the total loss. Owing to small insurance coverage, most losses caused by the disaster were not compensated. It was the result of weak social insurance awareness, incomplete disaster products, and an imperfect disaster insurance coverage and system.

Due to an imperfect social guarantee and remedy system, although the governmental organs carried out disaster evaluation, made relief and compensation for the stricken population after the "July 21" extraordinary rainstorm, there was still a deficiency in the scope of the compensation, standard, etc. For example, the non-local stricken population were beyond the scope of the guarantee, and the stricken population living in basements had to sleep in the plaza. To avert adverse effects by disasters such as a return to poverty, urban administrators should make constant efforts to improve post-disaster guarantees and insurance systems.

4.2.6 Cognition Deficiency of the Public in Climatic Risks

Before the "July 21" extraordinary rainstorm, the meteorological department had released a heavy rainfall prediction and early warning, but some citizens still went out, as usual, to participate in outdoor activities, and tourist attractions were still open to tourists. Some residents in the mountainous areas and tourists acted presumptuously instead of following the directions after receiving the flood notice, resulting in unnecessary losses. Some drivers carelessly decided to drive during the rainstorm. A media investigation showed that about 70 percent of the respondents were unaware or confused about the levels of the rainstorm early warning,[7] and most of the public did not understand the meaning of the disaster early warning signals. This shows that there was a deficiency in publicity, education, training and drilling for climate change risks, and that the public should enhance its awareness of and knowledge about climatic change risks.

---

7 *Color of Disaster*, http://www.legaldaily.com.cn/index_article/content/2012-08-01/content_3743262.htm.

4.2.7 A Lagging Behind of the Disaster Legal System

The Chinese *Flood Control Law* has explicitly stipulated that it is illegal to carry out road construction, building construction, sandy soil pile-up, breeding farm and agricultural planting on watercourses. However, such acts are not completely banned, resulting in the obstruction of flood channels. It was one of the main factors leading to the massive losses during the Beijing "July 21" extraordinary rainstorm. Disaster prevention and mitigation laws including the *Flood Control Law* have industrial and departmental features, which determine the boundedness in disaster management and departmental coordination. The complexity and systematization of urban meteorological disasters determine the importance of a multi-level participation, organizational coordination and linkage of the governmental organs, organizations, enterprises and citizens, as well as the difficulties and obstacles in the administrative organization and coordination. Therefore, it is a must to enhance the construction of a disaster legal system in order to adapt to the needs of the increasingly complex disaster management.

5 Countermeasures and Suggestions for Urban Climate Adaptability

Under climate change, the frequency and intensity of extreme climatic events may lead to an increase in urban climatic risks. Sustainable development is the challenge that cities must face. In order to govern the climate adaptability, it is necessary to re-recognize the relationship between human beings and the earth from the perspective of sustainable urban development, and enhance the building up of the climate adaptive capacity in ecology, engineering, technology and institutions, with an aim to strengthening urban climatic adaptability.

5.1 *Enhancing the Construction of Green Infrastructures and Ecologically Adaptive Measures*

The only means for flood drainage by engineering measures increases the flood drainage pressure of the limited drainage facilities and watercourses and wastes lots of valuable resources, which is not good for resolving the problems regarding natural urban ecology and water resource security. Ecological adaptation is capable of restoring the excessive climate vulnerability caused by human activities, to some extent, to reduce disaster impacts. Therefore, first of all, it is necessary to change the urban developmental planning ideas, protect the natural ecological environment, such as the urban water areas, forests and green spaces, and construct environmentally-friendly cities. Second, it is necessary to enhance the construction of green infrastructures, such as green roofs and gardens, and promote the function of climate protection of the ecological system.

5.2 *Enhancing the Urban Drainage and Water-holding Capacity with Adaptable Engineering Measures*

Apart from common low standards, the Chinese urban drainage network also suffers from disruptions, so that drainage network connectivity is poor, easily leading to water accumulation and waterlogging. Therefore, while improving the urban drainage standards, the connectivity of the drainage network should be strengthened to increase the overall capacity of urban drainage. In terms of full consideration of the relationship among the urban climate, water resources and floods, engineering measures, such as the construction of underground impounding reservoirs, seepage pavements and sinking gardens, should be taken to enhance the urban water-holding capacity and reduce surface runoff caused by rainstorms in cities.

5.3 *Enhancing Adaptable Technical Measures and Improving the Transmission of Meteorological Disaster Warnings and Information*

In terms of information management, an integrated disaster information management platform needs to be established and improved, which would be able to integrate the information about geography, precipitation, hydrology, disastrous situations, people and materials and provide technical support for an active and timely mechanism for disaster early warning, alarms, departmental linkage, emergency decision-making and information release for urban disaster prevention and reduction. In terms of meteorological science, it is important to intensify climate disaster monitoring and early warning, develop advanced meteorological satellites, weather radars and automated surface observation systems, and set up an all-weather, high-density and perfect disaster monitoring and early warning network. In terms of urban construction, small separated pieces of land should be used for development and low impact technology should be applied to accumulate water and relieve urban waterlogging, thus improving the urban landscape and ecology, and promoting the natural and ecological circulation of water.

5.4 *Enhancing the Institutional Adaptability and Improving the Disaster Risk Management System*

The construction of a Chinese climate adaptability system lags behind that of other countries, together with more weaknesses in its risk evaluation systems before and after the disaster, climate change protection systems, routine disaster management systems, insurance guarantee systems, etc.

5.4.1 Strengthening the Climatic Risk Evaluation

Climatic risk evaluation is aimed at recognizing risks, vulnerability and impacts, thus providing a basis for adaptive measures. It is an important part of the disaster management and climate adaptability, and can provide decision-making support for urban planning and construction. Governmental authorities should strive to promote urban climatic risk evaluation, prepare a climatic risk zoning map to act as the basis for urban construction, population and industrial layout, avoid climatic risk vulnerability and enhance the urban climate adaptability. Besides, it is necessary to enhance climatic risk evaluation on construction projects, such as urban roads, bridges and park areas, so as to have an understanding of the risk source, vulnerability and possible adaptive approaches, and enhance the management of the climate protection facilities according to the realities. Climatic risk evaluation should involve as many stakeholders as possible, and integrate the stakeholders' knowledge and the climatic adaptation needs, thus making urban planning and construction more scientific and reasonable.

5.4.2 Revising the Traditional Disaster Prevention Standards and Making a Scientific Planning for the Urban Waterlogging Prevention Facilities

Most of the Chinese urban planning and architectural standards are formulated according to the precipitation intensity of the previous dozen years. The changing pattern of distribution of the precipitation intensity due to climatic change posts new challenges to such traditional standards. Cities should lay down multi-level climate protection standards, including drainage design for urban infrastructures such as roads and bridges, climate protection standards for underground spaces, and construction standards for park water storage and leakage facilities. By the improvement of climate protection standards, the urban surface runoff would be reduced, and urban ecological environmental pollution would be mitigated.

5.4.3 Improving the Routine Disaster Management System

Occurrence of disasters is closely related to inefficient routine disaster management. In routine disaster management, hidden-danger checking and inspection should be enhanced to ensure that the drainage system is in good condition, that climate security hidden dangers, such as water or electricity leakage, is eliminated, and dangerous old houses are reinforced; river treatment should be enhanced, especially the small and medium-sized rivers, to ensure their smooth flowing; disaster or danger marks should be installed at risky points. Pre-arranged planning for urban "lifeline" systems such as urban traffic, water supply and electricity supply should be strengthened and refined. Relevant departments, enterprises and citizens should be organized to participate in the emergency drills. Climatic risk management should be included in the developmental planning of industries and districts. Furthermore, it is important to improve the public's awareness of disaster prevention and reduction and the adaptive capacity of climate change, and enhance the management on the groups that are most vulnerable to climate change.

5.4.4 Formulating a Basic Disaster Management Law

China should formulate a high-level basic disaster management law to define powers and responsibilities of all stakeholders in the disaster chain, regulate and coordinate the disaster management process by law. Disaster laws should stress the importance of "preventive measures" and upgrade the status of climate adaptability. Urban planning and construction should take into account the situation of climate change and the stakeholders' adaptive needs, and public participation should be encouraged, with the aim of providing a legal guarantee for scientific decisions, departmental linkage, disaster prevention and reduction, monitoring and management.

5.4.5 Developing Disaster Insurance Products and Refining Disaster Guarantee Mechanisms

The development of financing products integrated with meteorological disaster prevention and the application of financial instruments in the risk management of natural disasters, provides a new approach for the transfer of climate disaster risks,[8] and promotes the sharing of the risk mechanism among the insurance company, the government, enterprises and residents. Systems, standards and procedures relating to insurance compensation remain to be improved, and the timeliness, accuracy and efficiency of nuclear disaster loss assessment and compensation should be strengthened. Besides, efforts should be made to enhance the climate disaster publicity education, expand the scope of and the policy study for the insurance guarantee in an attempt to prevent poverty and the return to poverty due to the disaster, as well as promote the sustainable development of the group most vulnerable to climate change.

---

8 Tan Feng, Fang Yu, and Jiang Tong, *Impact, Trend and Reduction Policy of Global Climatic Disasters, Annual Report on Actions to Address Climate Change (2011)*, Social Sciences Academic Press, 2011, 186–197.

CHAPTER 14

Impact of Land Use Change on Low-carbon City Construction in Shenzhen

Liu Weiling, Zhang Linbo and Zhang Lili

Abstract

Carbon emission from land use change is a human carbon source, second only to the combustion of fossil fuels. As the Chinese urbanization process is radical, carbon emissions from land use change ought to be and will be the key factor for reaching the objective of carbon emission reduction. By taking the rapid urbanization of Shenzhen as a typical case, this research estimates carbon pools, carbon sink and carbon sink potential in different types of land use, and analyzes the characteristics of carbon pools in different types of land use. It studies the urban ecosystem carbon emission/sequestration caused by the transformation of different types of land use, reveals the carbon emission features of land use change and the driving mechanism at different stages in the course of urbanization, carries out carbon sink regionalization in terms of Shenzhen land use, and puts forward suggestions for low-carbon city construction, thus providing new research ideas and scientific guidance for other Chinese cities to achieve carbon emission reduction and sink increase.

Keywords

land use change – carbon emission – carbon source/sink – Shenzhen – low-carbon city construction

* Liu Weiling, an associate researcher at the Chinese Research Academy of Environmental Sciences, mainly engaging in studies on regional and urban ecology, and landscape ecology; Zhang Linbo, director and researcher at the Eco-environmental Research Office of the Chinese Research Academy of Environmental Sciences, mainly engaging in studies on theory and method of regional and urban ecology, and ecocity construction; Zhang Lili, a postgraduate researcher at the Chinese Research Academy of Environmental Sciences.

 | DOI 10.1163/9789004280021_015

Carbon emissions from land use change is a human carbon source, second only to the combustion of fossil fuels,[1] as well as one of the key factors affecting the realization of the objective of Chinese carbon emission reduction. In the past 150 years, carbon emissions from land use change accounts for one-third of the total emissions from human activities, and a half of that from industrial emissions over the same period.[2] According to the predication by the World Wide Fund,[3] the annual carbon emissions from land use, land use change and forestry (LULUCF) may reach 1 billion tons, equal to the one-year industrial emissions of Japan. In the past 50 years, Chinese carbon emissions from land use change may have reached 10.6 Pg C, accounting for 30% of national gross man-made carbon emissions and 43% of industrial emissions over the same period.[4]

China's urbanization process has been radical, with an unprecedented speed and scale. Its urbanization level increased from 20% to more than 50% from 1978 to 2012.[5] In the next two decades, China's urban population will increase by 185 million on the basis of the population in 2011, with an urbanization rate near to 65%. About 10 million people will shift from rural to urban areas annually,[6] equal to an annual increase of one mega city (with a population not less than 10 million) or five large cities with a population of about 2 million. The expansion rate of urban land is normally greater than that of the population.[7] The elastic coefficient of the scale of Chinese urban use is up to 2.28,[8] resulting in a more acute urban land use change caused by the urban

---

1 Quay, P.D., Tilbrook, B., and Wong, C.S., "Oceanic Uptake of Fossil Fuel CO_2: Carbon-13 Evidence," Science, 1992, 256: 74–79; Houghton, R.A., and Hackler, J.L., "Sources and Sinks of Carbon from Land-use Change in China," Global Biogeo-chemical Cycles, 2003, 17: 1034–1047.

2 Houghton, R.A., "The Annual Net Flux of Carton to the Atmosphere from Changes in Land use 1850–1990," Tellus, 1999, 51B: 298–313.

3 World Wide Fund (WWF), *Noticeable Annual Carbon Emission of One Billion Tons from Land Use*. http://www.wwfchina.org/wwfpress/presscenter/pressdetail.shtm?id=967, 2009.

4 The Ministry of Land and Resources of the People's Republic of China, *Low-carbon Emission: New Subject of Land Use Regulation*. 2009-12-25, http://www.mlr.gov.cn/xwdt/jrxw/200912/t20091225-130977.htm.

5 Xu Kuangdi, *China Development Forum 2013 Special Issue, Chinese Urbanization*, 2013, (4): 52–56.

6 Jiang Xiaolan and Cheng Biding, *Research on Chinese New Urbanization Developmental Stage Characteristics and Trend, Regional Economic Review*, 2013: 130–135.

7 Johnson, M.P., "Environmental Impacts of Urban Sprawl: A Survey of the Literature and Proposed Research Agenda," Environment and Planning, 2001, 33: 717–735.

8 Liu Wenjia, *Supporting City with Village: Much Behind of Construction Land Adjustment, The Chinese Newspaper of Land and Resources*, 2007-05-03.

population surge. It is anticipated that the new urban land required will be about 50,000 km^2 by 2030,[9] equal to 0.52% of the national land territory and the area of 20 medium-sized counties. The resulting carbon emission effect should not be neglected.

At the Copenhagen Conference, the Chinese government committed itself to reducing CO_2 emission per unit of GDP by 40%–45% by 2020, and included this as a control index in the medium- and long-term planning of the national economic and social development. Studies show that as the Chinese energy structure dominated by coal is hard to be changed in the short term, it is difficult to realize the emission reduction objective only by common low-carbon control policies such as the transformation of the energy structure, improvement of energy efficiency and industrial policies; in addition, enormous costs will be required, with an annual cost increase of over 100 billion USD.[10] The potential of carbon emission reduction by optimization of the land use structure is about one-third of common low-carbon policies (mainly energy and industrial policies), therefore, land use regulation measures may relieve the pressure of greenhouse gas emission reduction faced by China to some extent, which is of paramount practical significance and of application value for China to fulfill and keep its voluntary emission reduction commitments.[11]

As a Chinese special economic zone, Shenzhen has grown itself into China's youngest megalopolis from a frontier town with over two decades of development since its founding. It has created a miracle in world industrization, urbanization and modernization. Its condensed urbanization developmental process makes it an ideal experimental field for studies on the urbanization process. By taking Shenzhen as an example, this research analyzes the carbon emissions from the urban land use change in the past 30 years, reveals the relationship among the soil factor, climate factor and biotic factor in carbon emissions from urban land use change, highlights the features of carbon emissions from different modes of urban land use transformation and clarifies the impact rule of the land use transformation mode in carbon emissions, thus providing a scientific theoretical basis for urban land development and use,

9 Zhang Yongen, Chu Qingquan, and Wang Hongguang, *China Food Security Situation and Countermeasures during Urbanization, Research of Agricultural Modernization*, 2009, (3): 270–274.

10 Zou Ji, Fu Sha, and Wang Ke, *Cost for China's Realization of Carbon Intensity Reduction Objective, Environmental Protection*, 2009, (24): 26–27.

11 *Low-carbon Emission: New Subject of Land Use Regulation, The Chinese Newspaper of Land and Resources*, 2007-12-25.

greenhouse gas emission reduction and sink increase, and the formulation of a sustainable developmental strategy, etc.

1 Carbon Pool, Carbon Sink and Carbon Sink Potential Estimation of Different Land Use Types in Shenzhen

1.1 *Soil Organic Carbon Content and the Density Feature in Different Types of Land Use*

According to the spatial distribution characteristics of soil, the situation of vegetation and land use in Shenzhen,[12] a field investigation on soil organic carbon in main land use types was carried out on Shenzhen farmland, garden plots, forest land, mangrove forests, urban arbors, urban shrubs, urban grassland, the urban landscape belt, urban construction land and urban idle land. The investigation results show that the soil organic carbon content varies greatly with the type of land use. The mean organic carbon content of soil with a depth of 0–30 cm ranges from 0.72 to 40.52g/kg in different land use types, in the decreasing order of mangrove forest, wet land, forest land, urban landscape belt, urban arbor, urban shrub, garden plot, urban grassland, farmland, urban construction land and idle land. The soil organic carbon content of all types of land use and its amount of variation both decrease with an increase in the depth of the soil. Land use change has different impacts on soil organic carbon in different soil layers, with the greatest impact on the first layer of soil than the second and third layers, and in addition, soil organic carbon with a soil depth of 0–20 cm is mainly affected.

The organic carbon density of soil with a depth of 0–30 cm ranges from 0.27 to 13.36 kg/m^2 in different land use types, in the decreasing order of mangrove forest, wet land, forest land, urban landscape belt, urban arbor, garden lot, urban shrub, farmland, urban grassland, urban construction land and idle land. Soil organic carbon density shares almost the same order as the organic carbon content in different land use types, but there are also some differences, which is mainly due to the influence of soil bulk density.

The change in land use may lead to the change in the organic carbon storage and distribution in soil. As a whole, the transformation of mangrove forests, forest land, wet land, garden plots and farmland into urban construction land will bring about a decrease in both soil organic carbon content and density,

12 Liu Weiling, Zhang Linbo, Ye Youhua, Zou Jianfeng, Wang Yizhe, and Qi Yue, *Shenzhen Soil Organic Carbon and Density Feature in Different Land Use Types, Ecological Science*, 2011, (5): 486–492.

while the transformation of urban idle land or construction land into urban green land (urban arbor, urban shrub, urban grassland, urban landscape belt) will lead to an increase in both soil organic carbon content and density.

1.2 *Carbon Storage Features and the Spatial Distribution of Different Vegetation Forms*

Current forest vegetation in Shenzhen mainly includes coniferous mixed forests (such as the masson pine, slash pine and chinese fir), broadleaved mixed forests (such as eucalyptus, castanopsis fissa, acacia Crassicarpa, acacia confusa), theropencedrymion, bamboo forests, mangrove forests, shrubby vegetation represented by baeckea frutescens, hill gooseberry, and herbosa represented by dicranopteris, ferns and saccharum arundinaceum.

As a typical region with a rapid development of urbanization, Shenzhen's forest is mostly dominated by a young and middle-aged forest. In 2005, the total carbon from the Shenzhen forest was 225.04×10^4 Mg C, of which arborous layer carbon was 165.44×10^4 Mg C, accounting for 73.5% of the total carbon; the understory carbon reserve made up 16.3% of the total carbon; the litter layer carbon reserve was the least, only 22.94×10^4 Mg C, accounting for 10.2%. Shenzhen's mean carbon density was 25.63Mg C·hm-^{2}, far lower than the national mean density of 41 Mg C·hm-^{2} and the worldwide mean density of 86 Mg C·hm-^{2} in 2000, among which, the mean carbon density of arborous layer carbon, the understory and litter layers was 18.84 Mg C·hm-^{2}, 4.17 Mg C·hm-^{2} and 2.61 Mg C·hm-^{2}, respectively. The forest carbon stock reserve has a great variation in the spatial distribution among the different districts of Shenzhen, in the decreasing order of Longgang District, Baoan District, Yantian District, Luohu District, Nanshan District, Futian District and the Protection Area; the carbon density of various districts has no obvious correlation with the carbon reserve, with the carbon density decreasing order of Yantian District, Futian District, Luohu District, Longgang District, Protection Area and Baoan District.

Shenzhen's forest is classified into 10 types, i.e., arboreal forest, bamboo forest, open forest, shrubbery, afforested land, nursery land, non-forest land, mangrove wetland, economic forest and scattered forest, of which the arboreal forest and the economic forest cover a wide area, accounting for 54.1% and 28.4% of the total forest area, respectively. The arboreal forest carbon reserve is 178.68×10^4 Mg C, accounting for 79.4% of the total carbon; the economic forest carbon reserve is 20.03×10^4 Mg C, accounting for 8.9% of the total carbon; the scattered forest carbon reserve is 15.61×10^4 Mg C, accounting for 6.9% of the total carbon; the carbon reserve of other forest types are all small, with a proportion of the total carbon not more than 5%. Carbon density varies greatly with the forest type, with a variation range from 2.33 to

37.61 Mg C·hm-2, among which, the arboreal forest has the largest carbon density, that is, 37.61 Mg C·hm-2; the next is shrubbery, 30.9 Mg C·hm-2; then followed by the scattered forest, open forest, bamboo forest, afforested land, nursery land, economic forest, mangrove wetland and non-forest land.

Shenzhen's arboreal forest carbon is mainly stored in young and middle-aged forests, with a carbon reserve of 106.89×10^4 Mg C, accounting for 73.2% of the total arboreal forest carbon reserve; the mature forest (near-mature forest, mature forest and over-mature forest) carbon reserve is 39.22×10^4 Mg C, accounting for 26.8%. The mean carbon density of the arboreal forest was 30.76 Mg C·hm-2 in the whole city, lower than the national mean carbon density of 44.91 Mg C·hm-2 in 1998.[13] The mean carbon density of the mature forest is 38.36 Mg C·hm-2, 30% higher than that of a young forest (29.14 Mg C·hm-2) and a middle-aged forest (29.41 Mg C·hm-2).

According to the type of forest, and the area of forest, the total carbon and carbon density of the seven districts, the seven districts of Shenzhen may be divided into three zones. Zone I is a high carbon reserve area, including Longgang District and Baoan District. Zone I has the largest ecological public-welfare forest and commercial forest in Shenzhen, accounting for 52.7% and 27.2% of Shenzhen's total forested area, respectively. The carbon reserve of the two forests is 123.13×10^4 Mg C and 46.70×10^4 Mg C, accounting for 54.7% and 20.8% of Shenzhen's total forest carbon reserve, respectively. The carbon density of the two forests are 26.6 Mg C·hm-2, and 19.53 Mg C·hm-2, respectively. The type of forest is mainly other hardwood forests, eucalyptus, other soft broad-leaf forests, fast growing acacia, masson pine, broadleaved mixed forests and theropencedrymion. Zone II is a middle carbon reserve area, including Nanshan District, Yantian District and Luohu District, with an area accounting for 17.6% of Shenzhen's total forested area. The carbon reserve of this zone ranges from 12.79×10^4 Mg C to 20.49×10^4 Mg C, accounting for between 5.7% and 9.1% of Shenzhen's total forest carbon reserve, and the carbon density ranges from 18.18 to 46.18 Mg C·hm-2. In Zone II, the ratio of the area of ecological public-welfare forest and commercial forest is 3:2. The type of forest is mainly mixed coniferous-broadleaf forest, other soft broad-leaf forests, other hardwood forests and fast-growing acacia. Zone III is a low-carbon reserve area, including Futian District and the Protection Area, with an area accounting for 2.4% of Shenzhen's total forested area. The carbon reserve of the Futian District and the Protection Area is 5.63×10^4 Mg C and 1.57×10^4 Mg C,

13 Fang Jingyun, Guo Zhaodi and Bu Shilong, *Chinese Terrestrial Vegetation Carbon Sink Estimation from 1981 to 2000, Science in China (Series D): Earth Sciences*, 2007, 37(6): 804–812.

respectively, with the sum accounting for 3.2% of Shenzhen's total carbon reserve, and the carbon density ranges from 37.63 to 24.19 Mg C·hm-2. Zone III is mostly dominated by ecological public-welfare forests, and the type of forest is mainly mixed coniferous-broadleaf forest, fast-growing acacia, other soft broad-leaf forests and acacia confusa.[14]

1.3 *The Soil and Vegetation Carbon Sink in Shenzhen*

Shenzhen saw a total soil organic carbon pool of 78.74 Tg C and 41.07 Tg C in 1980 and 2010, respectively, which dropped by 37.67 Tg C during the period of 30 years, with a mean annual decrease of 1.22 Tg C. Due to the sharp decrease in the farmland soil area caused by rapid urbanization, Shenzhen's paddy land and dry land with high organic carbon content have decreased by 86,400 hectares, which results in a soil organic carbon pool loss of 29.77 Tg C. The forest land and tidal land have also decreased by 10.36 Tg C and 1.26 Tg C, respectively. The only land that owns the carbon sink function is garden soil; owing to the increase in organic carbon density and area, the carbon pool has increased by 3.72 Tg C in total. However, as a whole, Shenzhen soil organic carbon pool has been a 30-year process of carbon emission.

Shenzhen saw a vegetation carbon sink of 45.464×10^4 Mg C in 2005, equal to 3.58% of carbon emissions from energy consumption over the same period. Vegetation carbon sink exerts a small effect on carbon emission offset. The arboreal forest has the largest carbon sink, accounting for 59.2% of the vegetation carbon sink. Carbon sink varies greatly within the various districts in Shenzhen. Longgang District has the largest carbon sink, 24.763×10^4 Mg C/a, accounting for 54.5% of the vegetation carbon sink, followed by Baoan District and Nanshan District. Futian District and the Protection Area have the smallest carbon sink.

In the past three decades, Shenzhen's potential vegetation carbon sink has ranged from 43.52×10^4 Mg C/a to 55.82×10^4 Mg C/a, showing an insignificant variation. The annual potential carbon intensity has ranged from 0 to 624g C/m^2.[15]

14 Liu Weiling, Zhang Linbo and Ye Youhua, *Shenzhen Forest Vegetation Carbon Reserve Features and Spatial Distribution, Ecological Science*, 2012, (2): 144–154.

15 Liu Weiling, Zhang Linbo and Gong Bin, "IBIS model-based simulation study on variation of ecological carbon sink potential in Shenzhen City," Applied Mechanics and Materials, 2013, 260–261: 981–987.

2 Research on the Carbon Emission Effect from Land Use Change in Shenzhen

As one of the regions with the fastest urbanization development in China, Shenzhen land use has experienced a significant dynamic change. In the past 30 years, especially since the 1990s, the characteristics of land use have been sharply altered. Before the 1990s, the land use in Shenzhen was mainly farmland. However, due to the emergence of large-scale urban development and construction, the area and proportion of non-agricultural land constantly increased, and the proportion of Shenzhen's urban construction land was up to 46.7% by 2010, presenting the urban landscape of an emerging and rapid urbanization.

During the period of research, the type of land use in Shenzhen did not change, but the proportion dramatically changed. The area of all types of lands increased or decreased to some extent over time. The proportion of agricultural area in the total area declined sharply from 45.9% in 1979 to 1.8% in 2010, with a decrease of 86.42×10^3 hectares. As the original agricultural area enjoys good geographical and geomorphological conditions, it is the main source for the expansion of urban construction land. The area of forest land, tidal land, grass land and unused land decreased by 18.20×10^3 hectares, 2.77×10^3 hectares, 1.1×10^3 hectares and 6.26×10^3 hectares, respectively, while the area of garden plots, water body and urban construction land increased by 7.13 hectares, 8.5×10^3 hectares and 91.15×10^3 hectares, respectively. Before 2000, the expansion of urban construction land was mainly accomplished by the appropriation of agricultural land. After 2000, urban construction land started to appropriate ecological land. The development of urbanization and the adjustment of the agricultural structure are the main factors leading to land use change.

The effect of the process of urbanization on organic carbon content varies with the type of land use. The organic carbon emission caused by the decline in agricultural land is up to 256.93×10^4 Mg C. The carbon loss in forest land and tidal land is 148.97×10^4 Mg C and 30×10^4 Mg C, respectively. The increase in garden plots leads to an increase in organic carbon by 28.89×10^4 Mg C. No apparent change occurs to the organic carbon reserve of other types of land.

In the past 30 years, the rule of change in organic carbon reserve caused by Shenzhen's process of urbanization has been as follows (see Figure 14.1): from 1979 to 1989, the regional land use change led to a slight increase in regional carbon reserve, while from 1990 to 2010, the regional land use change led to a decline in the ecosystem carbon reserve with each passing year, with a total decrease of 377.87×10^4 Mg C and a mean annual carbon emission of 17.18×10^4 Mg C. The process of Shenzhen's urbanization has been a process of carbon emission as a whole.

Shenzhen's urban construction land has generated an obvious carbon emission effect. With the continuous increase of urban construction land, carbon emissions also constantly increased (see Table 14.1), from 7.73×10^3 Mg C in 1979 to $4{,}099.93\times10^3$ Mg C in 2010, with an increase of 529 times and a mean annual carbon emission of 127.88×10^3 Mg C. Compared with the carbon emission from energy consumption over the same period, the proportion of carbon

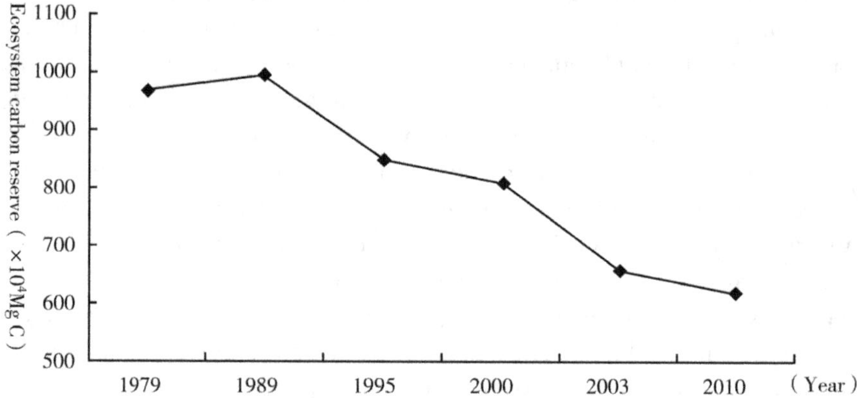

FIGURE 14.1 *Change in the ecosystem carbon reserve in Shenzhen from 1979 to 2010.*

TABLE 14.1 *Shenzhen's urban construction land area and carbon emissions from land use from 1979 to 2010*

Year	Urban construction land area (103 hectares)	Carbon emissions from land use (103 Mg C)	Carbon emissions from energy consumption (103 Mg C)	Ratio of carbon emissions from land use to carbon emissions from energy consumption (%)
1979	0.26	7.73	68.40	11.30
1989	13.43	391.54	2353.39	16.64
1995	55.16	1632.18	8758.80	18.63
2000	63.06	1837.31	16699.74	11.00
2003	70.37	2459.63	28680.12	8.58
2010	91.41	4099.93	50115.24*	8.18

NOTES: * IS 2009 DATA.

emission from land use shows the trend of first an increase and then a decrease; specifically, it increased from 11.30% in 1979 to 18.63% in 1995 and then decreased to 8.18% in 2010.[16]

3 Socio-economic Driving Mechanism Research on Carbon Emissions from the Change in Shenzhen's Land Use

3.1 *Quantitative Analysis for the Relationship Between Carbon Emission and Level of Urbanization*

Shenzhen's level of urbanization is positively correlated with carbon emissions from land use change or from energy consumption, and the correlation coefficient is 0.9106 and 0.9094, respectively. Co-integration test results show that the carbon emissions from land use or energy consumption have a long-term stable equilibrium relationship with the level of urbanization.

The short-term elasticity of carbon emissions from the change in Shenzhen's land use and energy consumption in respect to urbanization are 0.06 and 0.09, respectively, indicating that every 1% increase in urbanization brings a 0.06% increase in carbon emissions from land use and a 0.09% increase in carbon emissions from energy consumption. Carbon emissions from land use or energy consumption have a long-term equilibrium relationship with the level of urbanization. In the long run, the increase in carbon emissions from land use or energy consumption will always be accompanied by short-term improvement in the level of urbanization. Such conclusions demonstrate that for the long and short terms, Shenzhen must pay attention to carbon emissions, environmental protection and proper expansion while promoting the process of urbanization. Since their interactions have a "time delay," the relevant authorities should formulate short-term measures, and more importantly, develop a long-term strategic plan in the decision-making process.

In addition, Shenzhen's carbon emissions from land use or energy consumption have a one-way causal relationship with urbanization. The acceleration of the process of urbanization leads to the sustained increase in carbon emissions from land use, in which carbon emissions from energy consumption is the driving factor for urbanization. This means that in the course of urbanization, attention should be paid to the transformation of the economic developmental mode, the adjustment of the industrial structure, the

16 Liu Weiling, Zhang Linbo and Ye Youhua, "Study on Effects of Organic Carbon Emission of Land Use Change in Shenzhen City," Advanced Materials Research Journal, 2011, 304: 259–262.

development of low-carbon industries such as the service industry, and the promotion of resource-saving and an environment-friendly social construction. While promoting urbanization, strategic objectives of energy conservation and emission reduction should be fulfilled, thus realizing the sustainable development of China's urban economy.

3.2 *Quantitative Analysis of the Relationship Between Carbon Emissions and the GDP*

The economic growth of Shenzhen in the past 30 years is positively correlated with carbon emissions from the change in land use or from energy consumption, and the correlation coefficient is 0.9898 and 0.9971, respectively. The short-term elasticity of carbon emissions from the change in Shenzhen's land use with respect to the GDP is 0.45, indicating that every 1% increase in the GDP brings a 0.45% increase in carbon emissions from land use in the short term. The short-term elasticity of carbon emissions from energy consumption with respect to the GDP is 1.03, indicating that every 1% increase in the GDP brings a 1.03% increase in carbon emissions from energy consumption in the short term. The carbon emissions from Shenzhen's land use or energy consumption have a long-term co-integrated equilibrium relationship with economic growth, meaning that the short-term deviation between carbon emissions and economic growth will vanish, and will surely approach their path towards equilibrium in the long term.

In addition, Shenzhen's carbon emissions from land use or energy consumption have a one-way causal relationship with the GDP. Economic growth boosts the expansion of urban construction land, and further results in an increase in carbon emissions from land use; the GDP increase also leads to an increase in carbon emissions from energy consumption, revealing that Shenzhen's current economic growth mode is driven by high carbon emissions, which is the typical extensive mode of economic growth. Therefore, the realization of carbon emissions from energy consumption is sure to generate adverse effects on the economic growth of Shenzhen regardless of whether they are short or long term.

3.3 *A Convergence Study on Carbon Emissions in Shenzhen's Socio-economic Development*

Since 1979, the actual per capita GDP, per capita energy consumption carbon emissions and per capita construction land carbon emissions of Shenzhen all have shown an upward trend.

From Table 14.2, it can be seen that Shenzhen has a per capita land use CKC and per capita energy consumption CKC. When Shenzhen's per capita land

TABLE 14.2 *Estimation of the integral coefficient between per capita land use CKC and per capita energy consumption CKC*

	α	$\beta 1$	$\beta 2$	Inflexion point (RMB)	Sample number
Per capita land use	−1.069903	0.149663	−0.007290	$e^{10.26}$=28708.60	31
Per capita energy consumption	1.152962	0.841661	−0.038273	$e^{11.00}$=59604.89	31

TABLE 14.3 *Inflexion point predication of Shenzhen's energy consumption CKC and land use CKC*

	Mean annual growth rate of per capita GDP*(%)	Inflexion point (RMB)	Number of years for reaching inflexion point (year)	Year of reaching inflexion point (year)
Per capita carbon emissions from energy consumption	7	59604.89	17.7	2027 (since 2009)
per capita carbon emissions from land use	7	28708.60	6.9	2016 (since 2009)

NOTES: * BY REFERENCE TO THE *OUTLINE OF THE 12TH FIVE-YEAR PLAN FOR SHENZHEN'S NATIONAL ECONOMIC AND SOCIAL DEVELOPMENT*.

use carbon emissions are at the inflexion point, the actual per capita GDP is 28,708.60 yuan RMB. When per capita energy consumption carbon emissions are at the inflexion point, the actual per capita GDP is 59,604.89 yuan RMB.

Without considering the national carbon emission control objectives, when the actual per capita GDP is 59,604.89 RMB or 28,708.60 RMB, per capita carbon emissions from both land use and energy consumption reach their peak and then gradually decline according to the nature of the parabola and the inflexion point theory. Based on the mean annual growth rate of 7% of per capita GDP in the "12th Five-Year Plan," the time for per capita energy consumption and per capita carbon emission to reach the inflexion point is 17.7 years and 6.9 years, respectively. Then, the per capita carbon emissions from energy consumption or land use decline (see Table 14.3).

4 Shenzhen's Ecosystem Carbon Sink/Source Regionalization

According to the goals and principles of Shenzhen's ecosystem carbon sink/source regionalization and on the basis of the quantitative assessment of Shenzhen's ecosystem carbon reserve (density), carbon source/sink function and carbon sink potential difference, an ecosystem carbon source/sink regionalization was carried out through the method in the *National Ecological Function Regionalization* and the main type of land use. As a result, Shenzhen's ecosystem carbon source/sink area has been divided into six categories and 55 zones, mainly including 15 high carbon sink zones, 17 medium carbon sink zones, 10 weak carbon sink zones, four carbon neutral zones, five medium carbon source zones and four high carbon source zones (see Figure 14.2).

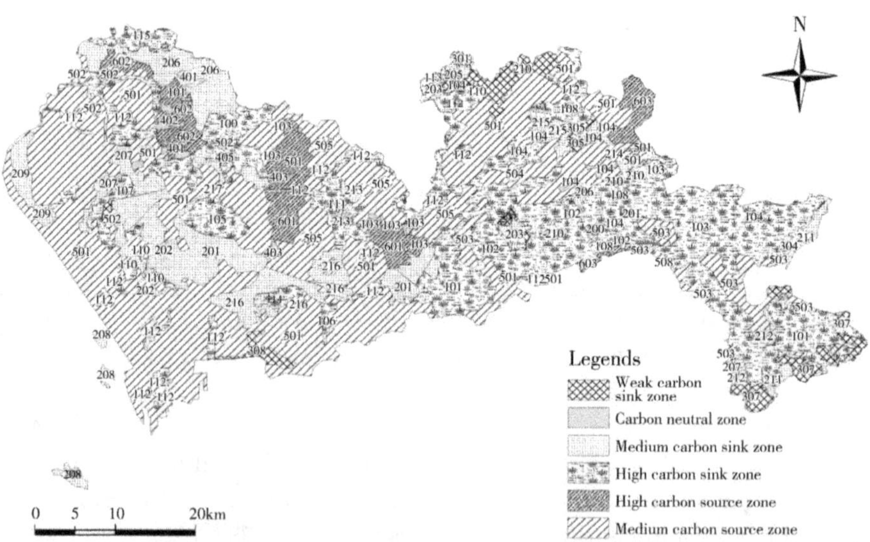

FIGURE 14.2 *Shenzhen's ecosystem carbon source/sink regionalization.*

The scheme provides technical support for realizing a reasonable use and development of urban land and a partitioned administration of the regional protection of the ecosystem carbon sink function.

5 Low-carbon Urban Construction

Low-carbon development is an inevitable choice for China to control its greenhouse gas emissions in the course of urbanization. The process of Shenzhen's industrialization and urbanization and the accumulation of social wealth require a lot of supply and consumption of land and energy. As a consequence, heavy carbon emissions are inevitable. Under the energy structure dominated by fossil energy, the emission reduction will surely impose limitations on the socio-economic development. Therefore, to construct a low-carbon city and promote a harmonious coexistence of the urban development and environmental protection, Shenzhen should explore a low-carbon land use mode and countermeasures for land use optimization, ecosystem carbon sequestration and sink increase, as well as a low-carbon economy based on its characteristics of land use, thus realizing sustainable urban development.

5.1 *Urban Land Use Optimization*

We should comprehensively enhance carbon sink capacity by the land use structure, scale, mode and layout to decrease or depress the excessive increase in carbon emissions. We should strictly implement land utilization planning to optimize the land use structure, control the surge of construction land and avoid urban sprawl, thus forming a compact, orderly and energy-saving urban land layout and improving the urban land use rate; efforts should be made to maintain a reasonable proportion of land structure, a regional carbon sink capacity and a good ecological effect; we should increase the area of forest land to relieve the downward trend of the area towards cultivated land, grassland, wetland and tidal land; we should strengthen the development of unused land, as well as land consolidation, improvement and reclamation, and improve land productivity and carbon sequestration by the reuse of land.

5.2 *Improvement in Ecosystem Carbon Sequestration and the Capacity for Sink Increase*

As a typical region with a rapid development of urbanization, Shenzhen's forest is mostly dominated by young and middle-aged forests. Shenzhen's mean carbon density was 25.63 Mg C·hm^{-2}, far lower than the national mean density

(41 Mg C·hm-2) and the worldwide mean density (86 Mg C·hm-2) in 2000. Therefore, it is necessary to energetically promote the construction of an ecological public-welfare forest and a man-made forest in the entire city in an attempt to improve the function of forest carbon sequestration.[17]

Shenzhen should work out a forest operational plan for each district, and perform classified operations and management. The main areas of contribution of Shenzhen's forest carbon reserve are Longgang District and Baoan District. The forest area of the main area of contribution accounts for 79.9% of the total citywide forest area, while the carbon reserve of the main contribution area accounts for 75.5% of the total citywide carbon reserve. Therefore, apart from the enhancement of forest management and protection to stabilize the carbon reserve of the main area of contribution, it is also necessary to enlarge the construction area of the ecological public-welfare forest to increase the density of the forest. Besides, more efforts should be made in tending the man-made forest in the main area of contribution, such as the stand improvement in the low-yield forest and in the low-efficient forest to improve the function of the forest carbon sink. The carbon reserve in the other districts is low, mainly coming from the ecological public-welfare forest. So, it is necessary to further enlarge the area of the ecological public-welfare forest, enhance the stand improvement and increase the density of the forest.

5.3 *Development of a Low-carbon Economy*

With the aim of building a "national low-carbon ecological demonstration city," Shenzhen should make a reasonable industrial plan based on its realities of short land resources, insufficient environmental capacity and ecological bearing capacity, to accomplish an efficient use of land resource, materials and energy and to avoid land waste and excessive land emission caused by repeated constructions; it should optimize the industrial structure, vigorously develop the low-carbon economic demonstration area and low-carbon ecological park, boost the circular economy, promote the transformation and upgrading of the industrial structure, carry out clean production, increase the carbon productivity of the per capita land area, and reduce the intensity of carbon emissions; besides, it should increase the ratio of the service industry in the economy, strive to develop energy-efficient industries such as finance, culture, logistics, tourism and exhibition, and hi-tech industries within the secondary industry, promote the transformation and upgrading of

17 Fang Jingyun, Guo Zhaodi, and Piao Shilong, *Estimation of Chinese Land Vegetation Carbon Sink from 1981 to 2000, Science in China Series D: Earth Sciences*, 2007, 37(6): 804–812.

labor-intensive industries and trade-processing enterprises, and make more efforts in low-carbon trade.

In the meantime, Shenzhen should enhance the cultivation of new-energy industries, with an emphasis on the acceleration of the process of industrialization of new energy dominant fields, and with a focus on the cultivation of new-energy industries such as solar energy, new energy vehicles, stored energy, biomass energy, wind power and nuclear power, so as to expand industrial scales. Moreover, to 2015, Shenzhen was expected to realize the total output value of new energy industries of over 250 billion RMB, and the total installed capacity of new energy power generation of over 8.40 million kW, accounts for over 50% of the citywide total installed capacity, so as to promote new-energy industries as its emerging pillar industry.

CHAPTER 15

Migration Planning and Its Practice for Adaptation to Climate Change: Ningxia Climate Migration Case

MENG Huixin

Abstract

The adaptation to climate change is closely associated with poverty elimination and sustainable development. Ningxia takes migration as an important measure for adaptation to climate change, which improves migrants' production and living conditions and the ecological environment to some extent. However, the approach adopted by migration engineering to address climate change and resource dilemmas also brings about new risks to migrants. Any planning for adaptation to climate change must be implemented through concerted actions at the grassroots level. The failure of migrants to adapt to new production and to a new lifestyle will make the adaptation objective of climate change hard to achieve. Therefore, to adapt to climate change efficiently, it is necessary to take into account the adaptive capacity of migrants, and attempt to explore a viable and valid governance pattern on the basis of providing social guarantees for migrants, respecting the voluntary adaptation will and actions of migrants, supporting and cultivating the mutual help of migrants' grassroots communities.

Keywords

climate change – adaptation – climate migration – synergetic development

Climate change exerts a far-reaching influence on sustainable economic and social development. IPCC pointed out in a 2001 report that, "The action of enhancing adaptability essentially means the promotion of sustainable

* Meng Huixin, a post-doctoral researcher at the Institute for Urban and Environmental Studies of the Chinese Academy of Social Sciences, mainly engaging in studies on adaptation, migration and sustainable development.

 | DOI 10.1163/9789004280021_016

development."[1] China is one of the countries that are greatly affected by climate change. Since the 1990s, extreme weather has brought about the direct economic loss of 200 billion RMB, and climate change has resulted in over 2,000 deaths annually. However, compared with the mitigation of climate change, China's climate adaptation has just started. Against such a background, the *National Strategy for Adaptation to Climate Change*, discussed and prepared by the National Development and Reform Commission, from 2011 has been basically completed. The *Strategy* includes adaptation to climate change in the regional developmental plan.[2]

In the *National Strategy for Adaptation to Climate Change*, Ningxia, located in a vulnerable climate area, is listed as an important national demonstration zone for adaptation to climate change. Ningxia has done a great amount of work in terms of adaptation to climate change. It is the first province that has included adaptation in the "11th Five-Year Plan," in which migration is an important measure for adaptation to climate change. As early as the 1980s, Ningxia has started to make plans for migration engineering, and at the end of 2010, the number of migrants has amounted to 660,000 in an attempt to relieve the aggravation of climatic drought, ecological imbalance and poverty in the central-southern area of Ningxia which is highly sensitive to climate. During the "12th Five-Year Plan" period, Ningxia has specially worked out the *Ningxia Eco-migration Plan for the Central-Southern Area during the 12th Five-Year Plan*, in which 346,000 people will be migrants as planned.[3] The research analyzes Ningxia migration polices and its practice to intensify the understanding of adaptation, which may help the practice of adaptation.

1 Ningxia Climate Risks and Climate Migration

Ningxia is located in a northwest inland area and is in a transitional zone of a semi-arid area and an arid area. The central-southern area of Ningxia has high mountains and deep rivers, where it is hard to draw water from the Yellow River, the rainfed agriculture on which peasants rely for their livelihood is at

1 IPCC, *Climate Change 2001: The Scientific Basis. Contribution of Working Group I to the Third Assessment Repot of the Intergovernmental Panel on Climate Change*, Cambridge and New York: Cambridge University Press, 2001, p. 165.

2 Zhao Chuan and Wei Wei, *Learning of Co-existence with Extreme Climates, 21st Century Business Herald*, 2013-07-04.

3 Shao Bin and Ma Zhongyu, *Ningxia Strategic Choice for Response to Climate Change, Top China*, http://www.zhyc.com.cn/inhalt.asp?id=732.

the mercy of the forces of nature, so that the area is very sensitive to climate change. In recent years, the area has suffered from drought and flood simultaneously; in particular, the drought has been further aggravated with a constant increase in frequency and area. According to the data from 1949 to 2008, the chance of drought in the central arid area is 71.9%, including a 14.0% chance of extreme terrible drought, a 15.8% chance of terrible drought, and a 42.1% chance of a slight drought. From 1991 until now, the frequency of drought has been increased, and the interval of drought at or above the terrible level (including extreme terrible drought) has been shortened, resulting in frequent occurrence. From 2000, both the frequency and the area stricken with continuous drought in spring, summer and fall have shown an upward trend. The central arid area and southern mountainous area, known as the "drought of nine years out of ten" suffers from "drought every year" now. Drought results in an ecological imbalance and a water shortage. The water resource available for use in the area is only 75.8 million m^3, so that it is one of the most arid and water-poor areas in the country. In the next 50 to 100 years, the water resources of the central-southern area will face a great challenge due to the decline in both precipitation and surface runoff, which makes the extensive cultivation of agriculture more severe. Since the 1950s, the maximum of mean crop yield per mu is only around 40 kg. With climate change, crop yield per mu presents a downward trend, only 20–30 kg, less than 1/10 of that in the Yellow River irrigation area. From the winter of 2003 until the spring of 2004, the central arid area did not see any effective precipitation for over 170 consecutive days, making it impossible to plant crops. In 2005, an extreme terrible drought occurred, the worst one in over 60 years, causing no harvest from summer and autumn grain crops in most of the central-southern area of Ningxia. As a consequence, both people and livestock lacked drinking water, which resulted in difficulty in solving the problem of food and clothing.[4]

Climatic disaster and water shortage aggravate the loss to the central-southern area of Ningxia. The annual direct economic loss caused by drought is 1.5 billion RMB at least, with 11 counties and districts in the central arid area and the southern mountainous area and with 30% of the population stricken. The region (Xiji—Haiyuan—Guyuan Region) is one of the 18 concentrated

4 Ma Zhongyu, the editor-in-chief of *Research on Ningxia Strategy for Response to Global Climate Change*, Sunshine Press of Yellow River Publishing Media Group, 2012; Chen Xiaoguang, Su Zhansheng, and Chen Xiaojuan, *Global Warming, Ningxia Climate Change and Impact, Ningxia Engineering Technology*, 2005, (4): 301–304; Shao Bin and Ma Zhongyu, *Ningxia Strategic Choice for Response to Climate Change, Top China*, http://www.zhyc.com.cn/inhalt.asp?id=732.

continuous poverty areas in China. Among the well-known "three-Xi poor regions" (i.e., Hexi, Dingxi in Gansu Province, and Xiji—Haiyuan—Guyuan Region in Ningxia Hui Autonomous Region), the region is weak in an economic and social developmental foundation, so that it is in extreme poverty, with the absolute poverty population accounting for 48% of the local population. The interaction among climate, environment and poverty intensifies the imbalance among the local population, resources, the environment and socio-economic development, further worsening the poverty. By the end of 2009, the per capita GDP of the region was only 29% of the average level of Ningxia, and 22% of that of Yinchuan; per capita net income of peasants in the region was only 72% of the average level of Ningxia, and 55% of that of Yinchuan. Based on calculations in accordance with the new poverty reduction standard, nearly a million members of the population in the region live below the poverty line, accounting for 1/6 of the total population of Ningxia. Due to ecology and population vulnerability, the region is listed by the UN World Food Programme as one of the "21 inhospitable regions."[5]

Climate change may lead to a large scale of population migration.[6] In other words, population migration is a mode for human beings to adapt to climate change—from a high climate-risky area to a low risky area, from rainfed agriculture sensitive to climate change to irrigation agriculture or non-agriculture adapting to climate change, with an aim to reducing climate risk to vulnerable populations and accomplishing climate poverty eradication.[7] As an important measure for adaptation to climate change,[8] Ningxia has organized and implemented a series of migration projects from 1980, such as village-hanging style migration, "1236 Engineering" migration, anti-poverty relocation and migration, and ecological migration in the central-southern area, with an aim to making peasants migrate from the central-southern climatic and ecological

5 Ma Zhongyu, the editor-in-chief of *Research on Ningxia Strategy for Response to Global Climate Change*, Sunshine Press of Yellow River Publishing Media Group, 2012.

6 IPCC, *Policymakers' Summary of the Potencial Impacts of Climate Change* (*Report from Working Group II to IPCC*), Geneva: IPCC Secretariat, 1990; Yu Qingnian and Shi Guoqing, *Environment, Climate Change and Population Migration, China Population Resources and Environment*, 2010 (7): 42–47.

7 Adger, W.N., "Social Vulnerability to Climate Change and Extremes in Coastal Vietnam," World Development, 1999, 27(2).

8 Of course, environmental and climatic factors may not be the necessary and sufficient conditions for population migration. In fact, environmental and climatic factors always interact with non-climatic factors such as economy, society and culture, generating a driving force for population migration. That is to say, adaptation is an important but not the only purpose for migration.

deterioration area to Yinchuan region in the southern mountainous area, and to the irrigation area of the Yellow River in the central-northern area.[9]

Migration engineering has relieved the livelihood of poverty populations effectively. Immigrant areas enjoy a relatively sufficient irrigation water source from the Yellow River, flat terrain, convenient traffic and unimpeded access to information. The good irrigation conditions enable peasants to use agricultural machineries for engaging in intensive irrigation agriculture, and cultivating more grain and commercial crops. In addition, it gives the peasants more time and allows for lower costs for migrant work. As a result, peasants not only have a stable agricultural income but also more opportunities to increase their non-agricultural income. Besides, the government puts a lot of resources into migration engineering and a centralized residence for migrants, allowing migrants to enjoy better public services including medical services, education and infrastructure. Therefore, the livelihood of peasants from the mountainous areas after migration, especially the migrants who have experienced a period of adaptation or have migrated to the northern Yellow River irrigation area, has been improved dramatically. For the migrants who migrated to areas near Yinchuan, their incomes from planting, breeding and labor are equal to eight times those in their native land.

At the same time, migration engineering improves the natural ecological restoration capacity and climate change adaptation capacity of the out-migration area. Migration relieves population pressure in the central-southern area, facilitates the natural restoration of the land that experiences excess reclamation, and the gradual restoration of vegetation, wild animals, the ecological environment and even the microclimatic environment. In 2003, three out of six groups of villages were migrated from Chanyao Village, Xiji County, with a total population of 901 in 153 households. The reclamation of barren hills and construction of terraced fields destroy the vegetation, further resulting in soil erosion and debris flow. After migration, the primeval forest, grass and wild animals, including aspen, willow, blackthorn, caragana microphylla, strawberries, bryophyte, boar and pheasant, etc., were recovered through eight years of returning farmland to forests, water resources preservation, and desertification prevention and control. Moreover, the microclimate has also been improved, manifested in more rain in summer and the formation of mountain springs.

9 Wang Chaoliang, *Village-hanging Style Migration Development—Research on the Establishment and Development of the Eco-migration Base of Hui Autonomous Region*, China Social Sciences Press, 2005; Zhang Minjian, *Practice Research on Eco-migration of the Southern Mountainous Area of Ningxia, Social Sciences in Ningxia*, 2006, (7): 66–69.

2 Resource Dilemmas and Migration Planning Governance

Migration means population relocation and resource redistribution. However, as Ningxia is mostly situated in an ecologically vulnerable area sensitive to climate change, the resettled areas still in Ningxia, even the central-southern area of Ningxia, also face the problems caused by climate change and water shortage in different degrees in spite of the flat terrain, access to water from the Yellow River and improved natural conditions. Studies show that the change in temperature and precipitation caused by global climate change exerts the largest impact on the arid areas.[10] According to feedback from residents in arid areas, precipitation has been unpredictable in recent years. Against the background of global climate change, Ningxia's temperature shows an upward trend in the past 50 years, with the mean annual temperature rising by 2.2°C, higher than the national mean level. In the past 10 years, the mean annual temperature has risen by 1.6°C, hitting the highest historical record. Especially in 2010, the mean annual temperature reached its peak. At the same time, in the past 50 years, the precipitation in Ningxia has presented a downward trend at the rate of 5.5 mm/10 yr. Besides, precipitation has great inter-annual and inter-decadal variations. Especially when water is required for crops in spring, the precipitation has generally declined. The variation in temperature and precipitation results in severe shortage of water resources in Ningxia. Climate drought leads to the decrease in inflow to the Yellow River, further resulting in a decrease in Ningxia's water supply from the Yellow River. Therefore, in spite of its location at the upper and middle reaches of the Yellow River, Ningxia has the poorest water resources in China, with per capita water resources of 198 m^3, only 1/12 the national level.[11]

So, the migration, especially intra-county migration in the central-southern area, intensifies the pressure of resources in resettled areas, and even destroys the local ecological environment. In the resettled areas in Yuanzhou and

---

10 Anderson, S., Morton, J., and Toulmin, C., "Climate Change for Agratian Societies in Drylands: Implications and Future Pathways," in Mearns, R., and Norton, A., eds., *Social Dimensions of Climate Change: Equity and Vulnerability in a Warming World*, Washington D.C.: The World Bank, 2010; Mearns, R., and Norton, A., "Equity and Vulnerability in a Warming World: Introduction and Overview," in Mearns, R., and Norton, A., eds., *Social Dimensions of Climate Change: Equity and Vulnerability in a Warming World*, Washington D.C.: The World Bank, 2010.

11 Ma Zhongyu, the editor-in-chief of *Research on Ningxia Strategy for Response to Global Climate Change*, Sunshine Press of Yellow River Publishing Media Group, 2012; Zhou Yuguo, Yang Beiqiao, and Ma Wenbo, *Ningxia Climate Change and Water Resource Sustainable Utilization, Ningxia Journal of Agriculture and Forestry Science and Technology*, 2012, (2): 70–71.

Tongxin, drinking water still mainly comes from the collection of rainfall in cisterns. In Haiyuan, villagers migrated from the mountain to the foot of the mountain. Before migration, villagers had a presentiment of drought, but the Qingshui River can still be used for irrigation. To accomplish migration, the government constructed an irrigation system. In the year of "dryland soil being reclaimed into paddy soil," the irrigation from the Qingshui River was cut off. Villagers living at the lower reaches of the irrigation system had to wait for water until irrigation in the upper reaches was completed, so that they suffered from more severe drought and water shortage. In their memories, what matters most is the drought, as they said that, "We encounter drought almost every year, and we are still in severe water shortage despite of the construction of the water channel." They have to dig wells, but the water from a deep well is salt water. What is worse, excessive well digging results in land salinization. "Before, we never dug wells. Now we have no choice but to dig wells. We have to dig wells with a depth of up to 100m and even more to find water. But the underground water is saline alkali water. Water drawn from those wells contains salt and alkali, which have a corrosive effect, resulting in a collapse of the water channel." All of these things show that, compared with the aborigines, migrants are in a more unfavorable situation.

Even in the Yellow River irrigation area in northern Ningxia, with climate change and worsening drought, the space available for migration is becoming smaller, supplemented by increasing difficulties and costs. As a consequence, the Ningxia migration system is facing the crucial problem of "how to migrate." In the climate and resource dilemmas, the Ningxia migration plan addresses climatic, environmental and resource pressures in the following ways.

2.1 *Root-cutting*

Unlike "village-hanging style migration" in the past, the "eco-migration" during the "12th Five-Year Plan" period is trying to use the root-cutting mode to restore the ecological and climatic environment of out-migration areas, develop and protect the ecological and climatic environment of the resettled areas. The word "village-hanging" refers to the traditional agricultural operational approach in the Loess Plateau dry farmland area, that is, the seasonal "flowing cultivation" or mountain reclamation production approach, in which village sites swing back and forth just like a pendulum.[12] Village-hanging migration

12 Wang Chaoliang, *Village-hanging Style Migration Development—Research on Establishment and Development of Eco-migration Base of Hui Autonomous Region*, China Social Sciences Press, 2005.

means migrants go back and forth between resettled areas and out-migration areas, and are not settled in one place, which enables migrants at the adaptation stage in early migration days, even those at the re-poverty stage, to overcome various difficulties by relying on resources in their native place. However, on the one hand, the village-hanging style migration is incapable of relieving the resource and environmental pressure of the out-migration area as soon as possible, and is not conducive to the restoration of the local ecological and climatic environment; on the other hand, the migrants' frequent return to the out-migration area makes it impossible for them to cultivate and irrigate the land in the resettled area in time, further resulting in land desertification and deterioration of the ecological environment of the resettled area. Moreover, migrants still occupy the land and water resources in the out-migration area, which is unfair to residents who did not migrate from the out-migration area. Thus, the root-cutting approach is applied in migration management, that is, water and electricity supplies are cut in the out-migration area to achieve complete migration.

2.2 *Land Concentration Operation*

The adjustment of planting pattern and operational pattern and the promotion of water-saving agriculture are the important strategies as a response to resource shortage. Water-saving technology such as drop irrigation can be applied in greenhouses, arched sheds and other planting patterns. The land is delivered to enterprises for intensive operations, which is conducive to adopting high-investment technology and equipment, and to yielding bigger harvests from scale operations. In migrant villages in the Yuanzhou District, solar greenhouses or large arched sheds are allocated to migrants, and efficient water-saving agricultural technology such as drop irrigation is promoted among the migrants by the government. In the "12th Five-Year Plan," 30,000 people have migrated to Hongsibao, and each person gets 1 mu of land, decreased by 1 mu compared with the former specified amount, which was not convenient for single household cultivation. For this situation, the government is attempting to manage land uniformly. Land is allocated in the name of the peasants, so that they have land ownership. Peasants may, by pooling the land as shares, share out bonuses each year, or earn money by labor force on the land. The delivery of land use rights to leading enterprises facilitates the adoption of the water-saving irrigation technique and equipment. Another operational approach is that peasants possess land ownership, land use rights and operational rights, but the planning pattern is under the uniform management of the government.

2.3 *Labor Migration*

Labor migration is a new approach to migration in Ningxia in recent years for adaptation to climate change under the limitation of climate and resources. Through deep development of the natural resources and socioeconomic resources, efforts are made to find any and all means of livelihood to transfer the population from the agricultural sectors that heavily rely on natural resources and are highly exposed to and sensitive to climatic risks to the non-agricultural sectors that have low exposure risk and sensitivity, thus relieving the population vulnerability in climatic risks and adapting to climate change. The 12th five-year migration plan of Ningxia plans to spend five years to complete the labor migration of 20,100 households and 86,500 people, accounting for 25% of the total task load—"Powerful measures are taken to guide migrants to work in the Economic Zone along the Yellow River, key cities and towns, industrial parks and bases. Labor service skills training will be launched for 118,000 people to strive to make one or two members of each household in the resettlement area become a migrant worker, so that the salary income of migrants increases and service income becomes the migrants' main source of income." In fact, before governmental planning, peasants had started to adapt to the climate change in the way of labor migration. According to data, limited by environmental resources, 600,000 peasants in poverty-stricken areas with a formidable ecological environment in Ningxia's central-southern area choose to leave the area to work, with an annual labor service income of more than 3 billion RMB. The proportion of labor service income in peasant per capita income is up to 40%. As a result, the labor economy drives voluntary labor migration. At present, 64,000 households of peasants in eight counties (districts) in mountainous areas have migrated to other places, accounting for 13% of all the agricultural households in the eight counties (districts). Especially in the central arid area with a formidable ecological environment, a great number of villages are now nearly devoid of inhabitants.[13]

3 Lifestyle Transformation and Guarantee Dilemma for Migrants

For migrants, response to climate change and resource dilemma governance mean full reconstruction of a lifestyle, production mode and social relationships, including the transformation of the production mode, consumption mode, social interaction and gender division of labor, such as from rainfed

---

13 Dong Ling, the editor-in-chief of *Poverty Alleviation of Xiji—Haiyuan—Guyuan Region*, Sunshine Press of Yellow River Publishing Media Group, 2012.

agriculture to intensive irrigation agriculture, even non-agricultural industries, from scattered residence to centralized residence, and from ethnic inhabitation to Hui-Han cohabitation, as well as the transformation of the reference group, social expectations and ideology due to cohabitation with aborigines in resettled areas. Life guarantee is the primary issue that migrants face.

Land concentration operation and labor migration make peasants, whose original survival foundation is land, bear market risks. Efficient water-saving agriculture poses a great challenge to migrants, who are from mountainous areas and are used to living at the mercy of the forces of nature, in the application of new agricultural technologies, such as planting of vegetables in the greenhouse and shed. The market for vegetables in the greenhouse and shed remains to be developed further. The delivery of land to enterprises for operation leaves peasants in a weak position in the face of powerful enterprises. Although labor migration planning stipulates that migrants should receive training, low-rent housing and comprehensive social guarantees to help them cope with the risks and adapt to the new work and life styles, the income and employment guarantees actually obtained by migrants are the result of the concerted efforts of three parties, namely, policy, enterprise and worker, rather than a simple written stipulation. In fact, the benefits provided by the government are mainly public services such as low-rent housing and education, while the working hours, income and basic guarantees mainly depend on the position of the migrants on the employment market. For example, new factory workers at the stage of internship or in the general operation of a new labor manner and technology have to work for a long time with low income, and even run the risk of unemployment at any time due to enterprises' production cycles or unsmooth development.

Therefore, migrants tend to take land as their survival guarantee. Peasants in Ningxia's central-southern area generally hold that the granting of land is the condition for migration. It is common for migrants to serve as labor worker and farmer simultaneously, which helps increase income to respond to climate change; they take land as a survival guarantee to cope with market risks. In spite of low agricultural gains from extensive cultivation, land, as a survival guarantee, is still heavily relied upon by peasants, including migrants. Migrants who have worked outside for many years with labor earnings as their major economic source are willing to spend money to purchase land. In their opinion, "Land means food. It is the solution to basic survival. When young, we can work for others, but when we are old, we cannot. To purchase land is to purchase insurance, which makes us feel secure." Moreover, farm work takes precedence over off-farm work. Only under the precondition of finishing farming work will peasants choose to go outside to work in case of surplus time and labor force. Peasants who have not migrated commonly maintain that, "No land, no migration. We cannot regard off-farm work as the only means of our livelihood. It may be alright when we are young. But when we are old, we cannot work outside. So, land is indispensable for us."

Migration between resettled areas and out-migration areas is an important way for migrants to adapt to a new life in early migration days. Neo-cultivated land needs a ripening process. As migrants are unfamiliar with new crop varieties and planting patterns in resettled areas, migrants may not have good harvests, so that they are worried about their livelihoods. Moreover, although the government invests a great amount of money in migration project, migrants still need to bear some expenses such as house construction costs, which is really a great burden for poor peasants from impoverished mountainous areas. Consequently, quite a number of migrants have to borrow money to complete migration. In such difficulties, a "native place" is an important resource available for new migrants to use—they can maintain their livelihood by stored grain or farming in their native place, or obtain help and support from relatives and friends in their native place. In the resettled area of the Hongsibao District, two adjacent villages have apparent disparities in living standards due to different distances from their native places. It is because migrants in the short-distance migration situation are capable of conveniently using resources from their native place while those in the long-distance migration situation are not.

Mobile employment is also a significant strategy for migrants who increase their income by off-farm work and regard land as a survival guarantee. Mobility includes the seasonal employment between farming and off-farm work, employment in different posts and regions, and the highly liquid, incidental employment such as odd jobs or casual labor at construction sites. In case of land as a survival guarantee, farming should be given the priority; in case of off-farm work to increase one's income in order to maintain one's livelihood, the primary objective of off-farm work is to maximize income instead of leading a peaceful, steady and comfortable work and life in the city. Mobility enables migrants to give consideration to both farming and non-farm work, further comparing and seeking maximum benefits.

However, the choice of the migrant is not fully accepted in the pattern of migrant governance in migration planning. Mandatory root-cutting will stop migrants from migrating between resettled areas and out-migration areas, and utilizing homeland resources and social support to get through the period of adaptation to migration. Through the governance patterns of the counterpart-assistance in out-migration areas and resettled areas and the labor migration industrial park setup, labor migration planning strictly stipulates that migrants must take up employment stably in a specific region and industry, with a disapproval of migrants' autonomous adjustment in farming and off-farm work in different regions, industries and posts. In addition, the planning specifically stipulates the scope of the out-migration area, to the accuracy of county, township, village and group. As a result, a small number of workers, who are willing to take up employment stably in a special region and industry after the individual-enterprise two-way selection and adaptation process, are not eligible for migration.

Consequently, both migrants and migration planning face a number of difficulties. Migrants without a fundamental survival guarantee live a difficult and miserable life. Though the government invests vast resources in migration project, it is still hard for migrants to settle down in the way as planned, not to mention the realization of the objective of adapting to climate change by population migration. In many new migrant villages, as lots of migrants fail to adapt to their new life, the colonization rate is low, and even not up to 30%. In labor migration resettlement areas, even the low-rent housing cannot attract migrants to settle down. Moreover, migrates who do choose to settle down are also not confident in their future urban life due to an unstable income and job. Very few people are willing to hand over their land in their native place five years later, and transfer rural household registration to urban household registration.

4 Promoting the Consistency between Migration Planning and Action

Ningxia uses migration project to adapt to climate change, which improves migrants' production, living conditions and the ecological climate environment to some extent. But, the approach used by migration project to address climate change and the resource dilemma forces migrants to undergo thorough reconstruction in social and life structures and face new risks. If migrants fail to adapt to new production and life modes, the objective of adapting to climate change will accordingly be hard to be achieved. Therefore, to efficiently adapt to climate change, it is necessary to consider migrants' adaptability and explore a more pertinent, viable governance pattern.

Agrawal has pointed out that as government and other public sectors have the power to command official actions and have the capacity to guide the investment directions of technology and fund in the course of adaptation to climate change, they are very likely to promote the adaptation strategies relevant to sharing and diversification. Therefore, national or governmental planning adaptation is indispensable and always dominant in practice. However, all external support will get results only by the adaptation actions at the grassroots level. Unfortunately, in the practice of adaptation, the adaptation actions at the grassroots level have neither received enough attention nor been included in the plans for adaptation. What is worse, the external support for the grassroots level to adapt to climate change is mostly limited to information and funds, with very little support in cultivating, forming and enhancing the adaptability of the local society.[14]

In the migration practice of Ningxia, the first issue faced by migrants is the guarantee of income and employment in the transformation of the production mode, which was ignored by social resources and support in the past. Therefore, to improve migrants' adaptability, it is necessary to intensify the governance pattern by carrying out specific actions, such as providing social guarantees for migrants, respecting migrants' adaptation will and action, supporting and cultivating mutual help among grassroots communities.

First of all, the first issue faced by migrants is the guarantee of income and employment. Migration planning should further intensify the pattern of social guarantees for migrants. From the history of world modernization and industrialization, social guarantee is used expressly to address the instability of income and employment caused by the macroeconomic situation, production seasonality, enterprise business performance, personal health and age.[15] Ningxia migration planning invests a large quantity of economic and social resources in migration project, and stipulates a comprehensive social guarantee to migrants. However, due to lack of specific policies (such as a guarantee policy for seasonal unemployment), no support is provided for the benign interaction between enterprises and migrants under market risks. So, migration planning needs further policy design to differentiate between universality and selectivity of the guarantee mode;[16] in terms of target orientation of selective guarantee (subsistence allowances such as low-rent housing), migration

14 Agrawal, A., "Local Institutions and Adaptation to Climate Change," in Mearns, R., and Norton, A., eds., *Social Dimensions of Climate Change: Equity and Vulnerability in a Warming World*, Washington D.C.: The World Bank, 2010.

15 Gilbert, Neil, and Terrell, Paul, *Dimensions of Social Welfare Policy*, translated by Huang Chenxi, Zhou Ye, and Liu Hong, East China University of Sciences and Technology Press, 2003.

16 In terms of target orientation, there is universality and selectivity in benefit qualification. The former refers to fundamental rights enjoyed by all citizens in common life issues, such as endowment, medical treatment, education and other social insurance or public services; the latter refers to the income guarantee or social relief offered to the poverty-stricken population who meets standards in accordance with income checks and other procedures. In terms of social control, physical distribution might be used to restrict the purpose of some reliefs, or the acceptance of education and training might be the precondition for receiving relief. Please refer to Gilbert, Neil, and Terrell, Paul, *Dimensions*

qualification based on the precise scope of out-migration areas is replaced by the guarantee standard based on reasonable poverty line of the off-farm work area, which is conducive not only to fair treatment of migrants from the central-southern mountainous area, but also concentration of resources to help poor migrants adapt to urban life; the fundamental social guarantees, such as endowment and education, which are supposed to be universal guarantees, is not granted based on beneficiary qualification, so that all migrants from the central-southern mountainous area can enjoy the treatment as urban citizens, and the migrant laborers, who voluntarily go outside to work for years and have the ability and intention to lead a stable work and life pattern in urban areas, can complete the last step of non-agriculturalization transformation.

Secondly, in terms of control of migrant labor actions, efforts are made to explore an effective way to manage population mobility. In the process of the reconstruction of production and life styles, the mobility between resettled areas and out-migration areas, and in different posts and industries in different resettled areas is an important approach and strategy for migrants to follow in order to seek adaptation space and social support, and is also a universal phenomenon in the process of population migration. In fact, population mobility is an important feature in modern society.[17] The regulation of Ningxia's migration planning on migrants' settlement and out-migration actions disapprove of the autonomous adaptation of migrants through mobility, intensifying the instability of resettlement and employment and the lack of social guarantees. Certainly, in the course of pilot labor migration, migration policies relax county-county restrictions, and implement the governance pattern of managing migrants based on a labor migration certificate, showing that migration planning is also undergoing constant adjustment in practice. On such a basis, the governance pattern intensifying the labor migration certificate and gradually relaxing the regulations on migrants' work place, occupation and industry, not only allow migrants to autonomously select and adapt through mobility in regions and occupations, but also introduce supporting policies to provide fundamental income and employment guarantees for migrants in occupational mobility.

At last, a variety of folk fraternity and NGOs are supported and fostered. Agrawal's analysis on 118 cases in 46 countries shows that most local social organizations involved in adaptation to climate change are non-official

---

of Social Welfare Policy, translated by Huang Chenxi, Zhou Ye, and Liu Hong, East China University of Sciences and Technology Press, 2003.

17 Wang Xiaoyi, *From Swing to Mobility: Adaptation in the Process of Population Migration*, *The Journal of Jiangsu Administration Institute*, 2011, (6): 50–54.

organizations, but they play a vital role in skill cultivation and capacity build-up for adaptation to climate change.[18] In Ningxia, migrants also have various modes of mutual help. As recalled by migrants who moved to Hongsibao ten years ago, in the early stage of migration, it is their relatives and friends who moved to Hongsibao earlier who teach new techniques to them, and then the new techniques are spread to others; it is their relatives and friends who provide social support such as grain and money to help them get through the transitional period. Therefore, while the "root-cutting" approach is adopted to maintain the ecological environment and utilize resources efficiently and fairly, substitutive measures should be taken to support and foster the development of other types of folk mutual help.

---

18 Agrawal, A., "Local Institutions and Adaptation to Climate Change," in Mearns, R., and Norton, A., eds., *Social Dimensions of Climate Change: Equity and Vulnerability in a Warming World*, Washington D.C.: The World Bank, 2010.

CHAPTER 16

Constructing a Low-carbon Urban Area, and Promoting Intelligent Development: Experimentation and Practice in Xiacheng District, Hangzhou City

Bao Jianqiang and Ye Ruike

Abstract

Low-carbonization and intelligentization of the urban areas of metropolises are two trends of future urban development, and experimentation with and promotion of a more low-carbon, more intelligent, cleaner and more convenient kind of urban development are epochal propositions. In recent years, when promoting low-carbon urban development, Xiacheng District, Hangzhou City has combined intelligent urban construction with exploring a new way of low-carbon and intelligent urban reciprocal development by creating new samples and new demonstrations of a "Beautiful Hangzhou"—"Beautiful China" as goals. Xiacheng District has taken the lead in experimentation and practice: first, it has carried out lots of basic work in the accounting field of carbon emission of the urban area; second, it has accumulated a lot of new experience in creating a low-carbon urban area; third, it has acquired a new way of thinking and new experimentation in low-carbon and intelligent urban reciprocal development.

Keywords

low-carbon urban area – intelligent urban area – carbon emission – digitization – intelligentization

* Bao Jianqiang, dean of the College of Politics & Public Administration of Zhejiang University of Technology, whose research direction is green low-carbon development; Ye Ruike, deputy director of the Green Low-carbon Development Research Center of Zhejiang University of Technology, whose research direction is low-carbon economy.

 | DOI 10.1163/9789004280021_017

In 2008, Hangzhou City took the lead in proposing the strategic idea of constructing a "low-carbon city." On December 28, 2009, Hangzhou held the mobilization meeting of constructing a low-carbon city; on December 29, 2009, the CPC Hangzhou Municipal Committee deliberated and passed the *Decision on Constructing a Low-carbon City* in order to construct a low-carbon city with the "six in one" concept, including low-carbon economy, low-carbon building, low-carbon transportation, low-carbon life, low-carbon environment, and low-carbon society.[1] In July 2010, eight cities, Hangzhou, Tianjin, Chongqing, Shenzhen, Xiamen, Nanchang, Guiyang and Baoding, became the first batch of low-carbon pilot cities. Xiacheng District is located in the central urban area of Hangzhou City, with a regional area of 31.46 km^2, a registered population of 403,700, and a permanent resident population of 526,100.[2] In 2012, the GDP in Xiacheng District was 59.88 billion RMB, with per capita GDP of 18,000 USD. In 2010, Xiacheng District formulated *the Development Plan for the Creation of a Low-carbon Urban Area in Xiacheng District (2010–2015)*; in 2012, Xiacheng District finished the investigative report: *Using Scientific and Technical Means, and Innovating the Social Management—Research and Practice of Digitization and Intelligentization of the "66810" Service-for-the-People System of Xiacheng*. When promoting low-carbon urban development, Xiacheng District combined with the intelligent urban construction and developmental trend to find a new way of creating a low-carbon and intelligent city.

1 The Urban Carbon Emission Accounting System: Quantitative Basis of the Low-carbon Developmental Decision in Xiacheng District

The first thorny problem in creating a low-carbon urban area is the accounting of the total carbon emission and the assessment of the carbon emission structure and characteristics in the region, and the mastering of the emphasis, difficulty, potential and direction of carbon emission in urban areas of metropolises. With the help of the Green Low-Carbon Development Research Center of Zhejiang University of Technology, based on the method of the formulation of a greenhouse gas emission inventory of the UN IPCC, according to the characteristics and laws of urban carbon emissions, and combined with the economic and social developmental conditions, the industrial structure

1 Hangzhou Municipal Government, *Decision on Constructing a Low-carbon City* (S. W. (2009) No. 37), 2009.

2 Hangzhou Bureau of Statistics, *Hangzhou Statistical Yearbook 2012*, Hangzhou Statistical Information Net, December 21, 2012.

and the characteristics of the statistical system of Xiacheng District, the Science and Technology Bureau of Xiacheng District carries out research on the statistical and monitoring methods of regional energy and carbon emission suitable to the existing statistical system, forming corresponding accounting methods and tools.

1.1 *The Main Fields of Energy Consumption and Carbon Emission in Xiacheng District*

As a "final energy consumption region" and a "consumption-type carbon consumption region," Xiacheng District mainly uses "primary energy" (gasoline/diesel, natural gas) and "secondary energy" (electric power), so carbon emission sources in Xiacheng District (hereinafter referred to as "carbon source") should include direct carbon emissions and indirect carbon emissions. Theoretically, carbon sources mainly include the following parts: (1) Industrial carbon source: mainly accounting for carbon emissions in the industrial field and in the industrial process; (2) Traffic carbon source: mainly accounting for carbon emissions from social vehicles and family cars in cities; (3) Household carbon source: mainly accounting for direct and indirect carbon emissions coming from consumption of water, electricity and gas by urban residents; (4) Building carbon source: building carbon sources excluding most hospitals, schools, hotels and shopping malls, but including those hotels, shopping malls and other buildings with office premises and office buildings. The carbon emission statistics of Xiacheng District normally cover only the related premises within the building area of more than 5,000 m²; other premises are classified into other carbon sources, mainly accounting for direct and indirect carbon emissions of office premises, office buildings and other buildings. (5) Other carbon sources: mainly including hospitals, schools, hotels, shopping malls, and roadside stores in the urban area. (6) Afforestation carbon sink: carbon sink accounting for an afforested area, greenbelt area and plants in the urban area. The total carbon emissions can be expressed as follows:

$$M_{\text{total carbon emissions}} = M_{\text{Industrial carbon source}} + M_{\text{Traffic carbon source}} + M_{\text{Household carbon source}} + M_{\text{Building carbon source}} + M_{\text{Other carbon sources}} - M_{\text{Afforestation carbon sink}}$$

1.2 *Accounting Method of Carbon Emissions in Xiacheng District*

The inventory or accounting method proposed by the IPCC is not suitable to the carbon emission accounting of cities or urban areas. Many large cities have no power supply systems (fixed carbon source) of independent and fully self-sufficient power plants, although cities are the major clients of energy consumption and carbon emissions. Carbon emission accounting of cities or urban areas needs to extend to the final energy consumption, from top to bottom. Meanwhile, due to the inconsistency between the statistical data of the final energy consumption and the statistical system of carbon emissions, Xiacheng District uses the method that combines the statistical data with direct investigation data, to carry out lots of investigations and measurements on the accounting of total CO_2 emissions of the region. The collecting method and accounting method of the measured data of total carbon emissions in Xiacheng District are divided into three parts: the first one is the statistical data, focusing on the statistical bureau, supplemented by departmental data, with reference to sampling survey data; the second is special investigation, carrying out fixed-point special investigations in shopping malls, buildings, power supply bureau (Nanbei), communities, etc.; and the third is the investigation via sampling questionnaires.

1.3 *Data Acquisition of Energy Consumption and Carbon Emission Accounting in Xiacheng District*

1.3.1 Industrial Department

Industrial carbon sources in Xiacheng District mainly involve two aspects, i.e. the energy activity and the industrial production process, and they are divided into direct energy consumption carbon source and indirect energy consumption carbon source. Method 1: direct carbon emission and primary energy data can be acquired from the statistical data of the statistical department, and the indirect carbon emission data, i.e. the electric power data, can be acquired through the electric power department (Nanbei Power Supply Bureau). Method 2: there are a few industrial departments in Xiacheng District, so the energy consumption data and the power consumption data of the related industrial enterprises can be acquired through the census method. Method 3: the two methods above can be adopted at the same time to check and correct the data.

1.3.2 Transportation Department

For the traffic carbon source—cars, Xiacheng District adopts the principle of territory to only count the vehicles registered in the jurisdiction (i.e. the ownership belongs to residents or related departments in Xiacheng District), without considering through traffic. This method acquires the vehicle data directly from the vehicle administration, the mileage through sampling surveys, converts the data to gasoline consumption, then to the average carbon emission according to related emission factors, and finally obtains the total carbon emissions of the transportation field (vehicles) by addition-multiplication.

1.3.3 Construction Department

The building carbon emission accounting has two main parts. Part 1 is the direct carbon emission produced in gas use, and there are three methods for the acquisition of data. Method 1: it acquires the related data directly from gas companies; Method 2: it adopts the data census method to research the data of related buildings since the number of buildings using gas is small and stable; Method 3: the two methods above can be adopted at the same time to check and correct the data. Part 2 is the indirect carbon emission produced in electric power use, and there are three methods of data acquisition. Method 1: it acquires the data regarding electric power consumption through the electric power department (Nanbei Power Supply Bureau), and then carries out the accounting according to carbon emission factors of electric power consumption; Method 2: by means of sampling survey and average accounting; Method 3: the two methods above can be adopted at the same time to check and correct the data.

1.3.4 Domestic Energy Consumption

Domestic energy consumption, and household energy consumption and carbon emissions are an important part of the energy consumption and carbon emissions in Xiacheng District. Household energy consumption is mainly composed of three elements. The first is household electricity consumption. Method 1: it acquires the related data directly from the electric power department; Method 2: since the electric power measurement and statistics of the electric power department are not divided by the administrative districts, and separate statistics are not carried out by residential areas, the data of household electricity consumption can be acquired by means of sampling questionnaire investigations, and then calculating the average accounting; Method 3: the two methods above can be adopted at the same time to check and correct the data. The second is household gas consumption. Method 1: it acquires the data directly from gas companies; Method 2: it adopts sampling surveys and then calculates the average accounting; Method 3: the two methods above can be adopted at the same time to check and correct the data. The third is the gasoline consumption of private cars; see "Traffic carbon source" accounting part for details.

1.3.5 Afforestation Carbon Sink

By referring to the average value of a mass of forest dry matter recommended by the IPCC and the afforestation condition of Xiacheng District, this method directly adopts the related statistical data to acquire the urban green coverage ratio of Xiacheng District, and calculates the afforestation carbon sink according to 850–950 kg/year for annual carbon sequestration capacity of 1 hectare of urban trees.

1.4 *Accounting Result of Energy Consumption and Carbon Emissions in Xiacheng District*

According to research results of the *Developmental Plan of Creating a Low-carbon Urban Area in Xiacheng District (2010–2015)*, after accounting research on carbon emissions in Xiacheng District (including direct emissions and indirect emissions), the total carbon emissions in Xiacheng District have slightly increased in recent years, as shown in Table 16.1, but the intensity of carbon emissions has shown a declining trend: the intensity of carbon emissions of ten-thousand-yuan output value (GDP) was 0.441 ton per ten thousand yuan in 2009, 0.409 ton per ten thousand yuan in 2010 and 0.351 ton per ten thousand yuan in 2011. Based on calculations with the permanent resident population of 526,000, the per capita carbon emissions in Xiacheng District was 3.41 tons/person in 2009, 3.58 tons/person in 2010 and 3.63 tons/person in 2011,[3] with a significant advantage compared with the USA, the EU and China's average value, respectively 17.2 tons/person, 7.3 tons/person and 6.6 tons/person.[4]

TABLE 16.1 *Accounting of carbon emissions through different ways in Xiacheng district*
Unit: 10,000 tons

Year	Industrial carbon source	Traffic carbon source	Building carbon source	Household carbon source	Other carbon sources	Afforestation carbon sink	Total carbon emissions
2009	15.3	51.6	39.2	24.3	49.6	−1.2	178.8
	8.6%	28.8%	21.9%	13.6%	27.7%	−0.6%	100%
2010	9.1	56.9	41.72	24.5	57.5	−1.2	188.5
	4.9%	30.6%	21.1%	13.2%	31.1%	−0.9%	100%
2011	8.3	58.8	43.1	25.6	56.8	−1.2	191.4
	4.3%	30.7%	22.5%	13.4%	29.7%	−0.6%	100%

3 The People's Government of Xiacheng District and Zhejiang University of Technology, *Developmental Plan of Creating a Low-carbon Urban Area in Xiacheng District (2010–2015)*, 2010.

4 UK Tyndall Center for Climate Change Research, *Challenge in Maintaining Global Temperature Rise Lower than 2°C*, Nature Climate Change, 2012-12-03.

2 Tripartite Linkage Among Scientific and Technological Innovation, Economy and Community: Summary of the Experience in Low-carbon Urban Area Construction in Xiacheng District

Xiacheng District has basically achieved three "low-carbon transformations" during the creation of a low-carbon urban area. First, the enhancement of scientific and technological innovation, the active development of strategic emerging industries conforming to urban developmental characteristics, such as intelligent power grids, the Internet of Things, intelligent city systems and new-energy industry, and the promotion of the low-carbon transformation of urban industries. Second, making modern service industries big, strong and fine so as to achieve the low-carbon transformation of the urban economy. Third, by propagandizing the concepts of "promoting low-carbon development, and creating a low-carbon urban area," guiding the public to participate in "low-carbon actions" to achieve the low-carbon transformation of the urban society, becoming a demonstration and example for national "low-carbon urban area" construction.

2.1 *Scientific and Technological Innovation: Source of Motivation for Low-carbon Development*

During the "11th Five-Year Plan" period, the comprehensive strengthening of science and technology in Xiacheng District was enhanced noticeably, becoming the national advanced district with scientific and technological progress for two consecutive times. The district introduced scientific research institutions to jointly build 19 technological innovation carriers, and signed 163 industry-university-research cooperation projects and 60 project financing activities, such as the national technological innovation fund, where 6 achievements won the national award for scientific and technological progress. In 2010, in the first batch of plan projects of scientific research on new product organized and implemented in Xiacheng District, five scientific and technological projects related to energy conservation and emission reduction (carbon) were approved with priority, with 475,000 RMB funding. In 2011, Xiacheng District increased its scientific and technological input in Xinghuo E-commerce Industrial Park and Zhejiang Guangyuan Technology Building, and promoted the construction of "two zones and multiple parks." Now, there are 81 hi-tech enterprises at the municipal level or above and 43 provincial-level software enterprises in the district; hi-tech industries of industrial enterprises above the designated size in the district realized a sales value of 4.408 billion RMB and a value added of 1.516 billion RMB over the whole year; the hi-tech output value accounted for 56.55% of the industrial output value of the district, and the value added of the hi-tech industry accounted for 74.62% of the industrial value added of the district. In 2011, the incubation area of the scientific innovation center of the district was expanded to 22,000 m^2, the number of incubating enterprises was up to 111, and seed funding was provided for 19 Internet of Things and incubation projects of energy-saving and environmental. In 2012, the district launched the creation of Zhejiang innovative pilot urban area, established the Postdoctoral Innovative Entrepreneurial Base of Zhejiang University of Technology, increased the scientific and technological incubation area by 5,000 m^2, newly cultivated six national hi-tech enterprises, 28 municipal-level hi-tech enterprises and R&D centers, with a ratio of output value of new products of the industrial enterprises reaching over 32.08%. The district was awarded the title of provincial intellectual property demonstration area.

2.2 *"2+1" Modern Industrial System: Economic Guarantee of Low-carbon Urban Area*

During the "11th Five-Year Plan" period, Xiacheng District further constructed a "2+1" modern industrial system, and the industrial structure was adjusted substantially. Through industrial structural adjustment and "suppressing the second industry and developing the third industry," 121 industrial enterprises above the designated size were moved out successively, reducing 379,000 tons of carbon emissions, accounting for 20.4% of the total carbon emissions in the district.[5] The industrial structural adjustment provides the developmental space for the third industry and modern service industry of Xiacheng District. By implementing the "service-oriented" urban economic development strategy, Xiacheng District has successfully created "4 centers" of commerce and trade, finance, exhibition, and culture, realizing the transformation from a "large district of service industries" to a "strong district of service industries." In 2012, Xiacheng District realized a value added of the service industry of 54.315 billion RMB, with a year-to-year growth of 9.35%, accounting for 90.71% of the gross regional domestic product, and the modern service industry accounted for 68.17%. The district won the comprehensive excellence award of modern service industries, and Wulin Commerce and Trade Service Industry Cluster was awarded the Zhejiang Modern Service Industry Demonstration Cluster title. The E-commerce Industrial Park was named Municipal Excellent Hi-tech Park for three years in a row (2010–2012), the demonstration experience center of

5 Bao Jianqiang, et al., *Develop Low-carbon Economy and Create a Low-carbon Urban Area—Exploration Practice and Theoretical Thinking of Promoting Low-carbon Development and Creating a Low-carbon Urban Area of Xiacheng District, Hangzhou City*, Proceedings of the 2010 China Sustainable Development Forum, 2010.

the Internet of Things was used to its fullest, and Heping Steel E-commerce Trade and Service Project was listed among the national E-commerce pilots. In the meantime, the building economy has become a new highlight and a new engine of the third industry and modern service industry of Xiacheng District, and its role of main platform of buildings community serving economy became obvious. In 2012, 114 target buildings realized a tax revenue of 6.7 billion RMB, with a year-to-year growth of 15.52% and tax contribution per square meter reaching 3,540 RMB; 67 buildings realized tax revenues exceeding 10 million RMB, 17 buildings realized tax revenues exceeding 100 million RMB, wherein two buildings realized tax revenues exceeding 500 million RMB,[6] becoming the first "China Headquarters Economic Development Practice and Research Base" of Hangzhou City.

2.3 *Construction of Low-carbon Communities: Foundation of the Masses for the Construction of a Low-carbon Urban Area*

The first important campaign for the construction of a low-carbon urban area in Xiacheng District is to construct "low-carbon communities." The district adopts a working mechanism of "government promotion, community subject, department links, participation by all people," and establishes the "low-carbon community" construction work steering group made up of the leaders of the Science and Technology Bureau of Xiacheng District, the Association for Science and Technology of Xiacheng District, the Environmental Protection Branch of Xiacheng District and all sub-districts. All pilot communities form a pilot work-leading group led by the main leaders of the communities, with specific responsibilities and a clear division of labor, so as to do a good job in the pilot projects. The Science and Technology Bureau of Xiacheng District, the Association for Science and Technology of Xiacheng District, the Sustainable Development Experimental Area Office of Xiacheng District and the Ecological Zone Construction Office jointly issued the *Notice on Carrying out a Pilot Low-carbon Community Construction*. Through investigation and survey, Xiacheng District has formulated "low-carbon community" construction reference standards and the "low-carbon family" reference standards of the district. By setting up 300,000 RMB special funds of low-carbon community construction for special purposes, Xiacheng District has made communities become the construction subjects; it has formed joint forces for the construction by department links; it has formed mass participation in low-carbon community construction by all-people participation activities such as low-carbon family

6 *Government Work Report*, the second session of the 14th People's Congress of Xiacheng District, Hangzhou City, 2013-02-02.

rating; and starts with low-carbon scientific popularization activities, such as holding low-carbon knowledge seminars and expert lectures, printing and distributing low-carbon publicity materials, and carrying out youth low-carbon actions, to allow the "low-carbon concept" to enjoy popular support, forming a complete low-carbon community construction atmosphere.

3 Interaction between Low-carbonization and Intelligentization: Trend of Creation of a Low-carbon Urban Area in Xiacheng District

Low-carbonization and intelligentization are two future developmental trends of urban areas of metropolises. Particularly, with the advent of the big data era, the era requires cities to be more low-carbon, more intelligent, cleaner and more convenient. In the experimentation in low-carbon urban construction, Xiacheng District especially focuses on the interaction between low-carbonization and informatization, and combines the construction of a low-carbon city and an intelligent city to adapt to changes in the new situation.

3.1 *Interactive Construction of Low-carbon Buildings and Intelligent Buildings*

According to statistics, the energy consumption of urban buildings accounts for 40% of urban energy consumption and carbon emissions. There are totally 103 commercial buildings above 5,000 m^2 in Xiacheng District, with the total building area of 2.59 million m^2 and the carbon emissions of buildings accounting for over 20% of the total carbon emissions in the whole district, and accounting for over 60% if the other buildings are included such as residential buildings, schools, hospitals, and government agencies.[7] Energy conservation and carbon emission reduction of buildings are imperative and have a great potential; promoting research and the formulation of energy-saving and low-carbon standards for buildings and combining the Internet of Things, cloud computing and other hi-tech means to create intelligent buildings are a basic approach and way to promote the energy conservation and the reduction of carbon emission in buildings. In recent years, Xiacheng District has made some achievements in the interactive field of low-carbon buildings and intelligent buildings.

7 Ye Ruike, et al., *Promote Building Energy Conservation and Carbon Emission Reduction, and Develop Low-carbon Building Economy—Typical Research Based on Targeted Commercial Buildings in Xiacheng District, Hangzhou City, Future and Development.*

First, Xiacheng District achieved rapid results in energy conservation and carbon emission reduction of buildings, and a batch of demonstration low-carbon buildings and projects have sprung up. Through scientific planning, Xiacheng District has promoted the green concept and low-carbon technology, and some enterprises and buildings have formed a relatively systematic working mechanism of energy conservation and emission reduction, with the appearance of a batch of typical advanced buildings and projects. Intime, Zhejiang Daily, Canhigh and other buildings have acquired obvious economic benefits and social benefits by virtue of new technology and new equipment after their technological transformation, such as low-carbon buildings represented by the Intime Department Store as well as energy conservation and carbon emission reduction projects represented by the ice storage of Biaoli Mansion, online energy consumption monitoring and the roof garden of Zhejiang Grand Hotel, energy-saving central air-conditioning and the underground water-free car washing of Canhigh Center, the electric balance test, the frequency conversion escalator and pump transformation at Intime Department Store. Low-carbon practices and technical improvement projects of energy conservation and carbon emission reduction of the buildings play a benchmarking and demonstrative role in the construction of low-carbon buildings in the district and even in the whole country.

Second, Xiacheng District has made progress in the interaction between the construction of low-carbon buildings and the intelligentization of buildings. In March 2011, Xiacheng District issued and implemented the *Financial Supporting Policies for Promoting the Economic Development of Buildings in Xiacheng District*, encouraging the creation of "green buildings" and intelligent buildings, and providing funds for investment subjects implementing the overall renewal and renovation of buildings according to 5% of the actual amount invested. Take Guangli Mansion as an example, a national demonstration building with the application of new technology and new materials. The mansion has a total site area of 8,820 m^2, 26 floors of main buildings, five floors of podium buildings, and three floors of basement, with the total building area of 61,280 m^2. With a storey height of 3.7–3.8m, the mansion uses the world advanced Hitachi layered central air-conditioning and double-pane hollow glass curtain wall to keep a constant temperature, wherein the Hitachi layered central air-conditioning has excellent energy-saving technology, and greatly improves the energy utilization rate by using a DC variable-speed compressor, stepless frequency conversion technology of the compressor, DC fan motor and "2 in 1" new exchanger with a subcooled loop. The 4,600-m^2 building is equipped with eight intelligent elevators, the most advanced in China; the parking proportion reaches the international standard of 150:1, which takes the lead in the office buildings of Hangzhou City.

Third, Xiacheng District has had a highly effective try in the field of development and utilization of new energy. Phase I project of 60 kW photovoltaic power generation of Zhejiang Daily Mansion in Xiacheng District is the first pilot and demonstration base of photovoltaic application in China. The project passed the acceptance of Hangzhou Electricity Bureau in March 2011 and has been added to the grid. The project was financed and constructed jointly by Zhejiang Daily Press Group and ERA Solar Co., Ltd.; the average annual generating capacity is about 60,000 kWh. By using the existing equipment to update, Hangzhou Electricity Bureau has the function of uploading the data of the user information collection system in a wireless way, saving millions of yuan. The photovoltaic power station runs with the grid connection but in the offline mode, and the electric energy produced is only used inside the Zhejiang Daily Mansion, without transmitting the power to the external power grid. The power station runs with the connection to the power grid of the electric power system and is the dynamic supplement of the power grid; the electric energy produced by the photovoltaic power generation system is used with priority.

3.2 *Interactive Development between Low-carbon Transportation and the Internet of Things*

Achievement of low-carbon transportation needs the technological support of the Internet of Things to form the "4 operation modes of low-carbon and intelligent transportation." The first is the low-carbon monitoring mode based on an intelligent environment; the second is the low-carbon consumption mode based on an intelligent energy network; the third is the low-carbon transportation mode based on intelligent transportation; and the fourth is the low-carbon operation mode based on intelligent management. In the process of establishing a preponderant position of urban public transportation in urban transportation and creating a low-carbon and intelligent urban transportation system, Xiacheng District has taken the following specific measures.

First, Xiacheng District created a "low-carbon and intelligent transportation," and adapted it to the "2 rails" new development. When constructing a bicycle and a walk system and further improving the main roads, secondary roads, branch roads and pedestrian facilities, Xiacheng District took full use of regional advantages to effectively connect the high-speed rails, subways and other low-carbon transportation hubs and public transportation systems, adopted the Internet of Things technology to dredge the transportation hubs of Hangzhou East Station and the passages of Xiacheng District, and effectively combined and seamlessly joined the various connections, to rapidly achieve a low-carbon collection and distribution as well as the "double zero-distance transfer" of time and space in the transportation hubs. In the meantime, Xiacheng District constructed a low-carbon and efficient transportation and logistics system to reduce the roundabout transportation, the repeated and empty transportation so as to reduce carbon emissions.

Second, Xiacheng District adopted measures such as a reasonable traffic light setting and traffic control of even- and odd-numbered license plates, realized vehicle distribution, reduced automobile exhaust emissions, relieved the traffic pressure of the central urban area, and promoted the "green-wave intelligent traffic light" technology at the crossroads. It also subdivided the multiple schemes according to traffic flow at different time periods, used "green-wave intelligent traffic light" technology at the crossroads, set "green wave tapes," and used "informatization" and "intelligentization" to increase the traffic speed, to solve traffic jams, improve environmental pollution, reduce traffic accidents and improve the transportation quality so as to achieve energy conservation and carbon emission reduction.

Third, Xiacheng District actively developed electric vehicles, and constructed the "electric vehicle—Internet of vehicles" model. To move toward the age of electric vehicles, the People's Government of Xiacheng District has made lots of pre-preparation work. (1) Formulate the development plans of electric vehicles for Xiacheng District, and implement preferential policies for the development of urban electric vehicles, e.g. the charging parking lots provide special stalls (with charging columns) for electric vehicles, communities provide parking space for newly-purchased electric vehicles, and the parking lots of units develop special charging parking stalls (with charging columns); take full use of the characteristics and advantages of the automobile culture boutique block on Shaoxing Road and by adding in low-carbon elements, make it become an urban promotion and demonstration area of electric vehicles. (2) Based on the demonstration operation of electric vehicles of Hangzhou City, apply the Internet of Things and cloud computing technology, and construct the "electric vehicle—Internet of vehicles" business model, i.e. electric vehicles and charging/electrical changing stations apply the sensor system of the Internet of Things to collect data and achieve unified fund settlement and intelligent monitoring on the electronic information platform.

3.3 *Simultaneous Promotion of a Low-carbon Community and an Intelligent Community*

Low-carbon community and intelligent community are general orientations of future community construction, and also important contents and foundation for low-carbon and intelligent urban construction. In a low-carbon community, the urban production mode, the living style and values advocate low-carbon economic patterns; the carbon emissions produced in all activities are minimized, helped by ecological green, the changing lifestyle of the people, and other measures as well, all of which help to achieve the goal of zero carbon emission. In an intelligent community, based on networks, intelligentization, informatization and digitization, full use of the digitization and related computer technology, network technology and control technology, carrying out all-around informatization processing and utilization of the infrastructures of community and various aspects related to the development of residents' life.

3.3.1 Use RFID Non-contact Automatic Identification Technology, and Promote Complete Awareness of Community Services

By arranging sensor networks in the community, an RFID wireless communication network made up of multiple sensor nodes is used to realize the automatic collection and transmission of information in the community, thus transmitting the data to each management control center of the community in a rapid, reliable and safe way through the ubiquitous network. After the practice of community managers over the years, the "66810" service-for-the-people system has gradually expanded from solving people's everyday life issues in all of the sub-districts and communities at the beginning, helping and supporting lonely old people and families with disabilities in the communities, to promoting professional, specialized and social services with the core content of "6 comes, 6 visits, 8 reports and 10 service-for-the-people lines" and applying the "66810" performance tracking information management system and other modern information technology community management and service systems. With the promotion of the instrumented awareness, it has been possible to master the real demands of the community residents more quickly and accurately, understand the developmental trend of all events in the community in real time, and achieve such functions of the community as active management, timely response, efficient problem-solving, e.g. intelligent family support, intelligent medical treatment and intelligent family prevention and control.[8]

8 The People's Government of Xiacheng District and Zhejiang University of Technology, *Use Scientific and Technical Means, and Innovate Social Management—Research and Practice of Digitization and Intelligentization of "66810" Service-for-the-People System of Xiacheng*, 2012.

3.3.2 Innovate the Working System and Mechanism, and Carry Out Low-carbon Community Construction

First, Xiacheng District has fully mobilized the initiative of all departments in the district to jointly carry out low-carbon community construction with unified actions. Low-carbon family creation by the Women's Federation of the district, promotion of energy-saving lamps by the Economy & Trade and Tourism Bureau of the district, classified garbage disposal by the Urban Management Office of the district, and construction of ecological garbage chambers by the Xiacheng Environmental Protection Branch form a unified force of "low-carbon community" construction through department linking up. Second, the People's Government of the district adopts resource integration strategies to explore the innovation of the system and mechanisms of low-carbon community construction. Currently, "4 mechanisms" were formed initially, including the multi-channel input mechanism, the co-construction mechanism of the government and communities, the participation mechanism for families and citizens, and the mechanism of striving for excellence by examination and assessment. Last, Xiacheng District has actively mobilized the masses, set a good example, advocated a low-carbon lifestyle, disseminated energy-saving tips, and paid attention to the saving of electricity, oil and gas in order to raise the concept of practicing a low-carbon style of living and working.

Xiacheng District has adopted the "grasping with two hands" strategy for the construction of low-carbon urban areas and intelligent urban areas. The goal of the low-carbon urban area better stimulates the innovative development of the related technology and means of an intelligent urban area; in return, a good construction plan of an intelligent urban area can better pave the way for the construction of low-carbon urban areas; the interactive development of low-carbon urban areas and intelligent urban areas will definitely be the theme of urban development in the future. Meanwhile, modern science and technology are a fundamental way to solve the increasingly severe problems of the ecological environment and of resource energy. During the "12th Five-Year Plan," Xiacheng District will accelerate its R&D and the demonstration and application of the Internet of Things, cloud computing and other new-generation information technologies, actively apply modern information technology, strengthen the comprehensive urban management, promote e-government construction, continuously improve the work efficiency of the government, intensify the construction of a digital community, improve the level of community service, accelerate the promotion of informatization of such social undertakings as education, science and technology, medical treatment, and health, and constantly improve the level of public service. By virtue of the pilot and demonstration projects of the Internet of Things, we should expand the technical application market of the Internet of Things, and develop the industry of the Internet of Things. We should actively apply new technology, to improve traditional industries, and promote the construction of the intelligent economic system, intelligent service system and intelligent management system in order to create an "intelligent Xiacheng."

CHAPTER 17

Positive Interaction between Low-carbon Urban Construction and Industrial Development: Taking Jiyuan City as an Example

Cong Jianhui, Liu Xuemin and Zhu Jing

Abstract

With the double restraints of low-carbon and economic growth, the realization of a positive interaction between urban construction and industrial development is a common new task for heavy industrial cities. This chapter discusses the interaction between low-carbon urban construction and industrial development, and defines the connotation of the positive interaction regarding two aspects, "industrial low-carbonization" and "the leading role of low-carbon urban construction in industrial development." Besides, taking Jiyuan City, Henan Province as a research case, based on the description of the condition of the industrial carbon emission of Jiyuan City, the chapter emphatically analyzes ideas and methods of its low-carbon urban construction as well as the measures adopted and results achieved in realizing a positive interaction between low-carbon urban construction and industrial development, in order to provide reference for the industrial and low-carbon development of other cities of the same kind.

Keywords

low-carbon city – industrial development – positive interaction – Jiyuan City

* Cong Jianhui, a doctoral candidate at the College of Resources Science & Technology of Beijing Normal University, whose research direction is resource economics and management; Liu Xuemin, a professor at the College of Resources Science & Technology of Beijing Normal University, whose research fields are regional economics, natural resources and environmental economics; Zhu Jing, a doctoral candidate at the College of Resources Science & Technology of Beijing Normal University, whose research direction is resource economics and management.

 | DOI 10.1163/9789004280021_018

As a gathering place of global physical capital and human capital, cities are a main spatial bearer of anthropogenic greenhouse gas emissions. It is a great urgency to promote the low-carbon development of cities, so low-carbon urban construction has also become an important policy focus of many national governments.

In lots of heavy industrial cities, the leading industry is mainly related to industrial departments which consume a large quantity of fossil energy, and greenhouse gas emissions of heavy industrial cities are much higher than those of other cities, so their low-carbon development arouses particular attention. For these cities, the industrial value added accounts for more than 60% of GDP generally, and the "path dependence" due to long-term industrial development makes it extremely difficult to adjust the industrial structure over a short period of time. Heavy industrial cities need to continue keeping a certain industrial growth to support the tax revenue and employment growth and continuously improve the level of residents' welfare, and they also need to control greenhouse gas emissions as much as possible and undertake corresponding responsibilities of emission reduction. How to find a way for a positive interaction between urban construction and industrial development under the double restraints of low-carbon and economic growth in such a dilemma is a new task. The solution to the problem is of significant importance for China which is at the middle stage of industrialization and has a lot of heavy industrial cities.

1 Interaction between Low-carbon Urban Construction and Industrial Development

In the definition of the connotation of low-carbon cities, although many scholars think that low-carbon cities pursue a "low-carbon" and "economical" development pattern which can realize the unification of ecological benefits, economic benefits and social benefits, whether there is a mutually-unified relationship in parallel between "low-carbon" and "industrial development" for heavy industrial cities has always been controversial.

The Environmental Porter Hypothesis with broad effects in the theoretical circle affirms that environmental governance can achieve win-win development of the environment and of the economy, which could be realized through the linking of two ways including "innovation offset" and "first-mover advantage of the international market."[1] Specific to the field of urban carbon

1 Chen Shiyi, *Win-win Development of Energy Conservation and Emission Reduction and China's Industry: 2009–2049*, Economic Research Journal.

emission, the hypothesis can be interpreted as follows: Enterprises make efforts to control greenhouse gas emissions; on the one hand, this will stimulate technological progress and improve the efficiency of production; on the other hand, low-carbon products made by the enterprises will be more recognized by the market, which will promote the improvement of competitiveness. There is plenty of research on empirical tests of this hypothesis, but the conclusions are inconsistent. Query of the hypothesis mainly focuses on two aspects.

First, the process of the enterprises' technological progress is slow, and the time that it takes technological progress to play its role is generally long, i.e. the Environmental Porter Hypothesis is more likely to be correct over a long period of time.

Second, there is a relatively perfect low-carbon market and consumers have a strong willingness to buy low-carbon goods, which is one of the preconditions for proving the correctness of the hypothesis, but in real life, this condition is difficult to meet.

What's more, the hypothesis is put forward centering on enterprises—microscopic subjects, so even if the enterprises' low-carbon behavior is indeed able to promote the improvement of their competitive advantages, it does not mean that there is a win-win relationship at the more macroscopic urban and national levels.

Generally speaking, urbanization and industrialization are symbiotic: industrialization is the cause while urbanization is the result. However, since industrial development will consume more fossil energy and will cause the growth of the total amount of urban carbon emissions, carbon emission control will produce adverse effects on industrial development at least in a short period of time; otherwise, the objective of emission reduction can be achieved only by relying on the autonomous market behavior of the enterprises within the cities. Exactly because the market mechanism fails in its performance in response to climate change problems, low-carbon urban construction stresses the leading and guiding role of the government. Therefore, the discussion of the interaction between low-carbon urban construction and industrial development needs to be understood through the meaning of "positive interaction" academically and theoretically.

It is undeniable that there is indeed a mode which can achieve a positive interaction between low-carbon with an absolute amount of emission reduction and industrial development theoretically, i.e. developing only low-carbon and zero-carbon industries (such as the new energy industry). This requires the core industries to have zero emissions and the whole industrial chain, and the lifecycle of all the products needs to have zero emission characteristics. The conditions for supporting this extreme mode are very strict: in addition to various factor endowments for the development of such industries locally, the product demands of domestic and foreign markets for these industries need to be far from saturated. For the current industrial cities, especially heavy industrial cities, it is difficult to meet these conditions. Besides, heavy industrial cities are generally national production bases for certain industrial products, and the output of these products will consume more fossil energy, but they have important strategic values and are indispensable for the country. Local governments cannot reduce their developmental space; on the contrary, they may even need to continue expand the production scales of such products.

According to the above analysis, it is unpractical for heavy industrial cities to adopt modes of accelerating the development of the tertiary industry and using a large quantity of new energy in a short term to achieve the reduction of the absolute amount of urban carbon emissions, but it is feasible to gradually reduce carbon emissions (i.e. the industrial carbon emission intensity) of the unit industrial added value by such methods as energy saving and technological progress. Furthermore, the development of low-carbon industries will optimize the regional developmental environment and make cities able to attract more external investments; the enterprises will also rely on advantages in low-carbon industrial products and technology accumulation to increase their market shares and improve their industrial output values in order to provide guarantees for further industrial development, thus producing a counteraction of low-carbon urban construction on industrial development.

Based on this, this chapter affirms that the positive interaction between low-carbon urban construction and industrial development includes two inseparable processes.

First, during industrial development, it is necessary to realize saving and efficient utilization of the traditional fossil energy by such methods as the improvement of management means and technological progress, optimize the structure of the energy utilization, increase the proportion of clean energy, and constantly reduce the carbon emission of unit industrial added value in order to achieve low-carbonization of industrial development.

Second, during the low-carbon urban construction, it is necessary to continuously optimize the regional ecological environment by low-carbon industrial development and the improvement of competitiveness, thereby obtaining more external investments and technology and selling more low-carbon products with a higher quality, so as to effectively promote the further development of urban industries.

What needs to be noticed is that the positive interaction defined in this chapter focuses on the synchronization and joint improvement of urban industrialization and low-carbonization in a medium- and short-term; the chapter also emphasizes that the "urban low-carbonization" of heavy industrial cities requires "industrial low-carbonization" first, and during a long period of time, the final direction of the development of the interaction is to urge urban carbon emissions to reach the peak as soon as possible and gradually achieve a continuous reduction of the absolute amount based on further optimization of the urban industrial structure. Besides, unlike the internal driving force in the interaction between industrialization and urbanization, the interaction between low-carbon cities and industrial development relies on the external force of government intervention.

In China's low-carbon urban construction practice, many heavy industrial cities have started exploring a development path conforming to the above characteristics, and we have carried out relevant research hereof. By taking Jiyuan City, Henan Province—a heavy industrial city in central China—as a research case, based on the description of the industrial development and carbon emission condition of the city, this chapter deliberately analyzes ideas and methods of its low-carbon urban construction as well as the measures adopted and results achieved in realizing a positive interaction between low-carbon urban construction and industrial development, in order to provide reference for industrial and low-carbon development of other cities.

2 Formation, Evolution and Industrial Development of Jiyuan City

Located in northwestern Henan Province, Jiyuan City has gained its name as the source of Jishui. It is the hometown of the Foolish Old Man in the legend and the birthplace of the story. Jiyuan City used to be a county of Jiaozuo City and experienced great changes in the administrative system in the past. The county was cancelled in 1988 and became a provincial city in 1997. Jiyuan City is the only city in Henan Province which has no districts or counties. The whole area of the city is 1,931 km^2, and the total population in 2011 was 680,000. With urban development, the status of Jiyuan City has gradually improved within the regional economy. The city was listed as one of the nine cities of the "Central Henan Urban Agglomeration" in 2003, playing the role of a bridge connecting the east and the west as well as the north and the south in the Central Plains Economic Zone.

As a typical heavy industrial city, Jiyuan City is the largest lead and zinc smelting base in China and an important base of steel, energy, chemical industry and machinery manufacturing in Henan Province. After the economic development over the years, the industrial departments of Jiyuan City have expanded gradually, forming 6 pillar industries mainly including steel, lead and zinc, energy, the chemical industry, machinery manufacturing, and mine-used electrical appliances. It achieved an industrial added value of 29.842 billion RMB in 2011, and the industry accounted for up to 76.7% of GDP. The city gathers a batch of leading national or regional enterprises of key industries, such as Henan Yuguang Gold & Lead Group Co., Ltd., Jiyuan Jinma Coking Co., Ltd., Qinbei Power Plant and Zhongyuan Special Steel Co., Ltd. The size of the main business income of the industrial enterprises mentioned above reached 111.833 billion RMB in 2011, the total profit exceeded 6.9 billion RMB, and the output of many industrial products ranked top in Henan Province and even in China, as shown in Table 17.1.

TABLE 17.1 *Output of the main industrial products of Jiyuan City in 2011*

Main industrial products	Output	Proportion in Henan Province (%)	Proportion in China (%)
Silver (ton)	156.03	—	12.54
Zinc (10,000 tons)	28.29	—	5.29
Lead (10,000 tons)	90.52	—	19.13
Generating capacity (100 million kW)	191.44	7.37	—
Coke (10,000 tons)	284.59	9.99	—
Caustic soda (10,000 tons)	35.73	21.43	—
Pig iron (10,000 tons)	244.85	11.38	—
Crude steel (10,000 tons)	273.52	11.54	—
Sheet glass (10,000 weight cases)	255.45	11.86	—

DATA SOURCE: 2012 *JIYUAN STATISTICAL YEARBOOK, HENAN STATISTICAL YEARBOOK* AND WEBSITE OF THE NATIONAL BUREAU OF STATISTICS.

It should be said that the industrial structure focusing on heavy industries drives the rapid development of the economic and social undertakings of Jiyuan City, and is an important motive force for the increase of the urbanization rate of the city from 40.94% in 2000 to 49% in 2010. Anyway we cannot ignore the fact that, with the rapid urbanization and industrial development of Jiyuan City, greenhouse gas emissions are increasing.

3 The Condition of Industrial Carbon Emissions in Jiyuan City

According to the format requirements of the *Global Protocol for Community-Scale Greenhouse Gas Emission* (GPC) (*1.0 Version*),[2] the accounting of the industrial carbon emissions in Jiyuan City from 2000 to 2010 is divided into two parts, which are industrial energy consumption and industrial production process.

3.1 *Method and Data*

3.1.1 The Accounting Method of Carbon Emissions for the Industrial Energy Consumption

Accurate accounting of carbon emissions is a worldwide problem. Due to the fluidity of greenhouse gases and the openness of urban systems, accounting for urban carbon emissions is quite complex. In defining the accounting boundary, this research used the scope established by the World Resources Institute (WRI) for reference and divides the industrial emission departments in Jiyuan City into three scopes;[3] but due to the limitation of the availability of the data, only the carbon emissions of Scope 1 and Scope 2 were included in the research.

The common accounting methods for carbon emissions of energy activities include: IPCC Guidelines for National Greenhouse Gas Inventories (including the department method and the reference method), the US Oak Ridge National Laboratory method, the method of the Energy Research Institute of the National Development and Reform Commission and the method provided by Xu Guoquan and others.[4] Comparatively speaking, the IPCC reference method is more meticulous in fuel classification and it is easier to get data than the department method, so it is suitable to carbon emission accounting research for domestic cities whose energy statistical systems are not currently

2 GPC is an agreement formulated jointly by international organizations such as the C40 Cities Climate Leadership Group, Local Governments for Sustainability (ICLEI) and the World Resources Institute (WRI); it aims unifying the formulation methodology and format of global urban greenhouse gas inventories, and it issued its first version globally in May 2012.

3 Scope 1 refers to direct carbon emissions of industrial departments within the administrative boundary of Jiyuan City; Scope 2 refers to indirect emissions caused by electricity consumption and heat consumption of industrial departments; Scope 3 refers to all other indirect emissions except Scope 2, such as greenhouse gas emissions hidden in imported/exported goods of the city.

4 Xu Guoquan, Liu Zeyuan, and Jiang Zhaohua, *Factor Decomposition Model and Empirical Analysis of Carbon Emissions of China: 1995–2004*, China Pollution, Resources and Environment, 2006 (6).

perfect. This chapter continues using the IPCC reference method in the selection of the accounting formula of the industrial energy carbon emissions in Jiyuan City, but it adopts the data of the *Guidelines for Provincial Greenhouse Gas Inventories of China* for the emission factors of fuels, significantly improving the accuracy.

The calculation formula is as follows:

$$E_{industry} = \sum F_i \times EF_i = \sum F_i \times \left(A_i \times B_i \times C_i \times \frac{44}{12} \right)$$

where, E_{industry} refers to the industrial carbon emissions in Jiyuan City; F refers to energy consumption; EF refers to emission factors of fuels; A refers to the net calorific value of energy; B refers to the carbon content of the unit heat value of energy; C refers to the oxidation rate of carbon; i refers to various energy types; and the rate of carbon sequestration is assumed to be 0.

For the secondary energy, an average is considered for the emission factors of electricity after being measured according to the actual energy consumption principle over the years, while the emission factors of heat are obtained by converting the emission factors of electricity according to 1 kWh = 3,600 kJ.

3.1.2 The Accounting Method of Carbon Emissions in the Industrial Production Processes

In addition to energy activities, chemical reaction processes or physical change processes in some industrial productions (mainly those of cement, lime, glass, the chemical industry and the metal smelting industry) can also produce greenhouse gas emissions, and there is a high proportion of this kind of emissions in some cities. According to the method provided by the *Guidelines for Provincial Greenhouse Gas Inventories of China*, this chapter gives an estimate of the carbon emissions from the production processes of glass, cement, lime, steel, lead and zinc smelting in the main industrial departments in Jiyuan City.

The specific estimate method is as follows: carbon emissions of glass production, lime production, and lead and zinc smelting processes are obtained directly by the product of the output data of industrial products and the respective emission factors. The emission factors in the lead smelting processes are calculated from the enterprise production data according to the material balance method, taking the value of 0.37. The value is less than the corresponding data of the IPCC, because the local lead smelting enterprises adopt advanced technology such as the "direct lead process" to greatly reduce the quantity of CO_2 production. Carbon emissions of cement production are obtained from the product of the output data of cement clinkers (the default rate is 65% of the cement output) and their emission factors. Steel production processes are determined comprehensively according to limestone consumption, dolomite consumption, the average carbon content of pig iron and the average carbon content of steel. The calculation formula of carbon emissions in industrial steel processes is as follows:

$$E_{steel} = AD_l \times EF_l + AD_d \times EF_d + (AD_r \times F_r - AD_s \times F_s) \times \frac{44}{12}$$

where, E_{steel} refers to carbon emissions in steel production processes; AD_l refers to the quantity of limestone consumed by steel enterprises as a solvent; EF_l refers to the emission factors of limestone consumed as a solvent; AD_d refers to the quantity of dolomite consumed by steel enterprises as a solvent; EF_d refers to the emission factors of dolomite consumed as a solvent; AD_r refers to the quantity of pig iron for steelmaking; Fr refers to the average carbon rate of pig iron for steelmaking; AD_s refers to the steel output; and F_s refers to the average carbon rate of steel products.

3.1.3 Sources of the Main Data

In this chapter, the activity data concerning the accounting of urban industrial carbon emissions are taken from the *Jiyuan Statistical Yearbook* over the years and the Energy Balance Sheet; the emission factors are mainly based on the *Guidelines for Provincial Greenhouse Gas Inventories of China*, where a small part of the emission factors have been revised according to local conditions.

3.2 *Analysis of the Condition of the Industrial Carbon Emissions of Jiyuan City*

With the combination of carbon emissions of the industrial energy consumption and industrial production process, Figure 17.1 shows the conditions of the industrial carbon emissions in Jiyuan City since 2000.

From Figure 17.1, it can be found that:

First, within the research period, the trend of the industrial carbon emissions of in Jiyuan City is obviously phase-based. In the first two years, at the beginning of the 21st century, the industrial carbon emissions in Jiyuan City basically kept a modest increase; from 2003 to 2007, the slope of the curve of carbon emissions gradually became bigger and the industrial carbon emissions rapidly increased, which is consistent with the historical background that heavy industrialization accelerated in that period; from 2007 to 2009, because of

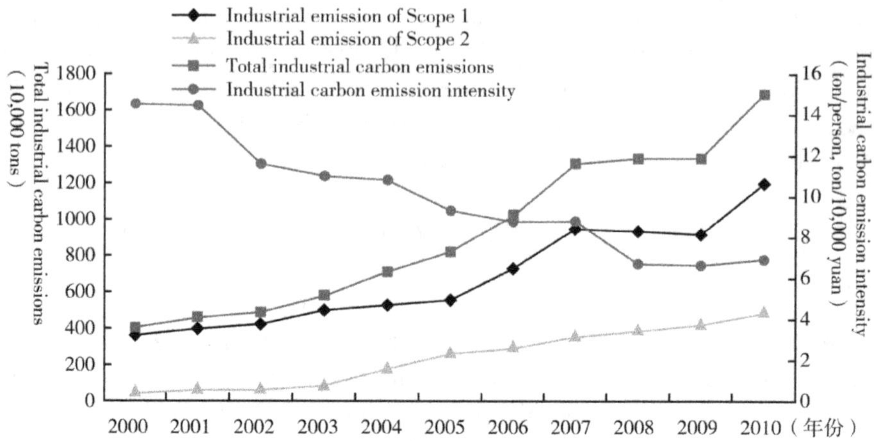

FIGURE 17.1 *The condition of the Industrial Carbon Emissions in Jiyuan City from 2000 to 2010.*

negative impacts of the financial crisis, the total industrial carbon emissions basically remained unchanged; and an increasing trend reappeared after 2009. From the angle of longitudinal comparison, the total industrial carbon emissions in Jiyuan City (Scope 2) reached 16,825,400 tons in 2010, with an increase of more than three times compared with the 4,020,000 tons in 2000, indicating that the rapid industrial development of Jiyuan City in that period gave rise to a significant increase in the absolute amount of carbon emissions.

Second, the intensity of industrial carbon emissions showed a gradual decreasing trend on the whole, and entered a stable fluctuation range after 2008. In terms of the average level of cities in China, although the absolute value of carbon intensity of the city was high, the decreasing rate of the intensity of industrial carbon emissions in Jiyuan City was relatively fast compared with some industrial cities in China, such as Shenyang and Wuxi.[5] This means that the low industrial carbonization of Jiyuan City had achieved phase-based results, which benefited from some specific and targeted measures of energy conservation and emission reduction adopted by the local government after 2000 under the macroscopic strategic guidance development such as the "new industrialization" and the "scientific outlook on development." See below for

5 Fengming Xi, Yong Geng, Xudong Chen, et al., "Contributing to Local Policy Making on GHG Emission Reduction through Inventorying and Attribution: A Case Study of Shen Yang," Energy Policy, 2011, 39.

the low-carbon urban construction measures which were carried out and have been greatly promoted with the new concept of ecological civilization development.

4 Promoting Positive Interaction between the Low-carbon City and the Industrial Development of Jiyuan City

According to measurements, the industrial carbon emissions of Jiyuan City account for 92.2% of the total urban carbon emissions. Since industrial development has a certain path of independence, it is unrealistic to change the industrial structure focusing on the industry of Jiyuan City in a short period of time. Therefore, the emphasis of low-carbon urban construction in Jiyuan City is on how to reduce carbon emissions in the existing industrial departments.

Only in terms of emission reduction efficiency, is the limitation of speed and scale of the industrial development undoubtedly a "shortcut" towards reducing carbon emissions in a short period of time. The per capita income level of Jiyuan City has a great gap compared with the developed cities in east China. People urgently pursue economic development, and as an important supply base of energy and non-ferrous metals in central China, Jiyuan City has a big responsibility for supplying safe of corresponding regional and national products, so industries will still be a core force in supporting the economic and social development of the city in the future. Therefore, low-carbon urban construction in Jiyuan City must consider emission reduction, development, employment, industrial security and other problems as a whole.

4.1 *Measures of Low-carbon Urban Construction*

With the pressure of the increasing emission reduction, Jiyuan City mainly adopts the following measures in the industrial field for the construction of a low-carbon city.

4.1.1 Eliminating Backward Production Facilities and Carrying Out Energy Conservation

In recent years, there has been a rapid increase in demands for characteristic industrial products of Jiyuan City by foreign and domestic markets, providing room for growth for a batch of small enterprises with low efficiency, backward technology, lax management and extensive operations. However, these enterprises have high energy consumption, poor economic performance and create severe environmental pollution. Moreover, some backward production lines and technologies also exist in enterprises with an overall good economic performance, which needs to be adjusted. The carbon productivity (the output value of unit CO_2 emissions) of these facilities with backward production is low, providing a entry point for government policies to reduce emissions of greenhouse gases from the point of view of "quantity control."

Jiyuan City formulated and strictly implemented a work program of eliminating the facilities with out-dated production during the "11th Five-Year Plan" and the "12th Five-Year Plan," where a batch of small enterprises with high energy consumption and poor economic performance were closed down, some small thermal power plants were shut down, mechanical shaft kiln production lines closed, solid clay brick-tile kilns were dismantled, and out-dated production lines, such as the sintering machine of lead smelting, were eliminated. Jiyuan City carried out an energy conservation program for key industries of energy consumption such as steel, non-ferrous metals, coal, electricity, chemical industry and building materials as well as for enterprises with an annual energy consumption of more than 5,000 tce; it implemented the improvement of the coal-fired industrial boiler, the utilization of waste heat and pressure, and the energy conservation transformation of electromechanical systems, strictly controlled the production scale of the electrolytic lead and electrolytic zinc within the city, and supported and developed the comprehensive utilization of secondary lead resources.

4.1.2 Developing and Cultivating Low-carbon Industries

Driving the urban development by low-carbon industries is the ideal operational mode of low-carbon urban construction. First, Jiyuan City has carried out low-carbonization-oriented upgrading and transformation of traditional industries, relied on existing competitive industries, lengthened the industrial chain, improved the added value of products, strengthened the independent innovation and industrial matching capacity, and advanced towards intensification, systematization, and intensive processing. Second, it grasped the opportunity of cultivating and developing the strategic emerging industries of the country, strengthened the planned guidance and policy support according to resource endowment, existing industrial basis and enterprises' capacity, actively developed a series of products of new energy electric vehicles, key parts of wind power equipment and other energy-saving and environmental equipment, and made efforts to cultivate new economic growth points characterized by low-carbon. At present, the proportion of emerging industries in the total number of industries has reached 30%.

4.1.3 Promoting a Circular Economy by the Combination of Government and Market

A circular economy consists of developing the economy based on circulation, regeneration and utilization of matters, taking reduction, recycling and recovery of resources as principles. Basic characteristics of the circular economy are low consumption, low emissions and high efficiency, which has many features in common with the concept of low-carbon development.

Some advanced enterprises in Jiyuan City, under the guidance of the market mechanism, have formed some circular economy chains with independent experimentation. For instance, the circular economy chain of "lead and zinc smelting—intensive processing—comprehensive utilization of waste materials—recycling of secondary lead" in the lead and zinc smelting industry was first established by Henan Yuguang Gold & Lead Group Co., Ltd. and other enterprises of the city motivated by market interests. This characteristic industrial chain has made full use of various waste materials, saving resources and reducing energy consumption, producing good economic benefits, and reducing carbon emissions to a certain extent.

The government summarized and acquired some advanced experience and successful modes in practices of different enterprises in developing a circular economy, disseminated and promoted them around the whole city, and encouraged the enterprises with potential needs to discuss the possibility of jointly establishing an industrial chain of circular economy. The application of the circular economy concept of Jiyuan Municipal Government was also reflected in its new round of industrial distribution, and particularly, it promoted the circular economy concept in Huling Industrial Agglomeration and Yuchuan Circular Economy Park which has played an important function in the economy of the city. The Huling Industrial Agglomeration focuses on the development of steel, coal, the chemical industry, non-ferrous metals, equipment manufacturing and other enterprises with a high energy consumption, and strives to establish two circular economy industrial chains of "coal—coke—coal tar and coke oven gas intensive processing" and "matching processing—equipment manufacturing—remanufacturing" in the agglomeration. The Yuchuan Circular Economy Park relies on the key enterprises in the park, focuses on new energy and leading industries such as the energy industry, non-ferrous metals and intensive processing, strives to establish three circular economy industrial chains of "chlorine alkali chemical industry—polycrystalline silicon—solar cell—solar photovoltaic power generation," "coal—thermal power—comprehensive utilization of pulverized fuel ash and slag" and "steel—intensive processing—comprehensive utilization of waste materials"; the park also promotes cyclic and efficient utilization of various resources in the agglomeration and achieves low emission of pollutants, up to zero emissions.

4.1.4 Paying Attention to the Technological Progress of Industrial Enterprises

Scientific and technological progress is the core and key to the response to climate change. Controlling the urban carbon emissions by virtue of science and technology tends to achieve "twice the result with half the effort," and the industrial enterprises are a hotbed of scientific and technological development. Based on resource-oriented industrial characteristics and focusing on energy conservation and consumption reduction, Jiyuan City has strengthened the leading role of science and technology in low-carbon urban construction from the point of view of three aspects.

First, it focuses on the application of hi-tech technology and appropriate advanced technology to upgrade and transform the traditional industries. A batch of inefficient industrial boilers, kilns, electric machines and transformers have been updated and transformed first. Currently, the energy efficiency indicators of the main energy-consuming equipment in the city have reached advanced levels in China. Second, it reinforces the policy support for low-carbon scientific and technological work, increases the scientific and technological subsidies of energy conservation and emission reduction year by year, encourages the enterprises to conduct technology research and development independently, and accelerates the transformation of scientific and technological achievements. Third, it attaches great importance to the introduction of colleges and universities and other external intellectual resources. In Jiyuan City, many well-known colleges and universities at home and abroad cooperate with the local enterprises in jointly tackling the scientific and technological difficulties.

4.2 *Results of Low-carbon Urban Construction and Effects on Industrial Development*

A series of measures in Jiyuan City regarding low-carbon urban construction have achieved obvious results: by eliminating the out-dated production facilities and carrying out energy conservation during the "11th Five-Year Plan," the energy consumption of unit industrial added value of the enterprises above the designated size reduced by 5.48%, 4.89%, 9.74%, 10.15% and 10.86% respectively; by developing strategic emerging industries with low-carbon characteristics, the city has improved the degree of industrial modernization in a cleaner way; the development of the circular economy and the accumulated technological progress of the enterprises have not only reduced carbon emissions, but have also produced extra economic benefits. Joint action of these measures makes the intensity of industrial carbon emissions in Jiyuan City decrease rapidly. At present, although the absolute amount of the industrial carbon emissions in Jiyuan City is increasing, the trend of a significant and rapid increase over a long period of time has been curbed preliminarily. Since 2010, the local municipal government has further carried out urban construction by focusing on low-carbon development, and as a consequence, the intensity of industrial carbon emissions has continuously decreased.

More importantly, the low-carbon urban construction in Jiyuan City has facilitated the local industrial development. Jiyuan City has successively applied for and become a "national sustainable development experimental area" (the Ministry of Science and Technology), a "national low-carbon transport system pilot city" (the Ministry of Transport), one of the second batch of national low-carbon pilot cities (the National Development and Reform Commission), and one of the first batch of US-Sino low-carbon ecological pilot cities (the Ministry of Housing and Urban-Rural Development), which has won for Jiyuan City a high reputation and has expanded the influence of the city nationwide. Many enterprises come to the city to invest in or start up a business, including traditional labor-intensive industrial projects brought by the industrial gradient transfer from the developed eastern areas of China (e.g. Foxconn) and a large number of low-carbon enterprises and projects, further expanding the benefits of scale and output of the industrial economy of Jiyuan City. Although we cannot calculate the proportion of the benefits brought by low-carbon urban construction from an urban level temporarily and we have no statistical data regarding extra profits obtained by increasing low-carbon technology investments from an enterprise level, the data are undoubtedly considerable.

From the analysis above, a positive interaction mechanism between low-carbon urban construction and the industrial development of Jiyuan City has been preliminarily formed. In the future, it will be necessary to comprehensively consider the existing carbon emission pattern and the economic and social developmental requirements, scientifically formulate an industrial emission reduction roadmap, strive to exceed the urban carbon emission peak as soon as possible, and achieve the reduction of the absolute amount of carbon emissions. In order to complete the task, in the short run, Jiyuan City needs to further improve its coal-based energy consumption structure, develop new energy and renewable energy, improve, perfect and promote a scientific developmental system and mechanism, and enhance its capacity for independent innovation; in the long run, the city needs to progressively adjust its industrial structure, put industry on the collaborative developmental track of the agriculture and the service industries, and reduce the proportion of the heavy industries in the national economy based on improving the benefits from industrial enterprises. In addition, as a hub undertaking industrial transfer in central China, Jiyuan City needs to focus on carrying out research concerning how to control the total emissions of greenhouse gases in the city during the dynamic process of undertaking industrial transfer, and how to strengthen the cooperation with the nearby counties and cities according to the fluidity of carbon emissions, so as to achieve a collaborative emission reduction.

5 Conclusion

The case of Jiyuan City has a certain applicability and referential significance to cities of the same kind in China. The low-carbon construction of the industrial cities cannot be achieved by restraining industrial development; on the contrary, we should achieve a positive interaction of the two during the process, take the reduction of the intensity of industrial carbon as the point of strength in the short run, and gradually adjust the industrial structure on this basis.

Key and difficult points of achieving a positive interaction between low-carbon urban construction and industrial development are finding some means and methods for reducing carbon emissions based on improving (not reducing, at least) the enterprises' benefits. By eliminating out-dated production facilities, paying much attention to industrial energy conservation, cultivating low-carbon industries, developing a circular economy and promoting technological innovation, Jiyuan City has experimented with some solutions to the problem, and has achieved remarkable results. The experimentation has also provided valuable reference for low-carbon construction of other domestic heavy industrial cities facing the same developmental restraint.

CHAPTER 18

Synergy Research on Overall Urban-rural Planning and Low-carbon Development: Taking Guangyuan City as an Example

Huang Liming, Zhou Yong and Liu Jianbo

Abstract

By taking Guangyuan City, Sichuan Province as an example, this research analyzes the synergistic relationship between the overall urban-rural planning and low-carbon development. Interaction exists between the overall urban-rural planning and low-carbon development, shown as a "synergistic effect" of the two in driving regional integrated development. Scientific planning and implementation of low-carbon development as well as the great promotion of a comprehensive, coordinated and sustainable urban-rural development are of great significance in breaking the urban-rural dual structure and achieving integrated regional development.

Keywords

overall urban-rural planning – low-carbon development – synergy theory

1 Introduction

The overall urban-rural developmental strategy is a significant strategic adjustment to correctly grasp the urban-rural developmental relationship under the new situation of China; it is also a concentrated embodiment of deeply implementing the Scientific Outlook on Development, gradually changing the urban-rural dual economic structure, fundamentally solving the "three agriculture-related issues" of China, and realizing a comprehensive, coordinated

* Huang Liming, president of the Guangyuan Low-carbon Economic Development Institute of Sichuan Province; Zhou Yong, full-time deputy director of the Guangyuan Low-carbon Development Bureau of Sichuan Province; Liu Jianbo, deputy secretary general of the Guangyuan Standing Committee of the National People's Congress of Sichuan Province.

 | DOI 10.1163/9789004280021_019

and sustainable development of the economy and society. For this reason, the 18th National Congress of the CPC proposed the overall urban-rural developmental thought of "promoting agriculture through industry, and driving rural development through urban development," which emphasizes the strengthening of the overall urban-rural development, reinforcing the rural developmental vitality, gradually reducing the urban-rural gap, and promoting urban-rural common prosperity. The low-carbon concept is a new developmental idea in response to climate change, a global revolution involving production modes, lifestyles and values, and a main mode of the economic development of human beings' future ecological society. Based on China's conditions with lots of coal, less oil and insufficient gas, developing low-carbon economy is the only way to increase energy efficiency and realize scientific and sustainable development; it is the key to adjusting the industrial structure and changing the growth pattern. The 18th National Congress of the CPC stressed sticking to the basic state policies of saving resources and protecting the environment, adhering to the guidelines of giving priority to conservation, protection and to a spontaneous recovery orientation, focusing on the promotion of green development, cyclic development and low-carbon development, forming a spatial pattern, an industrial structure, a production mode and lifestyle in order to save resources and protect the environment, and reversing the ecological environmental deterioration trend from the source, so as to create a good environment for production and the life of the people and to make contributions to global ecological safety. It is obvious that both the overall urban-rural planning and low-carbon development are scientific decisions strictly following the law of the development of the economy and society against the background of a Scientific Outlook on Development; they are of great strategic significance in promoting a comprehensive, coordinated and sustainable development of the economy and society.

Located on the border of Sichuan, Shaanxi and Gansu, Guangyuan City in Sichuan Province is a special area with "4 areas in 1," covering a remote mountainous area, an old revolutionary base area, a concentrated poverty-stricken area and the area severely hit by the "5·12" Wenchuan Earthquake. In the entire zone, there are many mountainous areas and little flat ground, and the mountainous areas account for 94% of the total area; the rural areas are overpopulated and the urban areas are less populated, and the population in the rural areas is up to 1,979,600, accounting for over 64% of the total population; the rural area is big while the urban area is small, so the driving force of "driving rural development through urban development" is insufficient; the economic aggregate is small and the level of industrialization is low, so the condition of "promoting agriculture through industry" is immature; its industrial foundation mainly consists of military industrial enterprises established in the 1960s, with little integration into the local economy and prominent dual economic characteristics. It is a typical example of "driving forth a big rural area through a small city." Since 2007, when Guangyuan City was determined as a pilot city of the comprehensive supporting reform of the overall urban-rural plan for Sichuan Province, and particularly after the Wenchuan Earthquake in 2008, as a severely-afflicted area, Guangyuan City has adhered to low-carbon reconstruction, seized the opportunity of an overall urban-rural pilot plan and post-disaster reconstruction, and has actively explored the path of synergistically promoting overall urban-rural planning and low-carbon development.[1] With the mutual promotion and interaction between the overall urban-rural development and low-carbon urban construction as well as complementing each other, apparent synergistic effects have been generated. The practices of Guangyuan City fully show that interaction exists between the overall urban-rural development and low-carbon development, an example of the "synergistic effect" of the two in furthering the regional development. Scientific planning and synergistic promotion of the overall urban-rural plan and low-carbon development are of great significance in breaking the urban-rural dual structure, solving the "three agriculture-related issues," and realizing a comprehensive, coordinated and sustainable development of the economy and society.

2 Analysis of the Synergy between the Overall Urban-rural Planning and Low-carbon Development

Overall planning means a unified plan, which is a fundamental method for practicing a Scientific Outlook on Development. The overall urban-rural development aims at grasping the overall situation, scientifically planning and coordinating development when dealing with the urban-rural developmental relationship, so as to promote urban-rural common prosperity and progress. The key to overall urban-rural development is reform. To break the obstacles of the system and mechanisms through reform so as to create the conditions for development, the essence is to break the production relationship that restricts the development of productivity and thus release the vitality of the emerging productivity. When dealing with the development of human and

1 Zhuang Guiyang and Yang Zhen, *Empirical Research on Low-carbon Development of Latecomer Regions—Taking Guangyuan City in Sichuan Province as an Example*, Guangming Daily Press, 2012.

nature, the goal of low-carbon development is to stick to respecting nature and conforming to nature, keeping energy consumption and CO_2 emissions at a low level, reducing the damage that humans do to nature, and promoting a harmonious coexistence of humans and nature. New social requirements will definitely bring about new technological revolutions, and a low-carbon economy represents the direction of future economic development and the track of the development of the emerging productivity. If we do not choose a new developmental road after breaking down the barriers of the overall urban-rural planning system and mechanism, industrial development will certainly repeat the same mistakes of township enterprises and extensive development, and the developmental quality of agriculture and even of modern agriculture will be affected due to the use of a large number of chemical fertilizers and pesticides. A new developmental road and a new developmental pattern are inevitable choices for overall urban-rural development, and low-carbon development emerges at the right moment and is organically integrated with the overall urban-rural plan.

Overall urban-rural planning is the road, and low-carbon development is the pattern. The development of overall urban-rural planning needs low-carbon development, while low-carbon development needs the support of the overall urban-rural planning system and mechanism. In spite of different logical categories of the relationships involved, in essence, the two have an obvious synergy in connotation, goal, approach and other aspects. The first is the synergy of connotation. The essence of a low-carbon economy consists of improving the efficiency of the energy utilization and the clean energy structure, and the core is to improve the efficiency of the utilization of resources and the degree of intensification through the innovation of the energy technology, the institutional innovation and the fundamental transformation of people's living style and developmental concepts, so as to promote the sustainable development of the economy and society. Overall urban-rural planning is not to simply transform the villages into cities and the rural residents into city residents, but to carry out the overall plan by taking the development of the urban-rural economy and society as a whole and taking all things into consideration, so its essence is to promote the transformation of the extensive economy of urban-rural areas into an intensive and efficient economy. The connotation of the overall urban-rural planning covers all aspects of economic and social fields, including the overall planning of systems, of the population, land, technology and other production factors, as well as that of the urban-rural relationship. In this sense, industry, land, resources, technology and other fields involved in a low-carbon economy are important aspects inevitable to overall urban-rural development. The second is the synergy of goal. Both overall urban-rural planning and low-carbon development are committed to solving serious contradictions and prominent problems in certain historical periods of the economic and social development; their common goal is to promote a more balanced and more harmonious development of the human society, and finally realize a comprehensive, coordinated and sustainable development of the economy and society. The third is the synergy of the approach. Overall urban-rural development aims at realizing an urban-rural synchronous development through the overall creation of a plan, industry, input, system and policy; low-carbon development also needs to synergistically focus on the plan, industry, input, system, policy and other aspects in order to promote the adjustment of the industrial structure, the energy structure and the consumption structure and produce changes in the developmental pattern and lifestyle. Therefore, the two are consistent in the approach to realization.

Seen from the practices of Guangyuan City, on the one hand, the earlier comprehensive supporting pilot reform work of the overall urban-rural planning has laid a solid foundation for the low-carbon development of the city. Since the implementation of the pilot work, by taking the promotion of urban-rural integration as the core, Guangyuan City has started with reform, carrying out reforms of the land, household registration, social security, rural finance and other aspects from the source of the system. Paid land circulation and the transformation of barren mountains and barren slopes opened a broad space for the construction of eco-agricultural parks and the development of economic forests; land and forest land were concentrated in the hands of large-scale growers and specialized growers, and specialized cooperative organizations sprang up like mushrooms; the reform of the household registration system promoted an orderly population mobility, bringing about vitality for the construction of small market towns in rural areas and new urbanization; the reform of the social security system, especially for seeking medical treatment, entrance into schools, endowment, and various trainings, allowed the rural residents to rid themselves of constraints and enjoy the same public services as the urban residents, stimulating the rural residents' enthusiasm for participation and effectively improving the laborers' quality; the reform of the rural financial system first established village banks to greatly activate land, funds, population and other production factors, effectively promoting economic construction, social development and the comprehensive progress of ecological civilization. Concerning the reforms, Guangyuan City has actively adjusted and optimized the industrial structure and promoted the transformation of the economic developmental pattern, and has preliminarily constructed a planning system and a modern framework for the industrial system conforming to the requirements of the scientific outlook on development and low-carbon economic developmental concepts.[2] It is without doubt that the overall urban-rural development has laid a solid foundation for Guangyuan City to construct a low-carbon industrial pattern and develop a low-carbon economy, promote a low-carbon lifestyle and construct a low-carbon city. Take the changes in the structure of energy consumption in the city as an example. By 2010, coal still occupied a large proportion, but it was reduced to 57%, and the proportion of such clean energy as natural gas, water and electricity increased to 36%; the energy consumption per unit of GDP had declined from 1.55 tce/10,000 yuan in 2005 to 1.24 tce/10,000 yuan, with a cumulative reduction of 20.11%; the energy consumption per unit of industrial value added had declined from 4.11 tce/10,000 yuan in 2005 to 2.85 tce/10,000 yuan, with a cumulative reduction of 30.67%. As a result, the significant improvement of the environment and other conditions won a warm praise from the citizens, rapidly forming a consensus of all people to construct a low-carbon city. On the other hand, low-carbon economic development and low-carbon urban construction provide constant power for the overall urban-rural development of Guangyuan City. In 2008, Guangyuan City first put forward the strategy of "low-carbon reconstruction, and low-carbon development" in the area severely afflicted by the 2008 Earthquake, actively promoted the construction of energy-saving technology and materials in post-disaster reconstruction, and strictly followed the standard of 50% of energy conservation to design and construct new housing and public building projects. The accelerated economic development highlights the low-carbon orientation of the city, and the unit energy consumption has declined year by year. During the "11th Five-Year Plan," the energy consumption per unit of GDP accumulatively declined by 20.11%, the energy consumption per unit of industrial value added accumulatively declined by 30.67%, and the economic structure was gradually optimized. The elimination of out-dated production facilities was strengthened and the rate of the output of the industrial value added of unit energy consumption gradually increased. Guangyuan City actively popularizes the environmental and energy-saving products, such as solar energy, comprehensively promotes soil testing for formulated fertilization, and constructs 10 new low-carbon rural complexes, with biogas covering rate reaching 50% and forest coverage rate up to 54%. It is anticipated that by 2020, the fixed carbon content of the ecological system of the city will reach 24,095,000 tons/year. Booming development of the low-carbon economy of Guangyuan City promotes the gradual optimization of the

2 Jiang Wei and Guo Jun, *Low-carbon Reconstruction—Case of Guangyuan City, a Severely-affiliated Area of "5·12" Wenchuan Earthquake*, University of Electronic Science and Technology of China Press, 2011.

economic structure, the enhancement of economic quality and the continuous improvement of people's livelihood; moreover, the whole economic society has continued to develop quickly and well with a double-digit economic growth in the last seven years, which also provides constant power for comprehensively supporting the pilot reform work of the overall urban-rural plan for Guangyuan City.

Above all, overall urban-rural development and low-carbon development are not isolated, but interconnected and integrated with mutual promotion where each one complements the other, which can produce strong synergistic effects.

3 The Measures and the Path towards Synergistically Promoting Overall Urban-rural Planning and Low-carbon Development

Based on carefully clarifying the synergistic relationship between overall urban-rural planning and low-carbon development, Guangyuan City boldly reforms and innovates developmental ideas, takes the lead in putting forward the strategy of "low-carbon reconstruction and low-carbon development" in the areas severely hit by the earthquake, gradually establishes the basic idea of "establish the city ecologically, become a powerful industrial city, enrich the city with culture and tourism, overall urban-rural planning." It sticks to using the low-carbon concept to promote overall urban-rural development and using the overall urban-rural planning concept to promote low-carbon construction, takes overall urban-rural planning and low-carbon development as "twin engines" to help the city catch up and leap forward, comprehensively promotes new industrialization, urbanization[3] and modern agriculture, and carries out fruitful beneficial experimentations.

3.1 *Promote Industrial Low-carbonization*

Low-carbon economic development is accompanied by industrial development. Particularly, underdeveloped areas, like Guangyuan City, must accelerate the pace of pushing the medium term of industrialization forward from the early term of industrialization. While promoting new industrialization, Guangyuan City carefully draws from the lessons it has learned and does not follow the old way of "first pollute and then control." At the same time that it accelerates the promotion of resource transformation and actively constructs

3 Qiu Baoxing, *New Urbanization: From Concept to Action*, Xinhua Digest, 2013 (4).

modern industrial parks, the city sticks to promoting industrial low-carbonization, vigorously develops new materials, electronics, food and beverages, water and electricity, natural gas development and utilization, tourism and other competitive low-carbon industries, promotes the cyclic and comprehensive utilization of industrial wastes and regrown raw materials; it strengthens the transformation of energy-saving technology to make comprehensive energy consumption of the unit product gradually close to the national advanced standard value, makes efforts to construct the "5+2+1" modern industrial system characterized by a high-end orientation, a cluster orientation and an orientation towards ecology, and enhances the engine of accelerating development and driving agriculture forward through industry powerful. Through these efforts, by 2012, the industrial park area of the city had increased from 9.9 km^2 in 2007 to 72.2 km^2 with an industrial concentration ratio up to 61.5% and an industrialization rate breaking through 40%, historically entering the medium-term developmental stage of industrialization; there were 347 industrial enterprises above designated size, including 141 enterprises with an output value of 100 million RMB, and the number was 4.5 times that in 2007; the output growth of the main industrial products was up to 66%, with 59.4% for the growth of raw iron ore and 33.5% for the growth of commercial ready-mixed concrete. The leading status of industry has basically been formed. Meanwhile, industrial energy consumption per 10,000 yuan of GDP of the city has declined year by year. The Guangyuan Economic and Technological Development Zone has been upgraded to national level, and Wangcang Industrial Concentrated Zone has been successfully built to be a provincial-level economic developmental zone and has been determined as the circular economy demonstration park of Sichuan Province.

3.2 *Develop Low-carbon Agriculture*

The key to overall urban-rural planning is urban areas, and the core is to solve the "three agriculture-related issues." According to the developmental requirements of modern agriculture, the old way of traditional agriculture must not be followed in planning and demonstration. During this process, Guangyuan City has always followed the low-carbon and efficient agricultural developmental concept, and improved its agricultural infrastructure according to the new idea of "industrial park supporting facilities and integration of parks and villages" and by taking the construction of industrial parks as the core and the construction of small ecological affluent villages as the point of strength; moreover, it has continuously enhanced the level of modernization of the industries in the parks and nearby regions, improved the production and living conditions of rural residents, and strengthened the fundermental position of agriculture. By the end of 2012, the city had totally constructed 51 modern agricultural demonstration parks with an area of the core zone up to 520,000 mu (1 mu = 0.0667 hectare), and there are more than 150 enterprises and special cooperative organizations in the parks, wherein Cangxi County has been identified as a national modern agricultural demonstration area. The city has established a standardized production base for national green food raw materials with an area of 790,000 mu, and a nationally-certificated "three products" production base with an area of 1,930,000 mu. Influence and market shares of "Seven Guangyuan Specialties," "Jianmen Pass Local Chicken" and other characteristic competitive brands have gradually expanded. By cooperating with the Environmental Defense Fund of the USA, the city has been carrying out rural soil testing for formulated fertilization and greenhouse gas emission reduction projects; by cooperating with World Expo and the Guangzhou Asian Games, the city has implemented "green travel" carbon-neutral projects, etc. The amount of carbon trading projects declared and accepted is about 200 million RMB. In the six years since the pilot work began, both agricultural production and income have increased continuously, rural per capital net income has reached 5,649 RMB, doubled compared with that of 2007, and endogenous power of rural development has been constantly enhanced.

3.3 *Create Small Green and Low-carbon Towns*

During the new urbanization process, small towns have the following obvious advantages: low cost, low threshold, easy to group together the local population and provide employment; little difference in urban-rural form, great industrial relevancy, strong integration of humans' life; sufficient capacity of the environment, and huge developmental potential. Research shows that, attaching importance to and accelerating the development of small towns from the strategic perspective is an important path towards the realization of local or nearby urbanization for more than 1/3 of the rural population of China in the future. The scientific planning and construction of small towns is a strategic choice for overall urban-rural development against the background of new urbanization. The administrative region of Guangyuan City is 16,314 km^2 in area, and its governance covers 248 towns in 4 counties and 3 districts, including 85 organic towns below county level and 163 townships. The city is located where 3 geomorphological units, including Micang Mountain, Longmen Mountain and Penbei Low Mountain. Based on its special geographic position and complex natural conditions, the development of the small towns of Guangyuan City must persist in scientifically dealing with the relationship between developing characteristic industries and accelerating urban construction as well as between preventing and controlling environmental pollution and strengthening environmental protection; this development must fully consider the carrying capacity of environmental resources, and stick to the economical and intensive, green and low-carbon sustainable developmental path. First, guide the rational layout and coordinated development of the small towns. According to the level of economic and social development, location characteristics, resource endowment and environmental basis, and combined with the sustaining capacity of resources and the environment, reasonably set goals for the development of urbanization in all regions, formulate urbanization strategies and relevant policies and measures according to local conditions, guide the adjacent small towns in certain regions to focus on industrial relevancy, and achieve functional complementation, industrial interaction and coordinated development. Second, pay attention to the organic unity of small town construction and comprehensive utilization of resources. Leave the traditional developmental pattern with capital drive, resource consumption and extensive management, actively develop new energy technology represented by solar energy and biogas according to local conditions, develop a circular economy, and protect the ecological environment, in order to gain maximum benefits with minimum resource consumption. Third, strengthen the construction of ecological infrastructures in small towns. According to the requirements and standards of national environmental policies, solve such problems as garbage disposal, sewage treatment and environmental health improvement of key towns, and realize the harmonious coexistence of human beings and nature. Strive for national and provincial ecological environmental compensation funds, coordinate downstream cities on the Jialing River to support and compensate ecological environmental protection of Guangyuan City, and promote the city to rapidly construct a batch of small ecological towns with certain low-carbon scale effects like Baolun, Zhaohua and Jianmen Pass. After 2008, based on post-disaster reconstruction and the construction of the Tingzikou Water Control Project, by consciously guiding the transfer and regrouping of the population group the people from disaster-stricken areas and reservoir areas, the level of urbanization of the city has rapidly increased with an average annual speed of over 1 percentage point. In 2012, the total urban population of Guangyuan City exceeded 710,000, and the level of urbanization reached 36.42%.

3.4 *Strengthen Ecological Construction and Environmental Protection*

First, improve environmental sustaining capacity. By taking the construction of an ecological and livable city and a low-carbon pilot city as a catalyst, implement projects of the ecological construction of Guangyuan City, strengthen forest land conservation and forestry build-up, vigorously return farmland to forests or grassland, create artificial ecological public welfare forests, develop plantations for industrial utilization, improve the coverage and quality of the vegetation, and increase environmental capacity. In 2012, the rate of forest coverage of the city reached 57%, and the annual carbon sequestration capacity of the ecological system reached 13 million tons; the rate of green coverage of built-up areas was 41.2%, forestry land area was 15.15 million mu, living wood growing stock was 52.12 million m^3, the number of days with good urban air quality was 364 or above for three successive years, and the quality of the water in all of the drinking water sources was up to surface water Class III water quality standards. Second, strengthen the construction of infrastructures. Rationally arrange and plan household garbage disposal facilities in the city, establish and improve the pattern of the integrated treatment of urban-rural household garbage, and promote the level of urban-rural environmental ecological hardware. Earnestly implement key ecological projects such as comprehensive treatment of water and soil loss and rural energy construction, reinforce the comprehensive treatment of mountains, water, forests, fields and roads, intensify the greening of the town, passageways, and new residential quarters, reinforce the treatment of rural non-point source pollution, and focus on the improvement of rural ecology, in order to build a human settlement with picturesque scenery, clean and tidy villages and a beautiful environment. It is expected that by 2020 more than 90% of the villages in the city will have been built into ecological affluent villages, and more than 95% of the rural households will have become ecological affluent households. Third, promote a low-carbon lifestyle. Rationally arrange urban space, and strictly carry out related national and provincial energy-saving standards and regulations. Optimize a low-carbon approach to transportation, implement the urban low-carbon transportation developmental concept with an orientation towards "public transport," and promote low-carbon energy-saving means of transportation. Set August 27 as the Low-carbon Day of Guangyuan City, actively advocate a low-carbon lifestyle as the new fashion for urban and rural residents, cultivate and name 36 low-carbon developmental demonstration units in two batches, energetically construct green transportation networks, actively promote energy-saving technology and products, and carry out the "5 entries" (enter organs and units, enter communities, enter schools, enter villages, and enter enterprises) activities of low-carbon production and life modes. Remarkable results have been achieved.

Through efforts of over seven years, the low-carbon urban construction of Guangyuan City has successively won the following titles: the China Habitat Environment Example Prize, the "Chinese City of Low-carbon Contribution," the "Outstanding Low-carbon Development Contribution City," the "Ecologically

Advanced Low-carbon City," the "Outstanding Green Ecological City," the "National Low-carbon Pilot City," etc.

4 Main Enlightenments

Practices of Guangyuan City in synergistically promoting overall urban-rural planning and low-carbon development are of great practical significance due to the following enlightenments.

4.1 *Clarify Thoughts with Scientific Guidance*

Thought determines the way out, and concept is also productivity. Urban-rural integrated development is a great revolution in urban-rural economy and society, and the development of a low-carbon economy is also a new undertaking without existing experience for reference. For such remote and backward areas as Guangyuan City, compared with developed areas, their industrial structure is at an early stage of formation; with the strong elasticity of their economic pattern and economic system as well as low cost, small resistance and fast speed for the low-carbon transformation of industries, they can grasp comparative advantages to catch up and leap forward during economic transformation, avoiding detours and seizing the initiative. Guangyuan City considers carrying out a pilot reform of comprehensive support to the overall urban-rural planning and low-carbon urban construction as important opportunities to emancipate the mind and improve the thought; it continuously deepens its understanding, firmly seizes the opportunities, calmly tackles the challenges, closely combines the actual developmental conditions, scientifically formulates a general plan of a comprehensive supporting reform of the overall urban-rural planning and the plan of implementation of the low-carbon pilot city work of Gaungyuan City; it continuously improves developmental ideas, and guides various tasks of the overall urban-rural development and low-carbon development with appropriate ideas.

4.2 *Strengthen Leadership with Overall Planning*

Overall urban-rural planning and low-carbon development cover a wide range, including various aspects of the economy, politics, culture, society and ecology. To do the work well, the city must strengthen its organizational leadership, integrate the powers of all those involved, and uphold with overall planning. Guangyuan City established a leading group for overall urban-rural development and for the pilot comprehensive supporting reform as well as a leading group to deal with municipal low-carbon development, and all counties and districts as well as municipal departments concerned also built corresponding working institutions. Moreover, Guangyuan City established the Guangyuan Overall Urban-rural Planning Work Council, and took the lead in China in setting up a municipal low-carbon development bureau and low-carbon economic development institute to guide the comprehensive supporting reform of overall urban-rural planning and low-carbon development work of the city, creating a good pattern with active participation and voluntary planning of all those involved within the city. Guangyuan City maintained a "board of chess" concept with overall planning and classified guidance, and prepared the *Urban System Planning, Overall Planning of Towns, Overall Planning of New Village Construction in County Areas, New Village (Agglomeration) Planning,* and *Low-carbon Economic Developmental Plan during the 12th Five-Year Plan,* etc., ensuring full coverage of all regions; the city completed the low-carbon developmental roadmap of Guangyuan City, and took the lead in Sichuan Province in preparing the *Low-carbon Economic Development Plan during the 12th Five-Year Plan* and the *Implementation Plan of the National Low-carbon City Pilot Work of Guangyuan City*, and formulated the *Opinions on Constructing a Low-carbon Circular Economic Industrial Park and Realizing Low-carbon Development* and other series of measures to support low-carbon development. Unified planning prevents short-term actions of urban-rural construction, effectively avoids repeated construction, prevents vanity projects and unplanned projects in low-carbon construction, and stops "first development and then treatment" activities.

4.3 *Adopt Measures According to the Local Conditions with Classified Implementation*

Overall urban-rural planning and low-carbon development must be based on actual conditions, with classified implementation and steady progress under the guidance of unified planning, and it must avoid "one size fits all" and "the same" actions. According to regional geographic characteristics and the state of its economic development, Guangyuan City is divided into three economic function areas, including the valley corridor economic zone in the middle, the low mountain and hill economic zone in the south and the mountainous economic zone in the north. The valley corridor economic zone in the middle has stuck to the concept of "industrial orientation and driving urban development through industries," actively developed emerging competitive characteristic industries, realized the agglomeration of such factors as population and resources, and built a characteristic industrial belt and a regional urban agglomeration. The low mountain and hill economic zone in the south has developed characteristic ecological agricultural and sideline products and a processing industry on a large scale, accelerated the promotion of natural gas exploration, and built a comprehensive development and utilization base for natural gas. The mountainous economic zone in the north has based its activities on environmental protection, it has strengthened the comprehensive development of forestry in mountainous areas, intensely developed ecological tourism resources, vigorously developed characteristic industries such as agriculture and forestry, and built a characteristic mountainous economic zone. Facts show that, only with principles of being practical and realistic and adopting measures according to local conditions, can overall urban-rural planning and low-carbon development produce maximum synergistic effects.

4.4 *Increase Investments with Strengthened Security*

Capital investment is a guarantee of low-carbon urban development and overall urban-rural development. On the one hand, Guangyuan City has been raising funds for construction through multiple channels, explored and established fiscal, financial and insurance systems serving urban-rural integration and low-carbon development, actively strived for great support from the provincial Party committee, the provincial government and all departments, and increased its capital investment in urban-rural integrated development and low-carbon pilot city construction year by year, providing a strong guarantee for overall urban-rural development and low-carbon urban construction, and for promoting coordinated and rapid development of the urban-rural economy and society. On the other hand, the city has been organically integrating various policies, and has been using post-disaster reconstruction funds for scientific development. With national continuous poverty alleviation work and the implementation of the policy of ecological compensation mechanism of the main functional area, based on the cooperation with the Institute for Urban and Environmental Studies of the Chinese Academy of Social Sciences, developmental planning and path design provide sufficient financial and intellectual support for Guangyuan City, thus expanding good space for development.

4.5 *Stick to Reform with a Drive towards Innovation*

Reform is an eternal power, and innovation provides permanent vitality. The most important thing in overall urban-rural development is to establish a practical system and mechanism for low-carbon development and overall urban-rural development, and the most fundamental thing is to innovate through reform in order to break through system obstacles and repair system flaws. Guangyuan City has always insisted on taking comprehensive supporting reform as the impetus and low-carbon system construction as a support, focused on the establishment of an urban-rural sustainable system and mechanism, and has stuck to independent innovation through innovation driving and strategy promotion, to release a strong vitality for overall urban-rural planning and low-carbon development. For instance, Guangyuan City has innovated its developmental pattern of "integration of industry and city," "integration of park and city" and "integration of industry and village," created many overall urban-rural planning complexes and low-carbon developmental demonstration sites such as the Sanjiang New Area, the 081 New Industry Area, Huilonghe in Lizhou, Lujiaba in Wangcang, Sanguo Township in Qingchuan, Jianping Village in Jiange and Sanjing Village in Cangxi, and carried out integrated development of industry and city, park and city, as well as industry and village, well avoiding such problems as city hollowness, park isolation and marginalization of rural areas.

In 2007, Guangyuan City started to carry out the overall plan for the pilot urban-rural comprehensive reform; in 2008, as an area severely-hit by the earthquake, Guangyuan City started low-carbon reconstruction, and has embarked on the low-carbon economic developmental path since then; in 2012, Guangyuan City was identified as one of the second batch of national low-carbon pilot cities by the National Development and Reform Commission. Practices in Guangyuan City show that reform is the core of and key to overall urban-rural planning; we must promote urban-rural development through reform, or there will be no new urbanization, industrialization and agricultural modernization. The key to a low-carbon economy is development. A low-carbon economy needs sound development, but not slow development, and seeks sustainable development in breaking the extensive economic pattern. With joint promotion of twin engines, i.e. overall urban-rural planning and low-carbon development, Guangyuan City has experienced significant changes in its society and economy and is achieving a new leap forward.

CHAPTER 19

Socio-economic, Energy and Carbon Emission Indexes of Various Countries, Regions and Cities

Data Handling: ZHU Shouxian

TABLE 19.1 *Human development index and relevant index of various countries and regions*

HDI ranking	Country (region)	Human development index (HDI)	Life expectancy (age)	Per capita gross national income (GNI) (USD)	Urbanization rate (%)	Dependency ratio (%)
Extremely high human development level						
1	Norway	0.955	81.3	48688	79.7	51.5
2	Australia	0.938	82	34340	89.4	49.3
3	U.S.A.	0.937	78.7	43480	82.6	50.7
4	The Netherlands	0.921	80.8	37282	83.6	50.6
5	Germany	0.92	80.6	35431	74.1	51.7
6	New Zealand	0.919	80.8	24358	86.3	51.4
7	Ireland	0.916	80.7	28671	62.5	50.8
8	Sweden	0.916	81.6	36143	85.4	55.5
9	Switzerland	0.913	82.5	40527	73.8	47.9
10	Japan	0.912	83.6	32545	91.9	59.6
11	Canada	0.911	81.1	35369	80.8	45.1
12	Korea	0.909	80.7	28231	83.5	38
13	Hongkong, China	0.906	83	45598	100	32.3
14	Iceland	0.906	81.9	29176	93.8	49.6
15	Denmark	0.901	79	33518	87.1	54.1
16	Israel	0.9	81.9	26224	91.9	61.6
17	Belgium	0.897	80	33429	97.5	53.3

* Zhu Shouxian, Ph.D., associate researcher of the Institute for Urban and Environmental Studies of the Chinese Academy of Social Sciences.

DOI 10.1163/9789004280021_020

HDI ranking	Country (region)	Human development index (HDI)	Life expectancy (age)	Per capita gross national income (GNI) (USD)	Urbanization rate (%)	Dependency ratio (%)
Extremely high human development level						
18	Austria	0.895	81	36438	67.9	48.1
19	Singapore	0.895	81.2	52613	100	35.4
20	France	0.893	81.7	30277	86.4	55.7
21	Finland	0.892	80.1	32510	83.8	53.5
22	Slovenia	0.892	79.5	23999	49.8	45
23	Spain	0.885	81.6	25947	77.6	48.4
24	Liechtenstein	0.883	79.8	84880	14.3	
25	Italy	0.881	82	26158	68.5	53.8
26	Luxembourg	0.875	80.1	48285	85.7	46.1
27	UK	0.875	80.3	32538	79.7	52.7
28	Czech	0.873	77.8	22067	73.4	42.9
29	Greece	0.86	80	20511	61.7	50.6
30	Brunei	0.855	78.1	45690	76.4	41.6
31	Cyprus	0.848	79.8	23825	70.7	41.4
32	Malta	0.847	79.8	21184	95	42.1
33	Andorra	0.846	81.1	33918	86.7	
34	Estonia	0.846	75	17402	69.5	50
35	Slovakia	0.84	75.6	19696	54.7	37.9
36	Qatar	0.834	78.5	87478	98.9	18.3
37	Hungary	0.831	74.6	16088	69.9	46.2
38	Barbados	0.825	77	17308	44.9	40
39	poland	0.821	76.3	17776	60.8	40.5
40	Chile	0.819	79.3	14987	89.4	45.2
41	Lithuania	0.818	72.5	16858	67.2	44.9
42	The United Arab Emirates	0.818	76.7	42716	84.7	20.9
43	Portugal	0.816	79.7	19907	61.6	50
44	Latvia	0.814	73.6	14724	67.7	47.3
45	Argentina	0.811	76.1	15347	92.7	54.4
46	Seychelles	0.806	73.8	22615	54	—
47	Croatia	0.805	76.8	15419	58.I	47.9

TABLE 19.1 *Human development index and relevant index* (cont.)

HDI ranking	Country (region)	Human development index (HDI)	Life expectancy (age)	Per capita gross national income (GNI) (USD)	Urbanization rate (%)	Dependency ratio (%)
High human development level						
48	Bahrain	0.796	75.2	19154	88.7	29.2
49	Bahamas	0.794	75.9	27401	84.5	40.9
50	Belarus	0.793	70.6	13385	75.5	40.5
51	Uruguay	0.792	77.2	13.333	92.6	56.2
52	Montenegro	0.791	74.8	10471	63.5	46.5
53	Palau	0.791	72.1	11463	85.1	
54	Kuwait	0.79	74.7	52793	98.3	41.1
55	Russia	0.788	69.1	14461	74	39.8
56	Romania	0.786	74.2	11011	52.8	43.6
57	Bulgaria	0.782	73.6	11474	73.7	47.3
58	Saudi Arabia	0.782	74.1	22616	82.5	49
59	Cuba	0.78	79.3	5539	75.1	41.8
60	Panama	0.78	76.3	13519	75.9	54.3
61	Mexico	0.775	77.1	12947	78.4	5.3.5
62	Costa Rica	0.773	79.4	10863	65.1	44.5
63	Grenada	0.77	76.1	9257	39.5	51.9
64	Libya	0.769	75	13765	77.9	55
65	Malaysia	0.769	74.5	13676	73.5	52.8
66	Serbia	0.769	74.7	9533	56.7	46.7
67	Antigua and Barbuda	0.76	72.8	13883	29.8	
68	Trinidad and Tobago	0.76	70.3	21941	14	38.6
69	Kazakhstan	0.754	67.4	10451	53.5	47.2
70	Albania	0.749	77.1	7822	54.5	46.1
71	Venezuela	0.748	74.6	11475	93.7	53.3
72	Dominica	0.745	77.6	10977	67.2	
73	Georgia	0. 745	73.9	5005	52.9	44.8
74	Lebanon	0.745	72.8	12364	87.4	45.1
75	Saint Kitts and Nevis	0. 745	73.3	12460	32	

HDI ranking	Country (region)	Human development index (HDI)	Life expectancy (age)	Per capita gross national income (GNI) (USD)	Urbanization rate (%)	Dependency ratio (%)
High human development level						
76	Iran	0.742	73.2	10695	69.2	38.7
77	Peru	0.741	74.2	9306	77.6	54.9
78	Macedonia	0.74	75	9377	59.4	41.2
79	Ukraine	0.74	68.8	6428	69.1	42.8
80	Mauritius	0.737	73.5	13300	41.8	39.6
81	Bosnia and Herzegovina	0.735	75.8	7713	48.8	40.5
82	Azerbaijan	0.734	70.9	8153	53.9	38.3
83	Saint Vincent and the Grenadines	0.733	72.5	9367	49.7	48.3
84	Oman	0.731	73.2	24092	73.7	42.8
85	Brazil	0.73	73.8	10152	84.9	46.8
86	Jamaica	0.73	73.3	6701	52.1	55.9
87	Armenia	0.729	74.4	5540	64.1	45.3
88	Saint Lucia	0.725	74.8	7971	16.8	46.9
89	Ecuador	0.724	75.8	7471	68	56.3
90	Turkey	0.722	74.2	13710	72.5	46.8
91	Columbia	0.719	73.9	8711	75.6	51.5
92	Sri Lanka	0.715	75.1	5170	15.2	50.6
93	Algeria	0.713	73.4	7418	73.8	45.6
94	Tunisia	0.712	74.7	8103	66.5	43.2
Medium human development level						
95	Tonga	0.71	72.5	4153	23.5	76.1
96	Belize	0.702	76.3	5327	44.5	60.8
97	Dominican	0.702	73.6	8506	70.3	58.3
98	Fiji	0.702	69.4	4087	52.6	51.7
99	Samoa	0.702	72.7	3928	19.6	72.7
100	Jordan	0.7	73.5	5272	83	66.9
101	Mainland China	0.699	73.7	7945	51.9	37.6
102	Turkmenistan	0.698	65.2	7782	49	48.4

TABLE 19.1 *Human development index and relevant index* (cont.)

HDI ranking	Country (region)	Human development index (HDI)	Life expectancy (age)	Per capita gross national income (GNI) (USD)	Urbanization rate (%)	Dependency ratio (%)
Medium human development level						
103	Thailand	0.69	74.3	7722	34.4	41.1
104	Maldives	0.688	77.1	7478	42.3	43.6
105	Suriname	0.684	70.8	7327	70.1	52.3
106	Gabon	0.683	63.1	12521	86.5	64
107	Salvatore	0.68	72.4	5915	65.3	60.6
108	Bolivia	0.675	66.9	4444	67.2	66.9
109	Mongolia	0.675	68.8	4245	69.5	46.8
110	Palestine	0.67	73	3359	74.6	79.5
111	Paraguay	0.669	72.7	4497	62.5	61.4
112	Egypt	0.662	73.5	5401	43.6	57.2
113	Moldova	0.66	69.6	3319	48.4	38.8
114	The Philippines	0.654	69	3752	49.1	62.4
115	Uzbekistan	0.654	68.6	3201	36.2	48.7
116	Syria	0.648	76	4674	56.5	65.2
117	Micronesia	0.645	69.2	3352	22.7	65.1
118	Guyana	0.636	70.2	3387	28.4	55.8
119	Botswana	0.634	53	13102	62.3	56.7
120	Honduras	0.632	73.4	3426	52.7	66.9
121	Indonesia	0.629	69.8	4154	51.5	47.3
122	Kiribati	0.629	68.4	3079	44	
123	South Africa	0.629	53.4	9594	62.4	52.9
124	Vanuatu	0.626	71.3	3960	25.2	70
125	Kyrgyzstan	0.622	68	2009	35.4	51.9
126	Tajikistan	0.622	67.8	2119	26.5	65.3
127	Vietnam	0.617	75.4	2970	31.7	40.9
128	Namibia	0.608	62.6	5973	39	64.8
129	Nicaragua	0.599	74.3	2551	57.8	61.2
130	Morocco	0.591	72.4	4384	57.4	49.2
131	Iraq	0.59	69.6	3557	66.4	84.3
132	Cape Verde	0.586	74.3	3609	63.4	55.8

HDI ranking	Country (region)	Human development index (HDI)	Life expectancy (age)	Per capita gross national income (GNI) (USD)	Urbanization rate (%)	Dependency ratio (%)
Medium human development level						
133	Guatemala	0.581	71.4	4235	50.2	82.4
134	Timor-Leste	0.576	62.9	5446	28.7	93
135	Ghana	0.558	64.6	1684	52.6	73
136	Equatorial Guinea	0.554	51.4	21715	39.6	72
137	India	0.554	65.8	3285	31.6	53.8
138	Cambodia	0.543	63.6	2095	20.1	53.2
139	Laos	0.543	67.8	2435	35.4	58.4
140	Bhutan	0.538	67.6	5246	36.4	49.7
141	Swaziland	0.536	48.9	5104	21.2	69.4
Low human development level						
142	The Republic of Congo	0.534	57.8	2934	64.1	79.3
143	Solomon Islands	0.53	68.2	2172	20.9	74.1
144	Sao Tome and Principe	0.525	64.9	1864	63.4	75.8
145	Kenya	0.519	57.7	1541	24.4	82.1
146	Bangladesh	0.515	69.2	1785	28.9	53
147	Pakistan	0.515	65.7	2566	36.5	63.4
148	Angola	0.SOS	51.5	4812	60	93.9
149	Myanmar	0.498	65.7	1817	33.2	43
150	Cameroon	0.495	52.1	2114	52.7	78.3
151	Madagascar	0.483	66.9	828	33.2	83.7
152	Tanzania	0.476	58.9	1383	27.2	92.6
153	Nigeria	0.471	52.3	2102	50.3	86.1
154	Senegal	0.47	59.6	1653	42.8	84.3
155	Mauritania	0.467	58.9	2174	41.7	73.1
156	Papua New Guinea	0.466	63.1	2386	12.5	70.3
157	Nepal	0.463	69.1	1137	17.3	64.1
158	Lesotho	0.461	48.7	1879	28.3	69.1

TABLE 19.1 *Human development index and relevant index* (cont.)

HDI ranking	Country (region)	Human development index (HDI)	Life expectancy (age)	Per capita gross national income (GNI) (USD)	Urbanization rate (%)	Dependency ratio (%)
Low human development level						
159	Togo	0.459	57.5	928	38.5	73.6
160	Yemen	0.458	65.9	1820	32.9	86.4
161	Haiti	0.456	62.4	1070	54.8	65.5
162	Uganda	0.456	54.5	1168	16	103.1
163	Zambia	0.448	49.4	1358	39.6	99
164	Djibouti	0.445	58.3	2350	77.1	62.8
165	Gambia	0.439	58.8	1731	57.9	83.8
166	Benin	0.436	56.5	1439	45.6	86.9
167	Rwanda	0.434	55.7	1147	19.4	84.2
168	Ivory Coast	0.432	56	1593	52	79.3
169	Comoros	0.429	61.5	986	28.1	82.8
170	Malawi	0.41 8	54.8	774	15.8	96.3
171	Sultan	0.414	61.8	1848	33.3	76.0
172	zimbabwe	0.397	52.7	424 t	39.1	71.6
173	Ethiopia	0.396	59.7	1017	17.2	77.3
174	Liberia	0.388	57.3	480	48.5	86
175	Afghanistan	0.374	49.1	1000	23.8	92.6
176	Guinea-Bissau	0.364	48.6	1042	44.6	79.7
177	Sierra Leone	0.359	48.1	881	39.6	80.8
178	Burundi	0.355	50.9	544	11.2	67.7
179	Guinea	0.355	54.5	941	35.9	85
180	Central African	0.352	49.1	722	39.3	78
181	Eritrea	0.35 1	62	531	21.8	78.9
182	Mali	0.344	51.9	853	35.6	97.3
183	Burkina Faso	0.343	55.9	1202	27.4	90.5
184	Chad	0.34	49.9	1258	21.9	92.6
185	Mozambique	0.327	50.7	906	31.4	89.1
186	The Democratic Republic of the Congo	0.304	48.7	319	34.8	94
187	Niger	0.304	55.1	701	18.1	104.8

HDI ranking	Country (region)	Human development index (HDI)	Life expectancy (age)	Per capita gross national income (GNI) (USD)	Urbanization rate (%)	Dependency ratio (%)
HDI group						
	Extremely high human development level	0.905	80.1	33391	81.2	50.3
	High human development level	0.758	73.4	11501	74.1	46.4
	Medium human development level	0.64	69.9	5428	43.7	47
	Low human development level	0.466	59.1	1633	33.6	75.5
Region						
	Arab countries	0.652	71	8317	57.2	59.7
	East Asia and Pacific region	0.683	72.7	6874	49.7	40.9
	Europe and Central Asia	0.771	71.5	12243	64.8	43.4
	Latin America and Caribbean area	0.741	74.7	10300	79.3	52.1
	South Asia	0.558	66.2	3343	32.9	54.6
	Sub-Saharan Africa	0.475	54.9	2010	37	83.4
	Least-developed countries	0.449	59.5	1385	28.9	75.5
	Small Island Developing States	0.648	69.8	5397	52.6	57.3
	The World	0.694	70.1	10184	52.6	52.0

NOTES:

(1) Human development index (HDI): it is a composite index used to evaluate the average achievements of the dimensions of three basic human developments (life expectancy, education and living standard). The classification standard of the human development index is a relative standard and is based on the quartile of 187, the number of all countries. It is classified into extremely high, high, medium and low human development index. There are 47 countries in the first three classifications and 46 countries in the last classification.
(2) Life expectancy: it is a statistical measure of the time a newborn is expected to live based on the unchanged death rate of all age groups at birth.

(3) Per capita gross national income (GNI): it is a figure obtained by the gross domestic product plus incomes earned through production factors, minus the income earned through foreign production factors, and then is converted into the international USD based on the 2005 PPP rate, and then divided by the total population. Data in the Table are calculated based on the 2005 PPP.
(4) Dependency ratio: it is a ratio of the population aged from 0 to 14 and at or above 65, and the population aged from 15 to 64. Data source: http://hdr.undp.org/en/reports/global/hdr2013/. http://datacatalog.worldbank.org/.

TABLE 19.2 *Energy consumption index of the main countries and regions in the world (2012)*

Country (region)	Carbon emission (10^6 t CO_2)	Energy consumption (10^6 toe)	Population (10^3)	GDP (current price, 10^6 USD)	Per capita carbon emission (t CO_2/person)	Carbon productivity (USD/t CO_2)	Energy emission coefficient (t CO_2/toe)
U.S.A.	5786.1	2208.8	313914	15684800	18.43	2710.77	2.620
Canada	619.6	328.8	34880	1821424	17.76	2939.68	1.884
Mexico	496	187.7	120847	1177271	4.10	2373.53	2.643
Total North America	6901.7	2725.3	469641	18683495	40.29	8023.98	7.147
Argentina	190.5	82.1	41087	474865	4.64	2492.73	2.320
Brazil	500.5	274.7	198656	2252664	2.52	4500.83	1.822
Chile	94.2	35.5	17465	268314	5.39	2848.34	2.654
Columbia	75.6	36.5	47704.43	369813	1.58	4891.71	2.071
Ecuador	35	14.4	15492.26	84532	2.26	2415.20	2.431
Peru	48.3	22.3	29987.8	197111	1.61	4080.97	2.166
Trinidad and Tobago	50.9	21.2	1337.439	23986	38.06	471.24	2.401
Venezuela	187.1	86.8	29954.78	382424	6.25	2043.96	2.156
Total Latin America	1182.1	573.5	581384	5348266	2.39	3853.22	2.086
Austria	65.1	33.1	8462.446	399649	7.69	6139.00	1.967
Azerbaijan	30.8	12.2	9297.507	67198	3.31	2181.75	2.525
Belarus	67.4	25.9	9464	63267	7.12	938.68	2.602
Belgium	141.5	60.6	11142.16	483709	12.70	3418.44	2.335
Bulgaria	45.2	17.9	7304.632	51030	6.19	1128.98	2.525

Country (region)	Carbon emission (10^6 t CO_2)	Energy consumption (10^6 toe)	Population (10^3)	GDP (current price, 10^6 USD)	Per capita carbon emission (t CO_2/person)	Carbon productivity (USD/t CO_2)	Energy emission coefficient (t CO_2/toe)
Czech	111.1	41.9	10514.81	195657	10.57	1761.09	2.652
Denmark	41.4	17	5590.478	314242	7.41	7590.39	2.435
Finland	45.9	26.4	5414.293	250024	8.48	5447.15	1.739
France	383.4	245.4	65697	2612878	5.84	6815.02	1.562
Germany	815	311.7	81890	3399589	9.95	4171.27	2.615
Greece	85.6	28.8	11280.17	249099	7.59	2910.04	2.972
Hungary	50.6	21.9	9943.755	125508	5.09	2480.40	2.311
Ireland	34.3	13	4588.798	210331	7.47	6132.10	2.638
Italy	406.3	162.5	60918	2013263	6.67	4955.11	2.500
Kazakhstan	197.9	58.1	16797.46	201680	11.78	1019.10	3.406
Lithuania	15.7	6	2985.509	42246	5.26	2690.83	2.617
The Netherlands	246	89.1	16767.71	772227	14.67	3139.13	2.761
Norway	44.8	48.1	5018.869	499667	8.93	11153.28	0.931
Poland	326	97.6	38542.74	489795	8.46	1502.44	3.340
Portugal	54.9	22.7	10526.7	212454	5.22	3869.84	2.419
Romania	81.9	33.6	21326.91	169396	3.84	2068.33	2.438
Russia	1704.3	694.2	143533	2014775	11.87	1182.17	2.455
Slovakia	36.2	16.9	5410.267	91619	6.69	2530.91	2.142
Spain	338.6	144.8	46217.96	1349351	7.33	3985.09	2.338
Sweden	50.5	52.9	9516.617	525742	5.31	10410.73	0.955
Switzerland	41.8	29	7997.152	632194	5.23	15124.26	1.441
Turkey	318.5	119.2	73997	789257	4.30	2478.04	2.672
Turkmenistan	64	25.8	5172.931	33679	12.37	526.23	2.481
Ukraine	321.7	125.3	45593.3	176309	7.06	548.05	2.567
UK	530.5	203.6	63228	2435174	8.39	4590.34	2.606
Uzbekistan	118.1	50.5	29776.85	51113	3.97	432.79	2.339
Europe and Central Asia	7037.5	2928.5	892490.2	21185280	7.89	3010.34	2.403
Iran	608.3	234.2	76424	514060	7.96	845.08	2.597
Israel	81.9	24.8	7907.9	242929	10.36	2966.17	3.302
Kuwait	100.6	36.4	3250.496	176590	30.95	1755.37	2.764
Qatar	79.9	31.6	2050.514	172982	38.97	2164.98	2.528

TABLE 19.2 *Energy consumption index of the main countries and regions* (cont.)

Country (region)	Carbon emission (10^6 t CO_2)	Energy consumption (10^6 toe)	Population (10^3)	GDP (current price, 10^6 USD)	Per capita carbon emission (t CO_2/person)	Carbon productivity (USD/t CO_2)	Energy emission coefficient (t CO_2/toe)
Saudi Arabia	615.3	222.2	28288	576824	21.75	937.47	2.769
The United Arab Emirates	233.2	89.3	9205.651	360245	25.33	1544.79	2.611
Middle East	2062.8	761.9	225378.3	2532143	9.15	1227.53	2.707
Algeria	116.5	44.6	38482	207955	3.03	1785.02	2.612
Egypt	223.8	87.1	80722	257286	2.77	1149.62	2.569
Africa	446.1	123.8	51189	384313	8.71	861.50	3.603
Total Africa	1156.9	403.3	1068444	2006779	1.08	1734.62	2.869
Australia	392.2	125.7	22684	1520608	17.29	3877.12	3.120
Bangladesh	66.1	26.3	154695	115610	0.43	1749.02	2.513
Mainland China	9208.1	2735.2	1350695	8227103	6.82	893.46	3.367
Hongkong, China	91.1	28.1	7155	263259	12.73	2889.78	3.242
India	1823.2	563.5	1236687	1841717	1.47	1010.16	3.235
Indonesia	495.2	159.4	246864	878043	2.01	1773.11	3.107
Japan	1409	478.2	127561	5959718	11.05	4229.75	2.946
Malaysia	218.8	76.3	29239.93	303526	7.48	1387.23	2.868
New Zealand	37.2	19.6	4433.1	139768	8.39	3757.20	1.898
Pakistan	165.9	69.3	179160	231182	0.93	1393.50	2.394
The Philippines	84.2	30.2	96706.76	250265	0.87	2972.27	2.788
Singapore	220.7	74	5312.4	274701	41.54	1244.68	2.982
Korea	763.7	271.1	50004	1129598	15.27	1479.1 1	2.817
Taiwan, China	326.5	109.4	23315.82	475036.6	14.00	1454.94	2.984
Thailand	332.3	117.6	66785	365564	4.98	1100.10	2.826
Vietnam	130.1	52	88776	141669	1.47	1088.92	2.502
Total Asia Pacific	15919.1	4992.2	3856679	22434758	4.13	1409.30	3.189

Country (region)	Carbon emission (10^6 t CO_2)	Energy consumption (10^6 toe)	Population (10^3)	GDP (current price, 10^6 USD)	Per capita carbon emission (t CO_2/person)	Carbon productivity (USD/t CO_2)	Energy emission coefficient (t CO_2/toe)
Total world	34466.1	12476.6	7046369	71666350	4.89	2079.33	2.762
OECD	13899.8	5488.8	1256610	46133956	11.06	3319.04	2.532
Non-OECD	20566.3	6987.8	5789759	25532394	3.55	1241.47	2.943
EU	3977.5	1673.4	504769.8	16653838	7.88	4187.01	2.377

DATA SOURCE: http://www.bp.com/en/global/corporate/about-bp/statistical-review-of-world-energy-2013.html. http://datacatalog.worldbank.org/.

TABLE 19.3 *Structure of energy consumption of the main countries and regions in the world (2012)*

Country (region)	Energy consumption (10^6 toe)	Structure of energy consumption (%)					
		Petroleum	Natural gas	Coal	Nuclear energy	Hydropower	Other renewable energies
U.S.A.	2208.8	37.12	29.61	19.82	8.3	2.86	2.3
Canada	328.8	31.72	27.56	6.65	6.59	26.16	1.31
Mexico	187.7	49.32	40.1	4.71	1.06	3.77	1.05
Total North America	2725.4	37.31	30.09	17.19	7.59	5.74	2.09
Argentina	82.1	34.34	51.84	1.19	1.71	10.19	0.73
Brazil	274.7	45.73	9.56	4.9	1.32	34.4	4.09
Chile	35.5	50.39	15.24	18.83	0	13	2.54
Columbia	36.5	34.76	24.25	10.97	0	29.54	0.48
Ecuador	14.4	76.16	4.12	0	0	19.25	0.47
Peru	22.3	42.91	30.28	3.57	0	22.39	0.85
Trinidad and Tobago	21.2	7.65	92.32	0	0	0	0.02

TABLE 19.3 *Structure of energy consumption of the main countries and regions* (cont.)

Country (region)	Energy consumption (10^6 toe)	Structure of energy consumption (%)					
		Petroleum	Natural gas	Coal	Nuclear energy	Hydropower	Other renewable energies
Venezuela	86.8	42.16	36.22	0.25	0	21.37	0
Total Latin America	665.3	45.42	22.33	4.24	0.76	24.91	2.35
Austria	33.1	37.6	24.59	5.95	0	26.91	4.95
Azerbaijan	12.2	34.4	62.22	0.02	0	3.36	0
Belarus	25.9	35.16	64.43	0.18	0	0.04	0.19
Belgium	60.6	50.49	25.14	4.92	15.05	0.15	4.25
Bulgaria	17.9	21.26	13.69	39.2	19.94	4.08	1.84
Czech	41.9	21.65	17.57	39.66	16.36	1.64	3.12
Denmark	17	44.84	20.69	14.6	0	0.02	19.85
Finland	26.4	34.43	10.48	10.92	19.88	14.33	9.96
France	245.4	32.97	15.58	4.65	39.22	5.37	2.21
Germany	311.7	35.77	21.72	25.41	7.22	1.54	8.34
Greece	28.8	53.52	13.12	25.96	0	3.46	3.95
Hungary	21.9	27.18	40.04	13.55	16.31	0.22	2.7
Ireland	13	47.88	31.01	11.44	0	1.75	7.92
Italy	162.5	39.5	38.04	9.97	0	5.77	6.71
Kazakhstan	58.1	22.04	14.7	60.14	0	3.11	0
Lithuania	6	43.14	49.96	3.16	0	1.04	2.69
The Netherlands	89.1	49.55	36.82	9.51	1	0.03	3.09
Norway	48.1	22.42	8.05	1.37	0	67.18	0.97
Poland	97.6	25.75	15.28	55.29	0	0.47	3.21
Portugal	22.7	48.09	18.61	12.82	0	6.65	13.83
Romania	33.6	26.24	36.11	19.83	7.72	8.23	1.87
Russia	694.2	21.24	53.96	13.52	5.81	5.44	0.02
Slovakia	16.9	20.9	31.82	18.98	20.71	5.86	1.72
Spain	144.8	44.06	19.5	13.33	9.59	3.2	10.31
Sweden	52.9	26.07	1.87	2.76	27.62	33.67	8.01
Switzerland	29	38.72	9.92	0.48	19.97	29.53	1.38
Turkey	119.2	26.41	34.97	26.27	0	10.99	1.36
Turkmenistan	25.8	18.78	81.22	0	0	0	0

Country (region)	Energy consumption (10^6 toe)	Structure of energy consumption (%)					
		Petroleum	Natural gas	Coal	Nuclear energy	Hydropower	Other renewable energies
Ukraine	125.3	10.52	35.63	35.57	16.28	1.89	0.1
UK	203.6	33.63	34.61	19.21	7.83	0.58	4.14
Uzbekistan	50.5	7.76	85.3	2.42	0	4.52	0
Europe and Central Asia	2928.5	30.04	33.29	17.65	9.11	6.52	3.38
Iran	234.2	38.25	59.99	0.38	0.14	1.23	0.02
Israel	24.8	54.75	9.42	35.41	0	0.03	0.39
Kuwait	36.4	57.52	42.48	0	0	0	0
Qatar	31.6	25.42	74.58	0	0	0	0
Saudi Arabia	222.2	58.36	41.64	0	0	0	0
The United Arab Emirates	89.3	36.56	63.44	0	0	0	0.01
Middle East	761.9	49.33	48.64	1.29	0.04	0.67	0.02
Algeria	44.6	37.33	62.36	0	0	0.2	0.12
Egypt	87.1	40.46	54.37	1.3	0	3.47	0.4
Africa	123.8	21.73	2.74	72.52	2.6	0.35	0.06
Total Africa	403.3	41.29	27.41	24.18	0.8	5.98	0.35
Australia	125.7	37.12	18.22	39.22	0	3.23	2.21
Bangladesh	26.3	21.37	74.41	2.69	0	1.46	0.06
Mainland China	2735.2	17.68	4.73	68.49	0.81	7.12	1.17
Hongkong, China	28.1	63.93	9.01	27.06	0	0	0
India	563.5	30.45	8.71	52.93	1.33	4.64	1.93
Indonesia	159.4	44.93	20.22	31.64	0	1.82	1.39
Japan	478.2	45.63	21.97	26.01	0.85	3.83	1.71
Malaysia	76.3	39.08	39.31	18.79	0	2.41	0.42
New Zealand	19.6	35.46	19.33	8.83	0	26.25	10.13
Pakistan	69.3	28.87	53.85	6.14	1.86	9.28	0
The Philippines	30.2	42.86	10.17	31.08	0	8.12	7.76
Singapore	74	89.5	10.11	0	0	0	0.39
Korea	271.1	40.11	16.61	30.18	12.55	0.26	0.29
Taiwan, China	109.4	38.53	13.4	37.55	8.36	1.12	1.05

TABLE 19.3 *Structure of energy consumption of the main countries and regions* (cont.)

Country (region)	Energy consumption (10^6 toe)	Structure of energy consumption (%)					
		Petroleum	Natural gas	Coal	Nuclear energy	Hydropower	Other renewable energies
Thailand	117.6	44.56	39.21	13.56	0	1.67	0.99
Vietnam	52	31.97	16.27	28.73	0	22.97	0.06
Total Asia Pacific	4992.2	27.83	11.27	52.26	1.56	5.79	1.28
Total world	12476.6	33.11	23.94	29.9	4.49	6.66	1.9
OECD	5488.8	37.76	26.12	19.19	8.1	5.75	3.08
Non-OECD	6987.8	29.45	22.23	38.31	1.66	7.38	0.98
EU	1673.4	36.53	23.87	17.55	11.94	4.42	5.68

DATA SOURCE: http://www.bp.com/en/global/corporate/about-bp/statistical-review-of-world-energy-2013.html. http://datacatalog.worldbank.org/.

TABLE 19.4 *Carbon emission indexes of the major cities around the world*

City	Country	Carbo emission (10^6 t CO_2)	Carbon emission reduction			GDP (billion USD)	Area (km^2)	Population (10^6)	Per capita carbon emission (t CO_2/person)
			(%)	Base year	Target year				
Addis Ababa	Ethiopia	0.3	75	2010	2020	12	504	2.9	0.10
Johannesburg	South Africa	19.54	30	2005	2025	58.1	1645	3.9	5.01
Lagos	Nigeria	17.2				43	3569	13	1.32
Jakarta	Indonesia	43.86	30	2005	2030	70.2	662	9.6	4.57
Seoul	Korea	49.75	40	1990	2020	224	605	10.6	4.69
Sydney	Australia	5.5	70	2006	2030	213.2	26	0.18	30.56
Tokyo	Japan	61.64	25	2000	2020			13.18	4.68
Yokohama	Japan	19.01	80	1990	2050			3.67	5.18
Amsterdam	Netherlands	5.09	40	1990	2025			0.75	6.79

City	Country	Carbo emission (10^6 t CO_2)	Carbon emission reduction			GDP (billion USD)	Area (km^2)	Population (10^6)	Per capita carbon emission (t CO_2/ person)
			(%)	Base year	Target year				
Berlin	Germany	20.17	40 85	1990 1990	2020 2050	123.3	892	3.44	5.86
Copenhagen	Denmark	2.12	100	2010	2025			0.54	3.93
London	Britain	44.3	60	1990	2025	565	1572	7.8	5.68
Madrid	Spain	12.04	20	1990	2020	124.85	607	3.2	3.76
Moscow	Russia	44.6	20.7	2009	2020			10.56	4.22
Oslo	Norway	2.34	100	1990	2050	67.7	454	0.62	3.77
Paris	France	7.4	25	2004	2020			2.26	3.27
Rome	Italy	10.01	7	1990	2012	157.5	1297	2.7	3.71
Rotterdam	The Netherlands	29.46	50	1990	2025			0.6	49.10
Stockholm	Sweden	3.19	100	1990	2050			0.86	3.71
Warsaw	Poland	13.3	20	2007	2020	74.29	517	1.72	7.73
Austin	U.S.A.	14.93	70	2007	2030	86	795	0.8	18.66
Chicago	U.S.A.	35.55	80	1990	2050	476.4	592	2.83	12.56
Los Angeles	U.S.A.	33.94	35	1990	2030	717	1216	4.1	8.28
New York	U.S.A.	53.36	30	2005	2030	1210.38	790	8.4	6.35
Philadelphia	U.S.A.	19.78	20	1990	2015	335.1	369	1.5	13.19
Portland	U.S.A.	7.77	80	1990	2050	124.7	376	0.56	13.88
San Francisco	U.S.A.	5.01	25	1990	2017			0.82	6.11
Seattle	U.S.A.	7.04	100	1990	2050			0.63	11.17
Toronto	Canada	20.66	30	1990	2020	120.86	630	2.62	7.89
Vancouver	Canada	2.69	33	2007	2020	35	114	0.6	4.48
Washington	U.S.A.	10.5	80	1990	2050	91.64	117	0.62	16.94
Bogota	Columbia	15.75	16	2007	2019	54.9	414	7.2	2.19
Buenos Aires	Argentina	9.81	33	2008	2030	75.1	203	2.89	3.39
Rio de Janeiro	Brazil	11.35	20	2005	2020			6.3	1.80
Santiago	Chile	19.59	20	2007	2020			6.7	2.92
Sao Paulo	Brazil	15.12	30	2005	2012	165.9	1509	11.4	1.33

DATA SOURCE: CDP Cities 2012 Global Report, http://www.c40cities.org/.

CHAPTER 20

Global Climatic Disaster and Loss Statistics

Data Arrangement and Calculation: Su Buda and Huang Jinlong

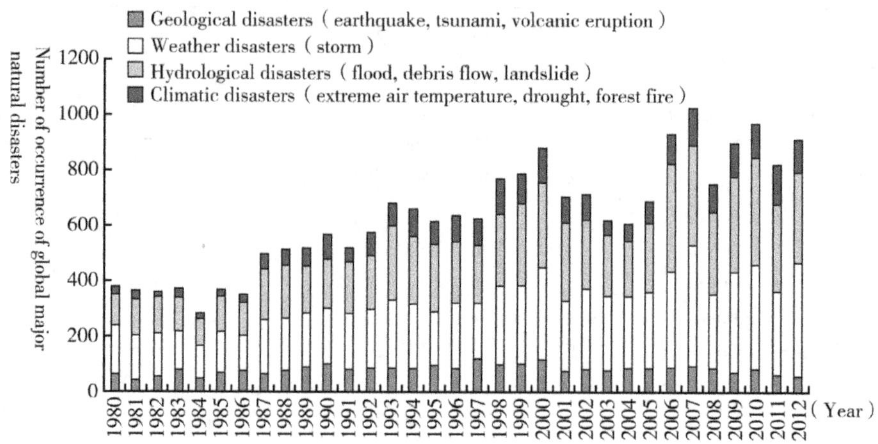

FIGURE 20.1 *Number of occurrence of global major natural disasters from 1980 to 2012.* DATA SOURCE: MUNICH RE GROUP AND NATIONAL CLIMATE CENTER.

TABLE 20.1 *Large-scale drought statistics of six continents from 1901 to 2010*

Region	Number of drought and proportion	Number of drought not more than 6 months	Number of drought not less than 12 months	Longest duration (month)	Maximum coverage (km^2)
Africa	76 (13%)	46	8	21 (1992–1993)	9.9 (33%, 1983.04)
Asia	185 (32%)	121	19	42 (1986–1990)	9.8 (22%, 1987.06)

* Su Buda, a associate researcher in National Climate Center, whose research field is climate change impact assessment; Huang Jinlong, a postgraduate of Nanjing University of Information Science & Technology, whose research field is climate change impact assessment.

 | DOI 10.1163/9789004280021_021

Region	Number of drought and proportion	Number of drought not more than 6 months	Number of drought not less than 12 months	Longest duration (month)	Maximum coverage (km^2)
Europe	81 (14%)	45	9	25 (1975–1977)	5.1 (51%, 1921.12)
North America	104 (18%)	65	15	41 (1954–1957)	7.5 (31%, 1956.10)
Oceania	45 (8%)	27	7	24 (1928–1930)	5.9 (77%, 1965.02)
South Africa	85 (15%)	58	10	19 (1982–1983)	10.8 (61%, 1963.10)
World	676	362	68	42 (1986–1990)	9.9 (33%, 1983.04)

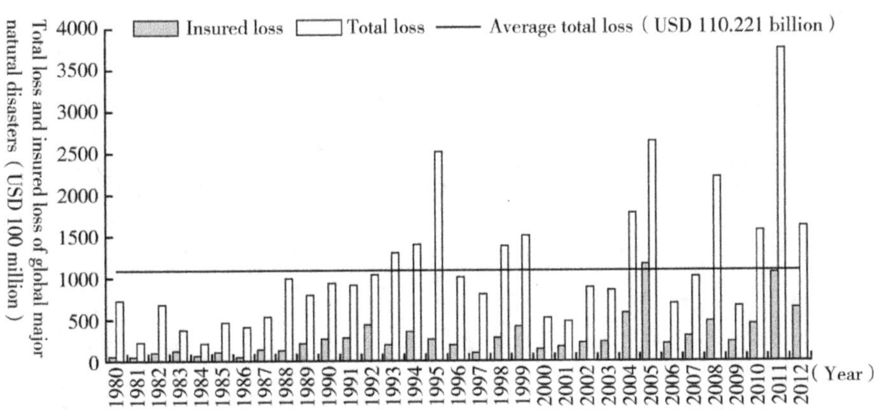

FIGURE 20.2 *Total loss and insured loss of global major natural disasters from 1980 to 2012 (Calculated at market value of 2012).*

DATA SOURCE: MUNICH RE GROUP AND NATIONAL CLIMATE CENTER. SAME FOR FIGURE 20.3 AND FIGURE 20.4.

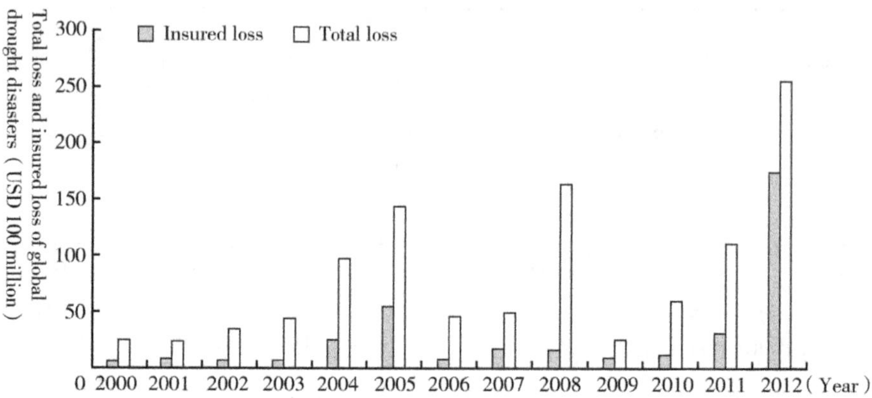

FIGURE 20.3 *Total loss and insured loss of global drought disasters from 2000 to 2012.*

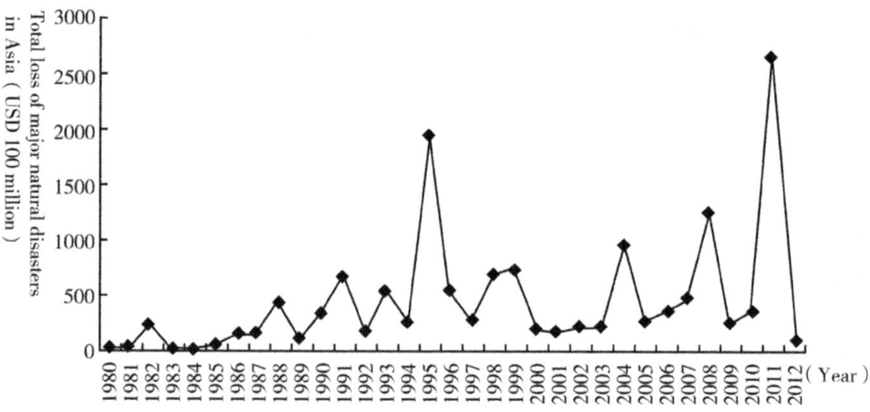

FIGURE 20.4 *Total loss of major natural disasters in Asia from 1980 to 2012.*

TABLE 20.2 *Summary of related major meteorological disasters in the US from 1980 to 2012*

Disaster type	Number of occurrence	Frequency (%)	Loss (USD 1 billion)	Proportion of loss in total (%)
Strong storm	50	34.7	110.9	11.2
Typhoon/ hurricane	33	22.9	485.2	48.8
Drought	17	11.8	240.1	24.1
Flood	16	11.1	85.1	8.6
Fire	12	8.3	23.2	2.3
Snowstorm	10	6.9	29.3	2.9
Freezing weather	6	4.2	20.5	2.1
Total	144	—	994.3	—

CHAPTER 21

Historical Statistics of Climatic Disasters in China

Data Arrangement and Calculation: SU *Buda and* HUANG *Jinlong*

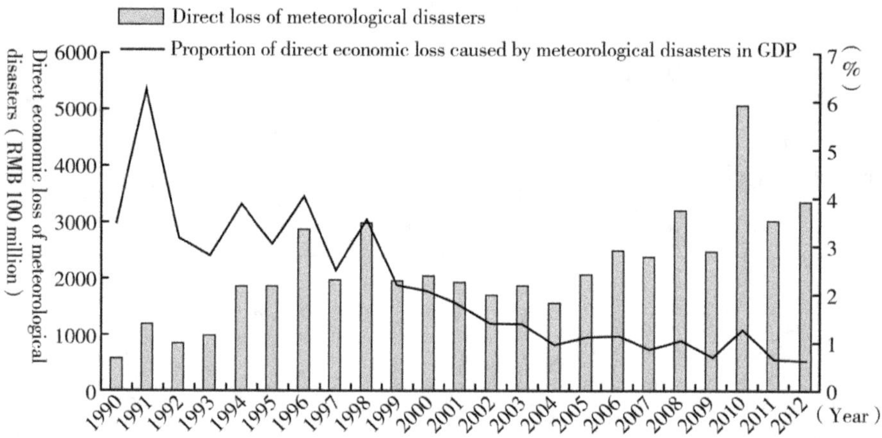

FIGURE 21.1 *Direct economic loss of meteorological disasters in China from 1990 to 2012 and its proportion in* GDP.
DATA SOURCE: *CHINA METEOROLOGICAL DISASTER YEARBOOK*, CHINA CLIMATE BULLETIN.

* Su Buda, a associate researcher in National Climate Center, whose research field is climate change impact assessment; Huang Jinlong, a postgraduate of Nanjing University of Information Science & Technology, whose research field is climate change impact assessment.

 | DOI 10.1163/9789004280021_022

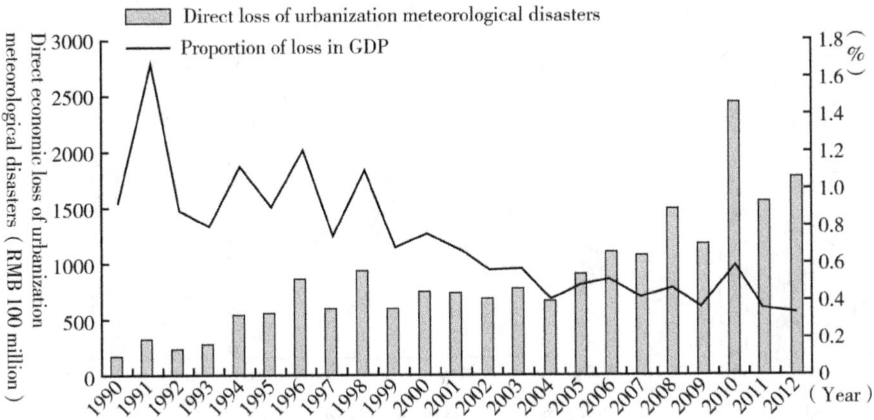

FIGURE 21.2 *Direct economic loss of urbanization meteorological disasters in China from 1990 to 2012 and its proportion in GDP.*
DATA SOURCE: *CHINA METEOROLOGICAL DISASTER YEARBOOK*, CHINA CLIMATE BULLETIN. NATIONAL BUREAU OF STATISTICS. SAME FOR TABLE 21.1.

TABLE 21.1 *China's meteorological disasters statistics*

Year	Crop disaster (10,000 hectares)		Population disaster		Direct economic loss (RMB 100 million)	Direct economic loss of urbanization meteorological disasters (RMB 100 million)
	Affected area	Crop failure area	Affected population (10,000 persons)	Population death (persons)		
2004	3765	433.3	34049.2	2457	1565.9	653.9
2005	3875.5	418.8	39503.2	2710	2101.3	903.3
2006	4111	494.2	43332.3	3485	2516.9	1104.9
2007	4961.4	579.8	39656.3	2713	2378.5	1068.9
2008	4000.4	403.3	43189	2018	3244.5	1482.1
2009	4721.4	491.8	47760.8	1367	2490.5	1160.3
2010	3742.6	487	42494.2	4005	5097.5	2421.3
2011	3252.5	290.7	43150.9	1087	3034.6	1555.8
2012	2496	182.6	27389.4	1390	3358	1766.3

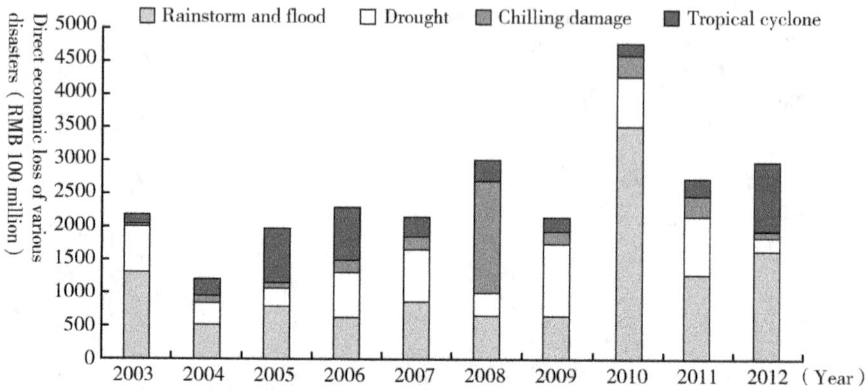

FIGURE 21.3 *Direct economic loss of various disasters from 2003 to 2012.*
DATA SOURCE: *CHINA METEOROLOGICAL DISASTER YEARBOOK*. NATIONAL BUREAU OF STATISTICS. SAME FOR FIGURE 21.4.

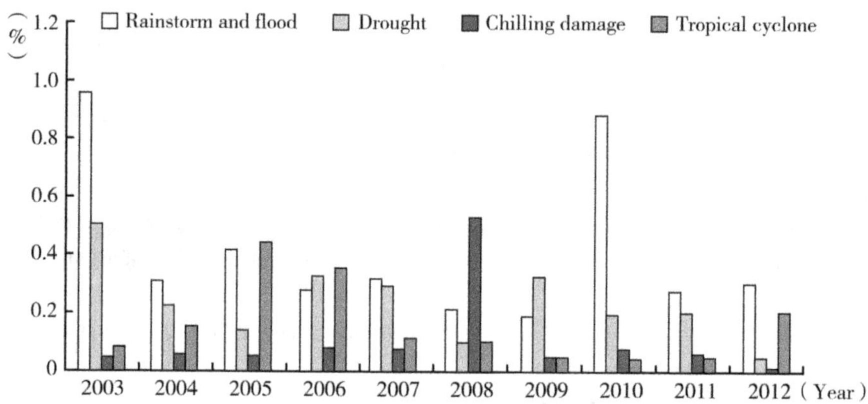

FIGURE 21.4 *Direct economic loss of various disasters from 2003 to 2012 and its proportion in GDP.*

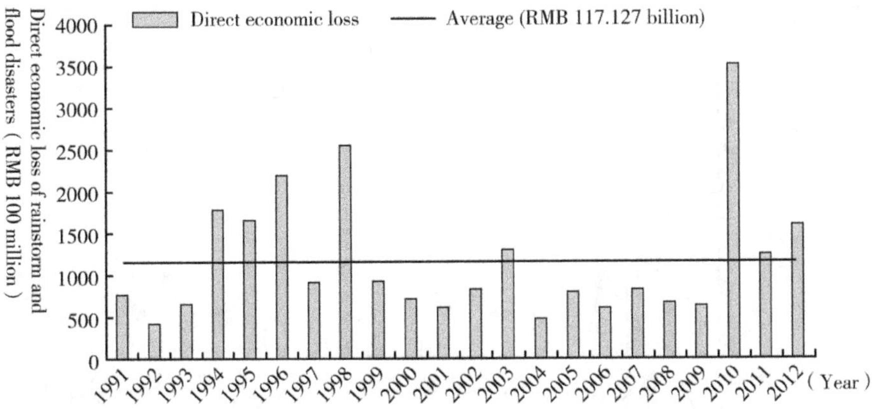

FIGURE 21.5 *Direct economic loss of rainstorm and flood disasters from 1991 to 2012.*
DATA SOURCE: *CHINA METEOROLOGICAL DISASTER YEARBOOK*, CHINA CLIMATE BULLETIN. SAME FOR FIGURE 21.6–FIGURE 21.8.

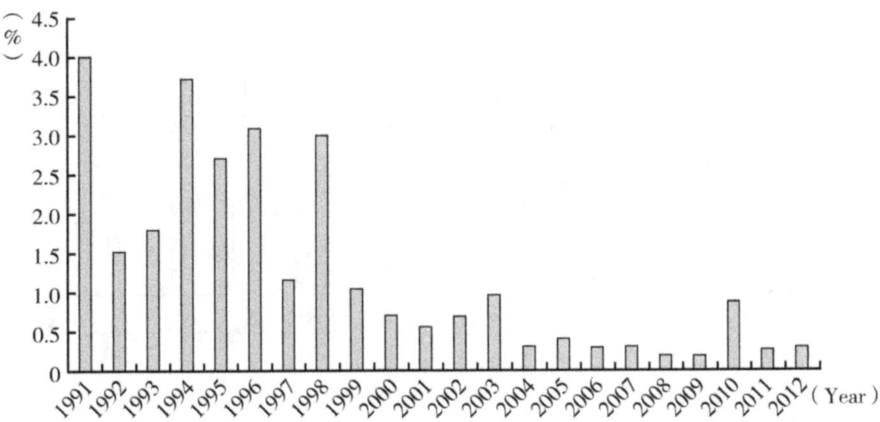

FIGURE 21.6 *Direct economic loss of rainstorm and flood disasters and its proportion in* GDP *from 1991 to 2012.*

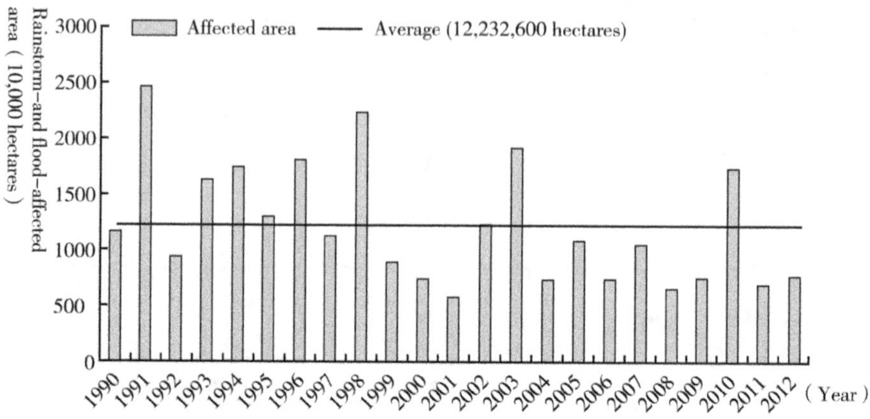

FIGURE 21.7 *Rainstorm- and flood-affected area from 1990 to 2012.*

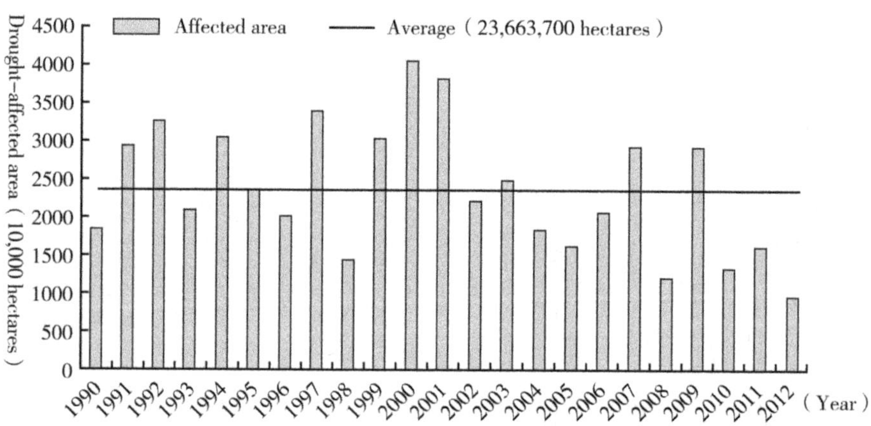

FIGURE 21.8 *Drought-affected area from 1990 to 2012.*

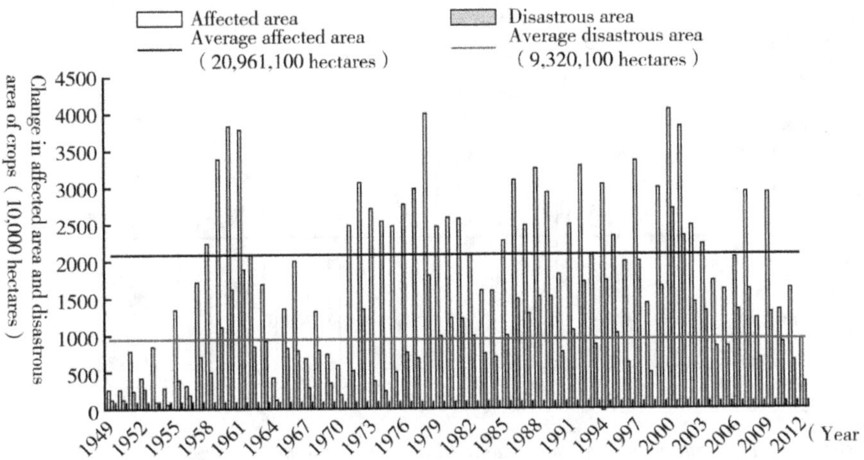

FIGURE 21.9 *Changes in affected area and disastrous area of crops in China from 1949 to 2012 (Drought Disaster).*
DATA SOURCE: ZZYS.AGRI.GOV.CN. NATIONAL FLOOD CONTROL AND DROUGHT RELIEF WORKING CONFERENCE.

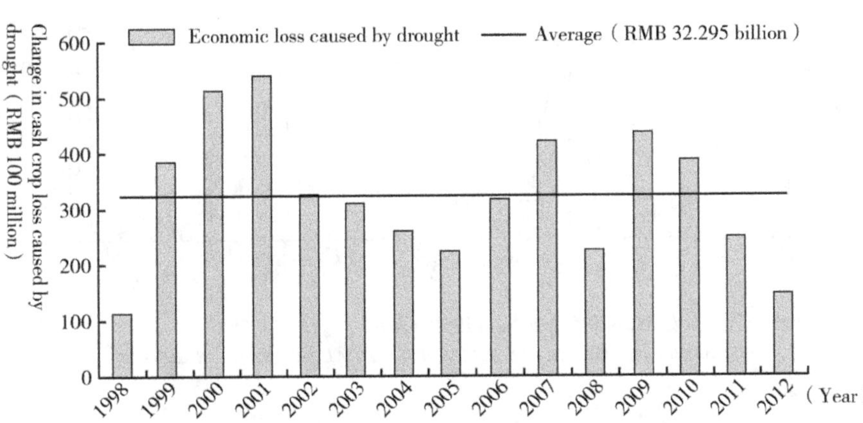

FIGURE 21.10 *Changes in cash crop loss caused by drought in China since 1998.*
DATA SOURCE: BULLETIN OF THE MINISTRY OF WATER RESOURCES. NATIONAL FLOOD CONTROL AND DROUGHT RELIEF WORKING CONFERENCE. SAME FOR FIGURE 21.11.

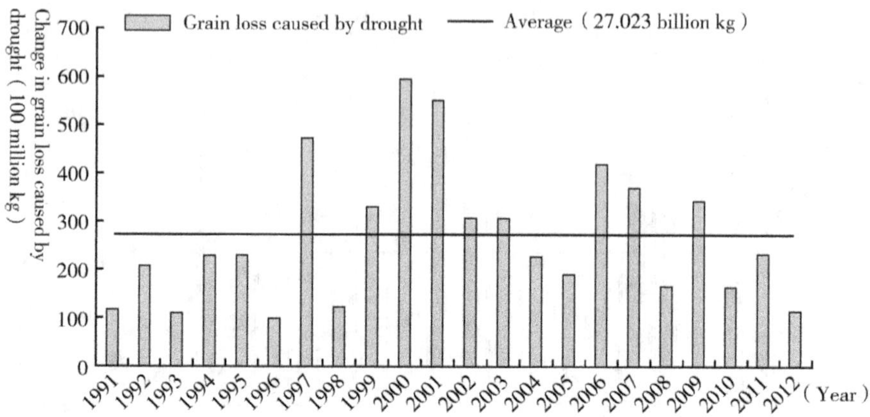

FIGURE 21.11 *Changes in grain loss caused by drought in China since 1991.*

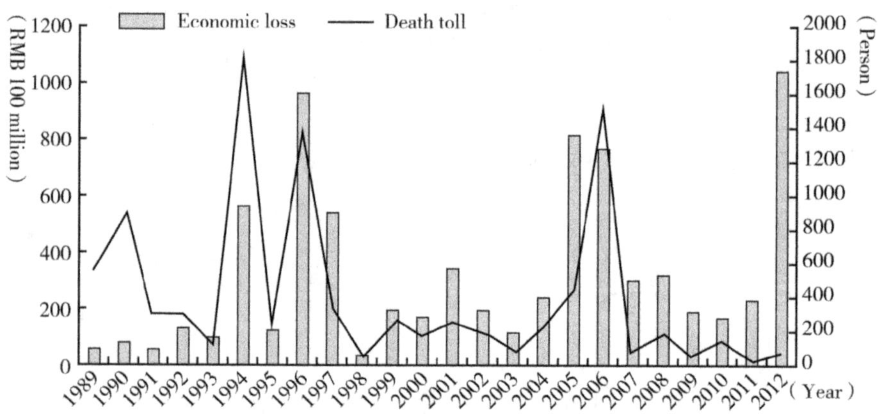

FIGURE 21.12 *Typhoon disaster loss from 1990 to 2012.*
DATA SOURCE: *CHINA METEOROLOGICAL DISASTER YEARBOOK.*

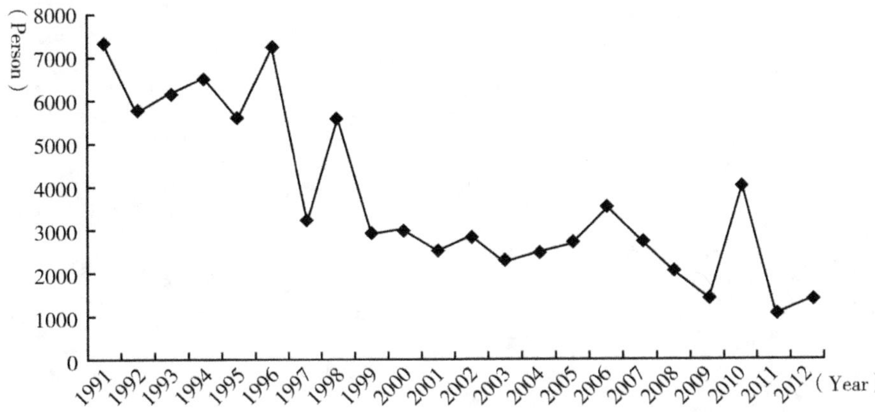

FIGURE 21.13 *Death toll caused by meteorological disasters in China from 1991 to 2012.*
DATA SOURCE: *CHINA METEOROLOGICAL DISASTER YEARBOOK*, NATIONAL CLIMATE CENTER.

Index

www.ingramcontent.com/pod-product-compliance
Lightning Source LLC
LaVergne TN
LVHW010624300926
849162LV00001B/55